W9-CJN-057

The Fats of Life

This book aims to fill the gap between unscientific comments about the hazards and benefits of high-fat or low-fat diets and weight control found in women's magazines and technical and medical reports about lipid biochemistry and obesity. It aims to explain in simple language the biology of feasting and fasting, fattening and slimming in wild animals as well as people. Topics include where fats come from and how animals and plants handle them, their natural roles in migration, mating, breeding and living in unpredictable habitats such as deserts and arctic regions, and their contributions to our cookery, paints and medicines.

The physiological mechanisms of digesting, transporting and utilising energy stores are discussed, along with the contribution of fatty tissue to body insulation and the protection of delicate organs. Archaeological, anthropological and physiological evidence is assembled to explore how, when and why people have become fat, and how evolutionary forces have determined the modern diversity of body size and shape. The book ends with a discussion of the contribution of dietary fats and obesity to health in the modern world.

Caroline Pond studied Zoology at Oxford University, completing a doctorate on insect flight in 1971, and taught comparative physiology there for 4 years. Her interest in adipose tissue began after she moved to the USA in 1975, and taught biology and veterinary anatomy at the University of Pennsylvania in Philadelphia. It seemed odd that textbooks of anatomy said so little about what was often the most abundant tissue in the body. She started as a lecturer at the Open University in Milton Keynes in 1979, and is now Reader in Biology. Since 1982, she and Christine Mattacks have been exploring the natural organisation and functions of adipose tissue, collaborating with scientists in Canada, Norway and elsewhere.

Also of interest in popular science

The Thread of Life: the story of genes and genetic engineering
SUSAN ALDRIDGE

Remarkable Discoveries
FRANK ASHALL

Evolving the Mind: on the nature of matter and the origin of consciousness
A. G. CAIRNS-SMITH

Prisons of Light – black holes
KITTY FERGUSON

Extraterrestrial Intelligence
JEAN HEIDMANN

Perils of a Restless Planet
ERNEST ZEBROWSKI

Dying to live – how our bodies fight disease
MARION D. KENDALL

The Clock of Ages: why we age – how we age – winding back the clock
JOHN J. MEDINA

Improving Nature? The science and ethics of engineering
MICHAEL J. REISS AND ROGER STRAUGHAN

Beyond Science: the wider human context
JOHN POLKINGHORNE

Reaching for the Sun: how plants work
JOHN KING

Figments of Reality: The Evolution of the Curious Mind
IAN STEWART AND JACK COHEN

CAROLINE M. POND

The Fats of Life

With drawings by Mat Cross and Sarah Sutcliffe

PUBLISHED BY THE PRESS SYNDICATE OF THE UNIVERSITY OF CAMBRIDGE
The Pitt Building, Trumpington Street, Cambridge CB2 1RP, United Kingdom

CAMBRIDGE UNIVERSITY PRESS
The Edinburgh Building, Cambridge CB2 2RU, UK http://www.cup.cam.ac.uk
40 West 20th Street, New York, NY 10011-4211, USA http://www.cup.org
10 Stamford Road, Oakleigh, Melbourne 3166, Australia

First published 1998

Printed in the United Kingdom at the University Press, Cambridge

Typeset in Monotype Ehrhardt 11/13pt, in QuarkXPress™ [WV]

A catalogue record for this book is available from the British Library

Library of Congress Cataloguing in Publication data

Pond, Caroline M. (Caroline Margaret)
The fats of life / Caroline M. Pond.
p. cm.
Includes bibliographical references and index.
ISBN 0 521 58321 7 (hc). – ISBN 0 521 63577 2 (pbk)
1. Fatty acids – Physiological effect. 2. Lipids in nutrition.
3. Fatty acids in human nutrition. I. Title.
QP751.P645 1998
612.3'97 – dc21 97–46515 CIP

ISBN 0 521 58321 7 hardback
ISBN 0 521 63577 2 paperback

Contents

Acknowledgements

The information quoted in this book comes from a wide variety of sources. Much of it is common knowledge, simply rearranged and reassessed to draw attention to the role of fats in the natural world. I have kept the notes to a minimum, and included only references to the original papers from which particular 'stories' are derived. I wondered whether to include a glossary of technical words, but eventually decided against it. Readers should use the index instead, and refer back to the text where scientific terms are first explained.

I am grateful to Hilary MacQueen, Christine Mattacks, Roger Moore, Eric Newsholme, Malcolm Ramsay and Irene and John Ridge who read early drafts of certain chapters and made helpful comments, and especially to Dick Colby of Stockton College, New Jersey, who read the entire manuscript very thoroughly, drawing attention to several errors and inconsistencies, making many suggestions for clarification and improvements, and advising on the subtleties of modern American English. I thank the colleagues who supplied the photographs and the artists who drew the illustrations.

I would particularly like to acknowledge the practical help and ideas provided by students and colleagues especially Christine Mattacks, with whom I have collaborated for more than fifteen years, and all the professional biologists, managers of domesticated, captive and wild livestock and ordinary people interested in science who have allowed me to participate in observing animals while alive and examining them after death.

Prologue

For much of this century, fats have suffered from bad press. 'Trim away the fat', 'fat and flabby' reflect the prevailing view that anything fatty is at best superfluous or ugly, often indicative of an indolent or intemperate lifestyle, and sometimes positively harmful. The implication of fats in causing common forms of heart disease was entirely in keeping with this view. Scientists explained how fats accumulated surreptitiously, gradually slowing life's traffic until the blockages they cause strangle it. Even more distressing is the fact that fats cannot easily be expelled by drugs, radiation, massage or even surgery; they are purged only by continual exertion and abstinence. More than any other biological materials, except perhaps genes, fats and fatness have acquired a moral dimension. Fats were seen as literally the agents of just retribution for the modern lifestyle, turning self-indulgence and sloth into debility and death.

Biochemists have been slow to undermine this attitude. They have understood the importance of proteins in living systems for more than 100 years. By the middle of the twentieth century, proteins were established as the primary gene product, and techniques for probing their internal structure and quantifying their activities were advancing rapidly. A score of different subunits can be assembled into thousands of different protein molecules, many of them precisely tailored to highly specific roles. The absence, misplacement or substitution of one crucial subunit can wreck a whole molecule. Similar structural variety in sugars and starches was also established in the nineteenth century, and it acquired new importance when the role of such molecules in cell adhesion and molecular recognition was identified in the 1960s.

Fats are as fundamental to life as proteins and genes, and living systems are the only natural source of the more complex kinds. Although it has long been known that all living organisms have some fats, and life as we know it could not continue without them, until very recently they were regarded as chemically rather boring, necessary only as fuel and for building barriers within and between cells. In the past 30 years, the intricacy and central importance of these roles have become clearer, and some fats have been found to have equally fundamental roles as messengers and regulators. Improved techniques for separating fats and determining the details of their chemical structures have revealed them to be much more diverse than previously believed. As in so many biological systems, the devil is in the detail. Far from being 'jacks of all trades', many fats turned out to have unique and specific roles. The incorporation of a single 'wrong' subunit into a complex fatty molecule can impair its function as seriously as comparable building faults in proteins or starches.

Anxiety about the sinister side of fatty activities is fuelled by the finding that humans make few fats themselves; most of the fats in our bodies come from our food. We cannot sit back and rely upon our bodies to organise their fatty components according to their own scheme, honed by evolution to be as efficient as possible. Fats stand uncomfortably between nature and nurture: far from asserting the omnipotence of genes to make us any proteins we need, fat requirements reaffirmed our dependence upon the rest of the ecosystem.

The Darwinian ideal of a fully functional, perfectly adapted organism implies that there is a place for everything and everything is in its place. Except of course for fatty tissue, which in many people's minds – biologists and physicians not excluded – has no rightful place of its own and expands untidily into parts of the body where it is not welcome. Fatty tissue is omitted from most anatomical drawings, or shown as a modest mass demurely making itself as small as possible. But as we all know, fat does not restrain itself in this way: it bulges in the wrong places, disappears from where it is required and generally behaves in an unpredictable and inconvenient way. Such intransigence has given rise to the belief that, alone among the major tissues, the distribution of fat does not follow any clearly identifiable basic plan so its arrangement is outside the axioms of comparative anatomy. Dispelling such notions is one of the main aims of this book: inheritance and function together determine the distribution and anatomical relations of fatty tissue as much as for any other tissue.

Describing a person as 'fat' carries more connotations than references to the selective expansion of any other tissue, such as 'muscular' or 'hairy'. The Romans had two quite different terms for this meaning of our overworked word 'fat'. The Latin adjective *pinguis* can mean plump or well fed, but the association between plumpness and certain traits of character led to its being used metaphorically to mean rich, fertile, dull, stupid or complacent.

So the verb *pinguescere* meant to prosper as well as to fatten, summarising perhaps the many processes that end in a contented, well-earned retirement. *Obesus*, from which we derive our word obesity, was not closely associated with fatness in the contemporary sense, and is more exactly translated as corpulent or, metaphorically, coarse, unsophisticated. It is perhaps significant that the English words derived from the former term, such as pinguedinous, pinguescent and pinguid, were in common use until the end of the eighteenth century but are now entirely replaced by those related to the more pejorative *obesus*.

For many organisms, being fat at least from time to time is indispensable, though far from easy: large concentrations of almost anything present biochemical complications. No organism accumulates large quantities of proteins without an immediate use for them, and only plants have perfected the art of laying down and maintaining large stores of carbohydrates. Some species of animals are professional fatties. They have evolved safe, efficient ways of accumulating, storing and releasing large quantities of potentially dangerous lipid fuel for hibernation, migration or as an adaptation to living in an unpredictable climate or habitat.

Instead of wasting time lamenting the tendency of fatty tissue to intrude where it is not required, expand when not desired and in other ways undermine health and happiness, this book extols the achievements of the naturally obese, animals that have mastered combining fatness with fitness. Unfortunately, scientists who study humans usually concentrate only on them, plus rats and, if necessary, mice. These species are chosen for their convenience and traditional role as laboratory animals, but neither naturally becomes obese, and can only be made so by drugs or other forms of artificial manipulation. Observations on naturally obese wild animals can provide insight into physiological adaptations to obesity that cannot be achieved by studying rats and mice.

Almost all biologists and physicians agree that many modern humans are exceptionally fat, fatter for a larger fraction of their lives than almost any other kind of land mammal or bird. Obesity itself is not really a disease – the existence of naturally obese wild animals shows that fatness does not necessarily exclude fitness – but in contemporary humans, obesity often leads to ill-health. While in some cultures, obese people are happy and healthy, in others, obesity leads to social and sexual disadvantages quite apart from any attendant medical problems. Many people dread the social consequences of obesity so much that they are willing to endure prolonged discomfort to avoid or correct the early stages of the condition, which by itself is medically harmless. Indeed, many readers of this book may be motivated at least in part by concern about being or becoming 'too fat'.

Commentators are unable to agree upon whether the human tendency towards obesity is adaptive, in the sense that it evolved in connection with some past or current function, or pathological, a deleterious spin-off from some other factor in our lives, possibly diet, or laziness. Part of the trouble is that the people doing the relevant basic research have contrasting approaches to biology and so are looking for quite different answers to the question of why people are fat.

To anthropologists and evolutionary biologists, 'why?' means 'for what reason?' What caused the feature to evolve? Why should such an apparently deleterious trait now be so prevalent? They shun the narrow complications of biochemistry, preferring to assume that if organisms have avoided extinction they must be adapted to their lifestyle. How an evolutionary will produces a physiological way is obviously interesting, not least because detailed knowledge of the mechanisms is essential to interventions such as drugs, diets and exercise regimes. Exploring the evolutionary 'will' helps us to understand the relationship between fatness, fecundity and longevity, and indicates the feasibility of adjusting our weight and body shape to modern tastes and habits.

To physiologists, 'why?' is always 'how?' 'by what biochemical means?' They avoid evolutionary issues, which they usually find boring and inconclusive, and regard historical 'justifications' for contemporary situations as irrelevant to the practicalities of diagnosis and therapy. Nonetheless, their investigations into the mechanisms of fattening and its metabolic consequences are essential to any discussion of why people become obese: they show us how far people are equipped to handle all that fat.

For many years, a lean figure and a low-fat diet have been extolled by the medical establishment, the fashion industry and show business. The glaring and persistent failure of most western people to attain anything approaching the ideal body shape or eating habits has led those naturally prone to stoutness to rebel against political and medical orthodoxy.[1] Other body features such as hair or muscle development do not arouse similar passions in those whose constitution and appearance are other than average. This book is not of that type. It is not a polemic for or against certain diets or lifestyles, nor does it condemn or condone obesity. It is not a slimming manual and does not offer dietary advice of any kind. It is strictly about lipids and fatty tissues and their roles in the natural world and human civilisation.

1

Introduction to fats

Fats and oils are a varied group of chemicals that do not readily dissolve in or mix with water. They dissolve in 'organic' liquids such as chloroform, benzene and acetone. These solvents are familiar as ingredients of nail polish removers and home dry-cleaning fluids: much of their efficacy in these roles derives from their capacity to dissolve the fatty components of stains caused by food, sweat and other biological materials. Fuel oil and its derivatives such as lubricating oils, paraffin (kerosene) and petrol (gasoline) share many of the properties of biological lipids, but they are chemically different, and, not being components of living organisms, are outside the scope of this book. So-called essential oils that are extracted from plants and used for perfumes, flavourings and aromatherapy are not oils in the chemical sense of the term.[1] They evaporate readily, forming 'essence' to which human noses are very sensitive.

By convention, 'oils' are liquid at room temperature, and 'fats', such as butter and lard, are solid or nearly so. Although impressive and important for their roles in living organisms, this distinction does not reflect a fundamental difference in chemical composition: it is merely a reversible change of physical state, just as ice, snow, and rain are different physical states of the same chemical substance, water. Fats are frozen oils, so to avoid confusion, hereafter the term 'lipid' will be used to describe both fats and oils, regardless of the physical state in which they are most usually found.

Basic discoveries

Until the end of the eighteenth century, living organisms, and most substances known to emanate from them, were regarded as too complex, or too sacred, to be amenable to chemical analysis in the same way as non-living 'minerals'. Almost all oils and waxes used as food, fuels or lubricants were derived from plants or animals, and so were rarely regarded as suitable subjects for scientific investigation. The distinction faded after the English nonconformist minister and schoolmaster, Joseph Priestley (1733–1804) observed that a candle burned and a mouse lived for longer if they were confined separately in sealed jars than if held together in a similar vessel, suggesting that mice and candles were competing for the same component of air.

A few years later, in 1783, the French civil servant and landowner, Antoine Lavoisier (1743–1794), in collaboration with the mathematician and astronomer, Pierre Laplace (1749–1827), demonstrated that the amounts of ice melted and 'air used' by a guinea-pig were equivalent to the heat produced and air 'consumed' by burning charcoal, and concluded that the chemical processes of respiration and combustion were fundamentally similar. Biological materials became 'fair game' as subjects for chemical analysis, although their complexity and their tendency to rapid decomposition made them technically difficult to study.

The basic structure of biological fats was elucidated by another enterprising and imaginative Frenchman, Michel-Eugène Chevreul. He was born in the Loire valley in 1786, when King Louis XVI and Queen Marie-Antoinette still ruled France, the aristocrats seemed safe in their nearby châteaux, and the basic design and construction of most kinds of machinery and vehicles had hardly changed since Roman times. As a young man, Chevreul moved to Paris, where his long and varied career included service at the Jardin des Plantes, directing the synthesis and applications of the dyes used in the manufacture of tapestries at the famous Gobelin factory, and being Professor of Organic Chemistry at the University of Paris.

The celebrations for Chevreul's hundredth birthday recorded over 500 publications between 1806 and 1885 on an immense range of topics in palaeontology, botany, zoology, physiology, chemistry and history of science. Such longevity was even more exceptional then than it is now, especially for a man (most centenarians are women), and suggests that a lifetime of exposure to dyes, oils and other 'chemicals' can't be that dangerous. By the time he died, which was not until 1889, exploitation of mineral oil (petroleum) was well under way, the electric motor, the internal combustion engine (used in cars, etc.) and the modern 'safety' bicycle had been invented, and would shortly revolutionise industry, homes and transport.

When Chevreul was young, what we now call organic chemistry was practised as an eccentric hobby by wealthy gentlemen such as Lavoisier, who funded his research out of his huge personal fortune until he was arrested and guillotined by revolutionaries at the age of 50. He lived to see it transformed into an important profession that serves major industries. His research on lipids underpinned the development of the protective coatings and the lubricants upon which all such machines depend.

During much of the first 30 years of Chevreul's life, France experienced political and social changes on a scale not seen for centuries, and, by the time Napoleon Bonaparte came to power in 1799, it was also embroiled in major wars with several other European states. The resulting loss of international trade, land reforms and the new Republic's active encouragement of rapid population growth[2] made food supplies a major issue during the first two decades of the nineteenth century. Napoleon was, from an early age, personally interested in many branches of science and archaeology, and was impressively knowledgeable on certain topics. He recognised the importance of science in military operations, civilian prosperity and national prestige, and befriended and financed its practitioners. Armies need good rations and hunger fuels discontent among civilians, so the newly established French Republic vigorously promoted research into what we would now call food technology, enlisting the help of its most promising scientists.

Starting in 1811, just before Napoleon's ill-fated invasion of Russia, and continuing until after his devastating defeat at Waterloo in 1815, Chevreul devoted himself to identifying the 'immediate principles' in mutton fat. Borrowing procedures long used in making soap (which was then still a luxury item, manufactured only on a small scale), he heated fats with alkalis (what we call potash or caustic soda) and purified the resulting mixtures. He named the clear, syrupy, sweet-tasting liquid that he extracted 'glycérine' (from the Greek word, γλυκυς, sweet).[3] In spite of its sweet taste, glycerine, now called glycerol,[4] is an alcohol, not a sugar, chemically similar to ethanol, the main active ingredient of beer, wine and other 'alcoholic' drinks. Glycerol contains three carbon atoms (and their associated hydrogen and oxygen atoms) while ethanol has only two, and it does not affect the nervous system and behaviour in the way that ethanol does. Both substances are produced in significant quantities by a wide variety of organisms, and can be broken down to release energy in a form that can be used by living cells.

Chevreul named the other major 'principles' he found in animal fat 'fatty acids', and showed that they occurred in the proportions of three fatty acids to each glycerol. When separated from the glycerol, fatty acids dissolve in alcohol and, by repeated extraction and precipitation with salts, could be purified sufficiently to form crystals. Chevreul noted that the fatty acids from mutton fat formed two distinct types of crystals, indicating that they were a mixture of at least two different chemicals.

Laboratory techniques and theoretical concepts about the structure of large biological molecules were then so limited that he was unable to take the analysis any further. He reported his finding in his classic book entitled *Recherches chimiques sur les corps gras d'origine animale*,[5] first published in 1823, and then turned his attention to dyes and pigments.

Fatty acids

Following Chevreul's pioneering work, scientists in France, Germany and elsewhere took up the study of natural lipids in a wide range of plants and animals, and their seeds, eggs and embryos. For a while in the middle of the nineteenth century, scientific knowledge of biological lipids was ahead of that of either proteins or carbohydrates (sugars and starches). Glycerol was found in the great majority of biological lipids (but not in mineral oils, including paraffin, motor fuels and lubricating oils), but the fatty acid components proved to be much more variable. By the end of the nineteenth century, dozens of different fatty acids had been described and named. Modern methods of extraction and separation have extended the list to hundreds that occur naturally in micro-organisms, plants or animals, and many more kinds can be synthesised artificially by chemical transformations. A few of the more abundant and physiologically important fatty acids are discussed in Chapter 3.

All fatty acids consist of a chain of carbon atoms, each attached to hydrogen atoms. The epithet 'acid' derives from the —COOH group on one end of the molecule: this H (hydrogen) atom (but not the many others elsewhere in the molecule) readily dissociates from the rest of the molecule, forming an H^+ (hydrogen ion), the hallmark of an acid. The other end consists of a carbon atom and three hydrogen atoms, and is called the methyl end, because of its resemblance to methane (CH_4), familiar as the main component of marsh gas. Figure 1 shows three alternative ways of illustrating the structure of a typical fatty acid with 16 carbon atoms that has the basic chemical formula $C_{16}H_{32}O_2$. A knowledge of the way in which carbon, hydrogen and oxygen combine with each other suggests the linear arrangement shown in Figure 1(a) as the most likely. It is often written as $CH_3(CH_2)_{14}COOH$ to show that the two ends of the molecule differ from each other and from the repeating chain of 14 similar units between them. Most of the carbon atoms are linked to two others and to two hydrogen atoms by a single bond, represented by a line. One of the oxygens is joined to a carbon by a double bond, shown as two parallel lines. The other forms the acid group.

During the first two decades of the twentieth century, Sir William (Henry) Bragg and his son, Sir (William) Lawrence Bragg, working at the Royal Institution in London used X-rays to determine the sizes of atoms and their spatial relationships to one another in various molecules. The common fatty acids were among the first molecules whose structure they and their colleagues worked out by such means. Fatty acids are genuine biological molecules, large enough to be interesting but small enough for the structure to be worked out without the need for the elaborate computing techniques that are used for similar analyses these days. The simpler fatty acids that they chose for study are chemically stable – they do not deteriorate when exposed to air and strong radiation – and, above all, as Chevreul discovered, they form regular crystals.

The model that the Royal Institution researchers built from wooden balls and wire to summarise their findings is shown in Figure 1(b). The carbon atoms (dark balls) link to each other in a zig-zag arrangement (called the bond angle) with their hydrogen atoms (light balls) on opposite sides. They form a repeating chain, terminating in the acid group on the last carbon ($COO^- H^+$), here shown on the right. The shape of molecules is so important to their properties and biological roles that organic chemists have devised ways of drawing them that emphasise it. Figure 1(c) shows the same fatty acid represented in this way: most of the hydrogen atoms are omitted, and the carbon atoms are understood to be at the bends of the zig-zag pattern revealed by the crystallographic discoveries. Only the acid group is shown in detail.

(a)

(b)

(c)

Figure 1. Three ways of illustrating the structure of a fatty acid molecule. (a) The basic arrangement of all the atoms. (b) A model incorporating information about bond angles derived from X-ray diffraction studies. (c) A modern scheme based upon (B), showing only the acid group in detail.

One way in which fatty acids differ from each other is in the number of carbon atoms in the main 'skeleton'. Fatty acids with from two to 36 carbon atoms have been found in higher plants or animals, and those of certain unusual micro-organisms can have up to 80, but the commonest have between 14 and 22 carbons. Those with fewer than eight carbon atoms are referred to, somewhat arbitrarily, as short-chain fatty acids, those with 8–12 carbons are called medium-chain fatty acids, and long-chain fatty acids have more than 12 carbon atoms. Fatty acids with even numbers of carbon atoms are generally more abundant in biological materials than those with odd numbers, although fatty acids with almost all possible numbers of carbons within this range have been found in small quantities somewhere in some organism.

Under special conditions such as thunderstorms, short-chain fatty acids with two or three carbon atoms can be produced from 'inorganic' sources of carbon such as methane or carbon dioxide, but so far as we know, medium- and long-chain fatty acids are made naturally only in living cells. All known self-propagating living cells (i.e. not viruses) contain some molecules that have a fatty acid component. They are thus key indicators of the presence, or former presence, of life. Together with amino acids (the basic components of proteins) and genetic material, long-chain fatty acids are sought in meteorites and in samples of rocks and dust collected by space probes, as evidence for the existence of life on Mars or other celestial bodies. So far, none has been found.

Larger lipids

Fatty acids are very weakly acidic compared with mineral acids such as nitric or sulphuric acid because the nitrate and sulphate groups repel H^+ ions much more strongly than hydrocarbon chains can, but the $COO^- H^+$ group is acidic enough for a high concentration of fatty acids to disrupt the biochemical workings of the cell. Fatty acids are therefore usually tidied away as part of larger, more complex molecules. They are often attached though their —COOH group of atoms to alcohols, characterised by having —OH groups, via a reaction called esterification. When an ester bond is made, a water molecule (H_2O) is formed from a hydrogen atom and an oxygen atom of one molecule (in this case, a fatty acid) and a hydrogen atom from the other molecule (in this case, glycerol).

Three such reactions form a triacylglycerol molecule (called triglyceride in older literature), as shown schematically in Figure 2. In the reverse reaction, water molecules are split and their components become part of the fatty acids and glycerol. This process can be called hydrolysis because water molecules are split, but it is also called lipolysis (i.e. breaking lipids). Esterification and lipolysis are usually facilitated in both directions by specific enzymes (of which more in Chapter 4). As biochemical reactions go, esterification and lipolysis are simple, involving only small quantities of energy,[6] and they can take place inside or outside cells, in the blood or in the gut.

At body temperature, the triacylglycerol molecules assume many more configurations than this static image suggests: the carbon atoms in both the

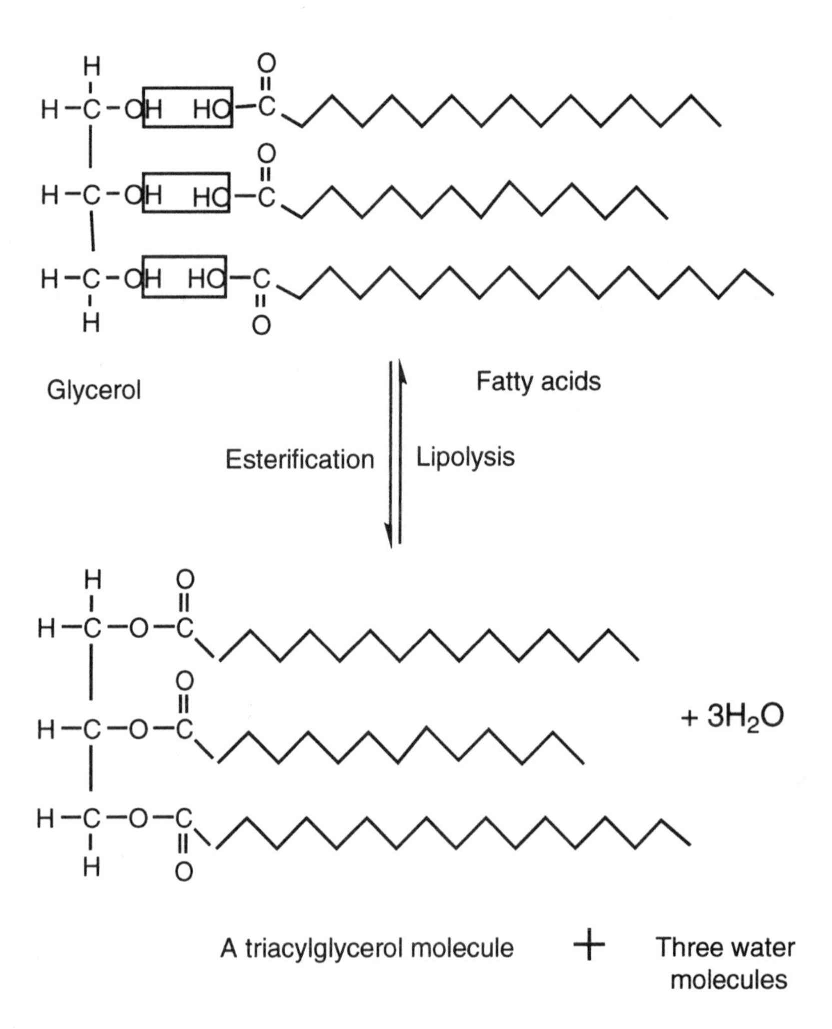

Figure 2. Esterification and lipolysis. Three fatty acids and a glycerol molecule (above) are joined by ester bonds to form a triacylglycerol molecule and three water molecules (below). The fatty acids have different numbers of carbon atoms.

fatty acids and the glycerol 'skeleton' can rotate relative to each other, producing a wide range of conformations. The three-pronged plan of triacylglycerols is fundamentally different from that of the other major category of energy-producing biological materials, the carbohydrates, in which most of the carbon atoms are arranged in rings. The simplest carbohydrates are the sugars, which have only one or two such rings; glucose, the predominant simple sugar in the blood of almost all vertebrates and many other animals, consists of a single ring, but sucrose, the commonest sugar used in cookery, has two rings.[7] Sugars are the basic units of so-called complex carbohydrates, such as starch and cellulose, which may contain thousands of rings, joined together in various different ways.

So-called free fatty acids move in animals' blood on their way between different kinds of cells, but in this state they are always bound to large protein carrier molecules such as albumin. As such, they are not really 'free', so 'non-esterified fatty acid' is a more accurate term than 'free fatty acid' although the latter is still widely used. In other circumstances, fatty acids seldom occur alone, except transiently during chemical reactions. In tissues other than blood, fatty acids are usually esterified to glycerol, forming monoacylglycerols if only one fatty acid is attached, diacylglycerols if two are attached, and triacylglycerols if three fatty acids are attached. All three kinds of acylglycerols occur in insects, vertebrates and probably many other kinds of animals. They have quite different roles, to be discussed in the following chapters.

In principle, triacylglycerols could include any kind of fatty acids, but in practice, we find that almost all the fatty acids in the major storage lipids in most animals and plants, including almost all the species that we use as food, are long-chain. The rows of carbon atoms pack beside each other so well that almost everything else, including water, is excluded. If left to themselves, triacylglycerols segregate from watery components of the cells to form droplets, giving a distinctive appearance to tissues in which they are abundant, as described in Chapter 2.

The triplets of fatty acids esterified to glycerol are not necessarily of the same type: the middle one is nearly always different from the other two, and often the molecule contains three different kinds of fatty acid. The triacylglycerols in edible oils and in lard and other animal fats consist almost entirely of long-chain fatty acids, but many of those in milk, especially that of cows, ewes and nanny goats, contain some medium-chain, and a few short-chain, fatty acids, from where they find their way into cheese, cream and butter. The origins and consequences of this situation are discussed in Chapter 4.

The other major kinds of glycerol-based lipids are phospholipids, which are basically similar to triacylglycerols in structure, but a phosphate group plus various other atoms (often including nitrogen) replaces one of the sites on the glycerol that in triacylglycerols combines with a fatty acid. This component has much greater affinity for water and water-based molecules than the fatty acids, so while they seek each other's company, the phosphate groups reach into the watery regions of cells. This combination of contrasting properties is the main reason why phospholipids are the major components of almost all biological membranes.

Membranes enclose whole cells, and define compartments within cells, keeping proteins, genes, carbohydrates and other large biological molecules together or apart, as required. They also maintain the appropriate concentrations of ions (including hydrogen, sodium and potassium ions), and other small molecules that create the chemical environment required for the larger molecules to function correctly. Although some cells sometimes use the fatty acid components of a few of their phospholipids for the synthesis of certain lipid-based messenger molecules, their primary role is to hold the cell together, so membrane lipids are known as 'structural lipids', in contrast to triacylglycerol 'storage' lipids.

Different enzymes make and break the ester bonds of the two kinds of fatty acid-containing lipids, so within the same animal, or even within the same cell, the mix of fatty acids in the phospholipids is not the same as that in the triacylglycerols. An immense range of different combinations of fatty acids can occur in phospholipids and triacylglycerols, and may change with time, forming membranes and storage lipids with slightly different physical properties such as melting temperature.

Waxes are another important class of biological lipids. They consist of a single long-chain fatty acid, which may have up to 34 carbon atoms, many more than are ever found in the fatty acid components of phospholipids or triacylglycerols, linked by an ester bond to an alcohol, also long-chain, with from 22 to 34 carbon atoms. The components are synthesised inside cells, as are other fatty acids and alcohols, but complete waxes are almost always found outside cells, where their properties of insolubility in water, chemical stability and resistance to decay are put to good use.

Waxes are most widespread as major constituents of the non-living coatings[8] on the outer surfaces of plants. The waxy layer forms the firm, shiny surface of many leaves, flowers and fruits that limits evaporation of water and sometimes, as in the shiny bluish needles of some conifers, protects the leaves from damage by reflecting the harmful components of strong sunlight. As the first point of contact between the plant and herbivorous insects and potentially dangerous fungi, its 'smell' and its resistance to mechanical or chemical attack are integral parts of the plant's defences against disease and destruction. Micro-organisms generally find such large hydrocarbons difficult to break down: butter, lard and suet eventually 'go mouldy', i.e. a fungus manages to establish itself and grow on them, but bees' wax 'keeps' for years, almost as well as hair, shell, tooth or dried bone.

Many animals also synthesise waxes, and as in plants, most are used externally. The skin of reptiles, birds and mammals, including ourselves, produces a mixture of lipids, including waxes (it can become concentrated as ear wax). Many insects, notably cockroaches, feel slightly greasy to the touch because they have a thin outer covering that contains waxes. Honey bees (and certain other insects) have glands that produce large quantities of wax, which they build into their honeycombs and the brood chambers that house the larvae. Some invertebrates, fish and a few mammals, notably sperm whales, synthesise large quantities of liquid waxes which are retained in internal tissues for various specialised roles.

Some properties of lipids

When pure, each triacylglycerol has sharply defined melting and boiling points, but lipids mix very well with each other, and most familiar fatty substances, including lard, butter and salad oil, are mixtures of dozens, sometimes scores, of distinct chemicals (though often one or two major types predominate). These mixtures of molecules do not readily pack neatly together to form crystalline solids as they cool so they freeze over a range of temperatures which differ according to their exact composition. Tallow at room temperature (about +20 °C), butter and lard kept in the fridge (at about +4 °C) or cooking oil stored in an ordinary deep freezer (at about −15 °C) form firm, crystalline solids that crack like ice when struck hard; when heated to about 50 °C, bees' wax and most culinary lipids are clear, free-flowing liquids. Short-chain fatty acids evaporate so efficiently that they are sometimes called 'volatile fatty acids', and people find the smell of many of the common ones unpleasant.

At the intermediate temperatures at which we normally encounter them, such mixtures of lipids are 'greasy', not quite solid, but not really liquid either. Most living processes take place between 0 and 50 °C, the range of temperatures at which most natural lipids are greasy or liquid. Many seeds, spores and even some adult organisms can survive many months at much lower temperatures but under such conditions, they do not actively feed, grow or breed.

By definition, lipids do not dissolve in water, but tiny droplets of liquid triacylglycerols can be dispersed in water to form emulsions, as in milk.

Untreated full-cream milk separates on standing at or below room temperature, to form 'cream', which is semi-solid if kept at about 5 °C in the fridge, and whey which remains liquid at this temperature. A few seconds of vigorous shaking re-emulsifies the cream into the whey, and it remains in this state for several hours, especially if the milk is fresh.

Nearly all lipids are less dense as liquids than water, so when not emulsified, they float. When soups and gravy separate, the lipids float on top of the watery phase. Small quantities of lipid, such as a few drops of oil spilt into a puddle, may form a layer thin enough to reflect light differentially, forming patterns of colours. However, while ice is less dense than cold water,[9] so it floats on top of it, frozen lipids behave as normal solids: their density increases as they cool and solidify. So fats always sink in chemically similar oils, though they usually do not become dense enough to sink in water.

The subjective experience of taste of foods that contain lipids depends upon their physical state as well as upon their chemical composition. Exactly how people perceive these properties is still not clear, but it has important implications for how we like our food to be prepared and stored. The taste of skimmed or semi-skimmed milk is less severely impaired by freezing (e.g. in a home deep-freeze) and thawing than that of full-cream milk, partly because there is less lipid in the former, and because it is thoroughly homogenised, i.e. emulsified with the whey.[10] Concentrated forms of milk lipids, such as heavy cream or cheese, should never be frozen: because each component has a different melting temperature, freezing and thawing disrupt the intimate association between the proteins, water and lipids, thereby altering the texture and the flavour. Some cheese connoisseurs even object to Camembert and other full-fat soft cheeses being chilled in a refrigerator.

Lipid roles

The most familiar biological lipids are the triacylglycerols that serve as the principal energy stores of many plants and all terrestrial vertebrates including mammals and birds. In this role, lipids are very efficient, much more so than the water-soluble energy store, glycogen, because large quantities can be packed into a small volume. Since much of the rest of this book concerns storage lipids, we will leave them for now and turn to other functions of lipids.

Skin lipids

Most other biological roles for lipids relate to their most distinctive property, the way they resist mixing with water. Skin, scales, feathers, hair and fur are normally coated with a thin but essential layer of lipids that act as anti-wetting and anti-bacterial agents. Many lipids on the skin of terrestrial vertebrates are acylglycerols, and they often include kinds of fatty acids that are never found in storage or membrane lipids: some have branched chains of carbon atoms, or additional atoms or groups of atoms.

The skin lipids of aquatic birds are particularly effective at repelling water, making it roll off as droplets 'like water off a duck's back', but all reduce the wettability of the body's outer covering. Skin, feathers and fur that are not easily wetted lose less heat in rain or spray, and may also be less susceptible to invasion from foreign bacteria and other tiny organisms that could penetrate the skin and cause disease. The lipids are secreted by special glands embedded in the skin and spread over the pelt or plumage by grooming or preening. Many birds spend a large proportion of their time preening, and if prevented from doing so, by an injury or impediment, their resistance to cold and wet is quickly impaired.

The durability of lipids in a watery world is also an advantage for the many mammals that identify each other by smell. Deer, foxes, large cats and many others have special skin glands from which they rub complex mixtures of odours derived from skin lipids onto trees and rocks. Such chemical messages can be 'read' weeks after they are laid down. Human perspiration consists mainly of water that dissipates body heat as it evaporates, leaving behind small amounts of salts and lipids including, in adults, derivatives of sex hormones. Spontaneous reactions with oxygen in the air, and the action of the many bacteria that live on the skin, convert these secretions into substances that feel and smell unpleasant to many people, so they are removed by frequent bathing and hairwashing.

Soap consists of molecules that form a bridge between lipid-soluble and water-soluble molecules: one end dissolves readily in water, and the other mixes well with lipids. By making lipid-associated particles soluble in water, soap is very effective in removing sweat, food stains and other biological 'dirt' from clothing. Washing the skin or hair with toilet soap removes only the outer layer of lipid, but prolonged contact with powerful laundry detergents makes skin 'dry' and prone to cracking. 'Drying' is a slightly misleading term because water is not the most important material that has been 'washed' away. Detergents remove the thin layer of lipids that holds the outer layer of dead cells in place and protects the living, sensitive tissues underneath. The application of additional lipids, such as Vaseline or hand-cream, immediately rectifies the situation and prevents further drying. Unless badly damaged, the skin continues to secrete its own lipids which build up to normal levels after a few days.

Fats for floating

The fact that biological lipids are less dense than water is of little consequence to land animals such as ourselves, but it is crucially important to aquatic animals and plants, because the type and quantity of lipid in relation to the watery tissues, and mineralised materials such as bone or shell, determines whether they sink or float. Many different kinds of invertebrates and fish use a wide range of lipids, including triacylglycerols and waxes, and various substances derived from lipids for buoyancy, but those of fish have been most thoroughly studied.

A gas-filled swimbladder is the main buoyancy organ of many bony fish, but sharks and rays never have such structures. So lipids often make a major contribution to their buoyancy and that of a few kinds of bony fish that have lost the swimbladder. Lipids that are involved mainly or exclusively in buoyancy accumulate in the liver, probably because of its central location inside the abdomen near the body's centre of gravity, rather than for any metabolic reason. In cod and related fish such as halibut, the lipids are triacylglycerols. Cod-liver oil has a density[11] of about 0.93, so its presence must make the fish only slightly lighter than water, and it may also serve as an energy store.

The liver of the deep sea shark *Centroscymnus* is huge, occupying 30% of the volume of its body,[12] and 80% of its mass is a unique lipid called squalene which has an exceptionally low density (0.86) for an organic liquid. Squalene, whose name derives from *Squalus*, a shark, is a hydrocarbon (a chain of carbon atoms each with two hydrogen atoms attached) similar to the backbone of fatty acids but without the acid group at one end. As far as we know, *Centroscymnus* cannot break down squalene (or any other hydrocarbon) to generate metabolic energy, so providing buoyancy must be the main function of the huge quantities of this material that the fish synthesises. This fish, and a few other kinds of slow-swimming sharks, notably the basking sharks, are neutrally buoyant, i.e. they can float motionless in water. As well as accumulating lighter-than-water lipids, they also reduce the heavier-than-water skeleton, which is poorly calcified in the vertebrae, the gill arches and fins, but not, of course, the jaws. They need the strength and hardness provided by calcium minerals to hold the huge mouth open, or, in the case of predatory sharks, impose a powerful bite.

Lipids in membranes

Although usually not given as much prominence as genes or proteins, membranes are equally fundamental to life. Phospholipids are the major structural components of the membranes of all kinds of cells, from bacteria (except a few very primitive and aberrant types) to mammalian tissues. The key to their biological role is the ability of the phosphate group to mix readily with water and water-loving molecules such as proteins, while the fatty acids prefer the company of lipids, particularly each other. Substances with an affinity for lipids, such as certain hormones and non-esterified fatty acids, can diffuse through the phospholipid membrane, as can water, oxygen and carbon dioxide. Oxygen from the air must pass through the membranes of the cells lining the lungs or gills before reaching the red blood cells that transport it deep into the tissues.

The majority of biological materials, including proteins and sugars, are water-soluble so their passage through cell membranes must be facilitated, and thus controlled, by an assortment of protein-based channels, receptors and transporter systems embedded in or on the membrane. The relative abundance of different types of such structures may confer unique properties on particular tissues or groups of cells within a single tissue. Some channels use chemical energy to 'pump' particular kinds of small molecules into or out of the cell, and the properties of others are transiently altered by the electrical voltage across them. Such 'policing' of molecular traffic into and out of cells is essential to the integrity of the cell, and to the role they play in the whole organism. Cells can swell and burst if too many small, water-soluble molecules and the water that accompanies them, are allowed to flow through the membrane and into the cytoplasm, the interior of the cell.

The cells of multicellular organisms co-ordinate their activities by a wide variety of chemical and electrical messages passed between them. Scores of messenger molecules have been characterised, and are known collectively as hormones, neurotransmitters, cytokines and other names depending upon their chemical composition, and between what kinds of cells they operate. They all have one thing in common: their action requires specific receptor molecules that are usually located on or in the membrane of the target cells.

As more and more different messenger molecules are discovered, and many are found to act on several different kinds of cells, it is becoming clear that the membranes of many complex cells harbour a wide range of receptor molecules that bind specifically to particular messenger molecules. The binding of messenger molecule to receptor molecule is signalled to the cell's genes and other parts of its interior by a relay of secondary messenger molecules that start at the receiving cell's outer membrane. The composition of the fatty acids in the membrane phospholipids influences their fluidity, where in or on the membrane the proteins are attached, and possibly how 'sticky' it is in attaching to other cells. The outer membrane is constantly turning over, with portions being reabsorbed into the cell, to be replaced by newly assembled membrane, thereby keeping the structure in good working order, and enabling frequent adjustments to the array of receptors, carrier molecules and channels that it supports.

These roles alone make membranes vital to all living cells. Damage to the cell membrane is often even more rapidly lethal than poisoning protein synthesis or altering genes. One of the earliest antibiotic drugs to be developed, gramicidin, kills bacteria by chemically punching holes in their membranes: the cell contents leak out, substances normally excluded or pumped out by the membranes seep in, and the bacteria quickly become so disordered that they die. Even if the membrane itself remains intact, cells can starve to death within minutes if essential nutrients do not get into them because their molecular escorts through the membrane barrier are poisoned, or specific channels are blocked. The main snag with such drugs is that bacterial and mammalian cell membranes are so similar that both are injured almost indiscriminately, so side-effects can be serious. The more modern anti-fungus drug, polymixin, works like a soap, disrupting phospholipids in the cell membranes. Fungi rarely get further than the skin, which can withstand the damage inflicted by the drug very much better than internal organs can, because the outer layers of cells are constantly replaced.

As well as the outer cell membrane, plant and animal cells and more complex unicellular organisms have several distinct kinds of internal membranes that segregate compartments, assemble enzymes in functional order and form surfaces on which biochemical reactions can take place. The layers and chambers formed by internal membranes of chloroplasts and mitochondria are clearly visible in electron micrographs such as that in Figure 3. They are as important to cell function as the external membrane, and even minor changes in their structure or arrangement can severely disrupt essential biochemical processes. In many cases, key enzymes are attached to

Figure 3. Internal membranes in mitochondria in an animal cell. This electron micrograph is magnified 9000 times, so the field of view is about 4 μm (0.004 mm) across. (Courtesy of Heather Davies.)

internal cell membranes, thereby enabling series of reactions, such as those involved in respiration, to take place in the right order. In brief, genes contain the information to build the cell, proteins catalyse the necessary chemical reactions, but phospholipids act as the marshals, holding the biochemical machinery together and helping to maintain the right chemical environment.

The maintenance of structural order in lipid membranes is critically dependent upon temperature. In many animal and plant cells, death from cold is due mainly to irreversible disorder of the lipid membranes. Almost all structural proteins and the majority of enzymes are perfectly preserved at temperatures far below zero. Biochemists routinely freeze tissues to as low as −80 °C for weeks before thawing them for further study. The blood of humans and other mammals also survives freezing, at least as far as its role in carrying oxygen is concerned, but, except in the cases of some species that are specially adapted to survive extreme cold, most cells and whole organisms are killed by freezing.[13] Their membranes have lost their selective permeability to different kinds of small molecules, and no longer maintain sequences of enzymes in the correct alignment in which they can work together to form a metabolic pathway. Being too hot, which in mammals and birds can mean just a few degrees above normal body temperature, is also lethal, because it disrupts the ordered structure of both proteins and lipids.

In 1816, Chevreul, motivated by little more than simple curiosity, extracted a lipid from gall stones (taken from human or animal cadavers) that would crystallise readily, but could not be broken into glycerol and fatty acids. He named this atypical lipid 'cholestérine', from the Greek words for bile, χολη, often used metaphorically to mean angry or embittered, and στερεος, which means stiff, firm or solid and, metaphorically, stubborn. Later research showed that cholesterol and other sterols consist of several contiguous rings of carbon atoms and thus have a fundamentally different structure from triacylglycerols (see Figures 1 and 2, pages 10 and 12).

Chevreul's choice of name, now altered to cholesterol, is apt in one way: cholesterol is more stable than most other cell components and endures for a long time in living tissues and after death. However, the name is misleading in other ways: cholesterol is most concentrated in bile, but that is by no means its only biological role. Sterols are essential components of the membranes of all kinds of nucleated cells, though their exact function there is still not very clear. Sterols probably affect membrane fluidity, which in turn determines what and how much molecular traffic enters or leaves the cell.

In most animals including all vertebrates, the membrane sterol is cholesterol, while that of insect cells is slightly different. The cell membranes of green plants contain yet another sterol. These differences between plants and groups of animals are probably very ancient: certain lineages of organisms 'got started' on particular sterols and the structure and roles of their membranes are adapted to that composition. For reasons that are still not fully explained, the outer membranes of almost all kinds of bacteria manage without any sterols.

Research into the origin and early evolution of life indicates that lipid membranes were essential to cell integrity and basic biochemical processes right from the beginning. Membranes were almost certainly the earliest roles of lipids in the evolution of primitive cells, billions of years ago. We have little direct information about the structure or habits of the first kinds of cells, since very few ancient rocks in which their fossils might be found have survived and are accessible to us now, but laboratory experiments can suggest how lipids assumed their fundamental role in primitive cells.

Pure phospholipids isolated from cells (or even synthesised artificially) assemble themselves into the double layer observed in natural cells when they are spread over water. The molecules line up side by side in two layers with their fatty acids meeting end to end in the middle, forming two sheets, one upside down relative to the other. The tendency of phospholipids to remain associated in the typical bilayer array is so strong that fragments of membrane remain after excess heat or chemical attack has rendered them useless to the cells that made them. The great, grey, green and greasy appearance of the Limpopo River (and other warm, slow-flowing waters) is due to the membranes and membrane fragments of millions of tiny algae, living and dead. Lakes and reservoirs often acquire a similar appearance after hot, sunny spells lead to 'blooms' of algae, which die a few weeks later.

Membranous tissues

The properties of membranes become so important to the function of some cells that their surface is enormously expanded. Many kinds of immune cells, including those that form pus after they have completed their job of killing invading bacteria, have an elaborately frilled outer membrane. The greatly increased surface area supports more receptors that enable these cells to detect and track their minute prey, and is also essential to the killing process: the cells engulf bacteria, or any other debris, dead or alive, in a pocket of cell membrane which then moves into the interior of the cell, where digestive enzymes break it down. A severely infected wound can produce literally kilograms of pus over several weeks: supplying the fatty acids necessary for the synthesis of enough phospholipids for all these cells that eventually die and are expelled as pus can impose a significant drain on the body's lipid reserves.

The outer surface of the cells that line the small intestine, the part of the gut that absorbs nutrients from the digested food into the body, is also deeply folded. The area of membrane that actually makes contact with the food is thus enormously extended, greatly speeding up the rate of absorption. Like some immune cells, these intestinal cells have short but busy lives. Even in mammals as large as ourselves, they are functional for less than a day before being shed and replaced. Many of their constituents are digested and reabsorbed, but some are lost in the faeces, representing one of the few ways, other than being oxidised to produce energy (see Chapter 4), or expelled as pus or as ingredients of eggs or sperm, that the body can lose lipids.

By far the most extensive and chemically sophisticated membranes are those of the nervous system, whose functions include transmitting signals in the form of electric currents. Metals, graphite and most watery solutions including those inside and outside of cells, conduct electric currents, but paper, rubber, polythene and oils do so only weakly or hardly at all, so they can be used as insulators. Phospholipid membranes also pose a barrier to the movement of electrically charged ions, which pass, if at all, though channels specialised for the purpose. Controlling the magnitude, timing and location of the flow of currents carried by different kinds of ions, notably sodium, potassium and calcium, is a major function of biological membranes, especially, but by no means exclusively, those in the nervous system.

Peripheral nerves that innervate muscles and send back information from sense organs are usually greatly elongated, branching cells, so a high proportion of their mass is membrane, just as a fair proportion of a twig is bark. Brains consist of a huge number of tiny cells, most of which are branched, though not necessarily elongated. So the whole nervous system, and major sense organs such as the eye and the inner ear, contain far more membrane than tissues such as muscle or bone, that consist of fewer, much larger cells.

In gross composition, the mammalian brain is around 60% lipid, almost all of it various kinds of phospholipids and cholesterol. In lean animals, the nervous system may be by far the largest concentration of lipids in the body, and so from a predator's point of view, one of the most nutritious parts. Fatty tissues are often regarded as the simplest, dullest, most expendable tissue, and the brain as a fabulously complex, glamorous and inviolate organ, but chemically they have a great deal in common. Indeed, one of the reasons for studying how the body handles lipids is its implications for the growth and functioning of the brain and the eye.

The fatty nervous system has little affinity for the water-soluble stains and dyes that attach readily to proteins and thereby make the internal structure of tissues like muscle, liver, gut and skin visible under the light microscope. The brain and the eye were among the last of the major tissues to yield to microscopical examination, after the Italian biologist Camillo Golgi developed (in the 1880s) a lengthy and never very reliable staining process involving potassium bichromate and silver nitrate. Although 'silver staining' remained standard for almost a hundred years, its chemical mechanism has never been fully explained, and it did not work well enough to reveal the outline and internal structure of entire cells. Until well into the twentieth century, it was not even certain that all nerve cells were discrete entities, each entirely enclosed in its own membrane.

The composition of cell membranes means that lipid-soluble molecules, even quite large ones, such as fatty acids, usually need less assistance in getting into cells than water-soluble molecules such as glucose. This property is one reason why alcohol so readily traverses the stomach wall and acts on nerves and muscles (whose membranes are rich in lipids) within minutes of being ingested. Anaesthetics such as chloroform, ether and halothane work fast and effectively because they are lipid-soluble, passing quickly through the membranes of the cells lining the lungs, and thence through further membranous barriers and into the brain, where they disrupt nerve function enough to induce unconsciousness. The liver immediately sets about extracting the foreign substances from the blood and inactivating them, eventually removing enough for the brain to recover, and the patient wakes up again. The solvents in glue and varnish alter mood by similar processes, and, if inhaled repeatedly, cause liver damage.

Potentially toxic foreign substances can easily follow the same route. One class of lipid-soluble substances that is causing concern are organophosphate insecticides that work by penetrating the fatty outer layer of insects' cuticle and disrupting their nervous systems. Unfortunately, they are chemically stable, so they persist for years in and on plants and animals and in soil and ground water, where other animals take them in. They accumulate in fatty tissues, including the nervous system and liver of vertebrates such as fish, birds and mammals, where they can cause mental disorders. These and perhaps other foreign substances have been implicated in the development of bovine spongiform encephalopathy, or mad cow disease, which might cause neurological disorders in people who eat contaminated beef. The main cause of the disease is a prion, which is a protein, but disturbances of lipid membranes may assist, or hinder, an infection becoming established and producing serious symptoms.

Final words

Biochemistry has long emphasised specificity: only one enzyme or receptor or antibody will do, and just one wrongly placed amino acid out of thousands is enough to spoil the whole molecule. Uniquely arranged, highly specific molecules such as proteins and those that form genes tend to be regarded as 'more important' than lipids, whose strengths are their ability to mix with each other, in the case of triacylglycerols to the exclusion of almost everything else. Phospholipids have 'a foot in both camps', enabling them to marshal themselves and proteins into arrays, and to control the passage of other molecules. There is no single 'right' composition for either triacylglycerols or phospholipids, even within a single kind of cell: their fatty acid composition and with them their biological properties, change from time to time, according to temperature and other factors. As will be described in the following chapters, they are often built from whatever happens to be available to the organism. The impression that emerges from the study of the biology of lipids is of plasticity, variability and adaptability.

2

Introduction to fatty tissues

As well as referring to a type of biological molecule, the term 'fat' is also applied to a living tissue, known to biologists as adipose tissue, or, to be specific, white adipose tissue (although it is never pure white, often cream-coloured, deep yellow or pink in colour) to distinguish it from 'brown' adipose tissue, discussed later. The name comes from the Latin word *adipatus* which means greasy or, in a culinary context, fried.

With the notable exception of nerve cells, most animal cells store some energy-providing materials for their own use when demand for energy is high or supplies of glucose and other fuels run low. The most widespread storage material is glycogen, a large, insoluble molecule formed by linking ('polymerising') thousands of glucose molecules, but tissues such as muscle and liver often contain small droplets of lipid. In some fish, especially those without adipose tissue, the liver and muscles stores may become massive, making these tissues look, feel and taste oily. The swimming muscles of fish such as salmon taste rich and delicious because they contain storage lipids as tiny droplets intimately associated with the muscle proteins, not segregated from them in separate cells, as in proper adipose tissue. Salmon taste best when caught on their way upstream to breed; the flesh of so called 'spent' salmon is much less appetising because most of the lipids (and some of the proteins) have been withdrawn to provide fuel for swimming and for the production of eggs and sperm.

Proper adipose tissue is unique to vertebrates. It occurs sporadically among fish, often in a greatly modified form, but is most extensive in mammals, birds, reptiles and amphibians. Its distinctive feature is a unique type of cell called an adipocyte, which can accommodate much larger quantities of triacylglycerols than any other kind of cell. Almost all the adipocytes' stores of lipid are not for their own use, but for export to other tissues as required.

Fat cells

In view of the importance of adipose tissue in human sexual relations (see Chapter 7), it is perhaps appropriate that its detailed study should be the product of one of the first husband and wife collaborations in science.[1] During the 1870s, Frances Elizabeth Hoggan, MD and her husband George studied the formation, internal structure and size changes of adipocytes using the most advanced staining and microscopical methods then available. The results[2] of their studies on human, mouse and guinea-pig tissues were reported to a meeting of The Royal Microscopical Society at King's College, London in March 1879. George read the paper, though the chairman noted that his wife had also contributed to the research, at the time a thoroughly modern state of affairs, since the Act of Parliament admitting women to medical organisations had been passed only 3 years before.

As animal cells go, replete adipocytes are huge, up to 4 nl in volume[3] in well-fed whales, though nearer 0.1–1 nl in rats and people, thousands of times larger than red blood cells, most brain cells and the cells of the immune system that protect the body from disease. In Figure 4, only the top layer of adipocytes is in the plane of focus and, because the tissue is not stained, the nucleus, cytoplasm and lipid droplet cannot be distinguished. This picture reveals the most characteristic features of adipocytes: unless severely depleted of lipid, they are spherical, or very nearly so, which is an unusual shape for functionally mature animal cells (other than eggs).

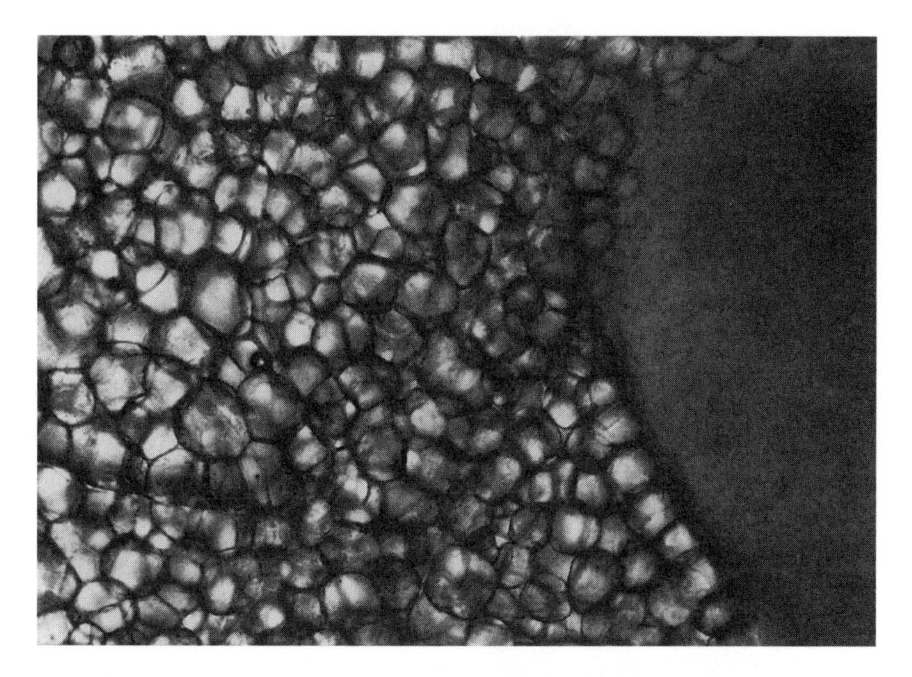

Figure 4. A thick, unstained section of part of the popliteal adipose depot of a young, lean rat, photographed with light shining through it from below. The elongated shadows are fine blood vessels that permeate the whole tissue, but are tiny compared with the adipocytes. The structure on the right is part of the popliteal lymph node that is embedded in this depot. It contains large numbers of cells of several different kinds, all too small to be visible at this magnification. The field of view is about 2 mm across (just under a tenth of an inch), so these adipocytes average around 70 μm (0.07 mm) in diameter, or about 0.2 nl (2×10^{-10} litres) in volume. (Courtesy of Hilary MacQueen.)

Adipocytes are packed closely in a tight mesh, so together they occupy most of the volume of the adipose tissue. This tissue comes from a young adult rat that was eating as much as it wanted but was not abnormally obese. Adipocytes of fat specimens would be much bigger, and those of animals subjected to prolonged starvation would be small and misshapen.

Molecular traffic into and out of adipocytes is regulated by a variety of receptors, carrier molecules, enzymes and other substances, most of which are produced by the adipocytes themselves. The genetic instructions for synthesising such proteins are located in the nucleus, which is essentially similar to that of other animal cells, but is relatively small, and is displaced from its usual position near the centre, as shown in Figure 5. The lipid droplet has no internal structure (the tiny blobs and shadows are staining artefacts) – it is almost pure, liquid triacylglycerol. The thin rind of cytoplasm between the outer membrane and the lipid droplet contains the biochemical apparatus for building and storing receptors, messenger molecules and enzymes plus a few mitochondria where the fuel that powers their synthesis is produced. Although occupying only a small fraction of the whole cell, especially in replete adipocytes, the nucleus, cytoplasm and outer cell membrane are very important in determining what substances are taken into and released from the cells.

In people and sedentary domestic animals, the proportion of lipid in whole adipose tissue is rarely less than 40% and it increases with fattening, reaching as much as 85% of the total mass of the tissue in middle-aged men who were 44% by weight adipose tissue.[4] The rest of the tissue is mostly protein and water. In people, changes in adipocyte volume seem to be due almost entirely to changes in the size of the lipid droplet. The same does not seem to be true of naturally obese wild animals: the proportion of lipid in adipose tissue is almost constant (at 42–66% depending upon the depot) in arctic foxes living wild in the high Arctic.[5] Their body composition ranges from 3–33% by weight visible adipose tissue, and the adipocytes enlarge and diminish in proportion to fatness, so we can only conclude that both the aqueous and the lipid components are changing as the adipose tissue expands and shrinks. Dwarf hamsters (*Phodopus*) naturally become very fat indeed during the summer, but chemical analysis shows that their adipose tissue never exceeds 46% lipid, even when it amounts to 35% of the body mass and they are eating as much as they like.

The lipid droplets are the most conspicuous and distinctive feature of mature adipocytes, and their role in lipid metabolism is by far their most thoroughly studied property, but it would be a mistake to conclude that

Figure 5. (opposite) Electron micrograph of fixed, stained thin section of white adipose tissue from a Djungarian hamster (*Phodopus campbelli*) magnified 9000 times. The adipocytes are about the same size as those in Figure 4, but the field of view is only 10 μm (0.01 mm) across, so only small parts of four adjacent adipocytes are visible. Each adipocyte has a thin rind of cytoplasm containing a few mitochondria, that appear as darker round bodies, and a huge lipid droplet. The nucleus of the upper right adipocyte is near the centre of the picture, surrounded by its own membrane. Collagen is visible as tiny black spots and streaks on the outside of the cells and is particularly clear towards the bottom of the picture. (Courtesy of Heather Davies.)

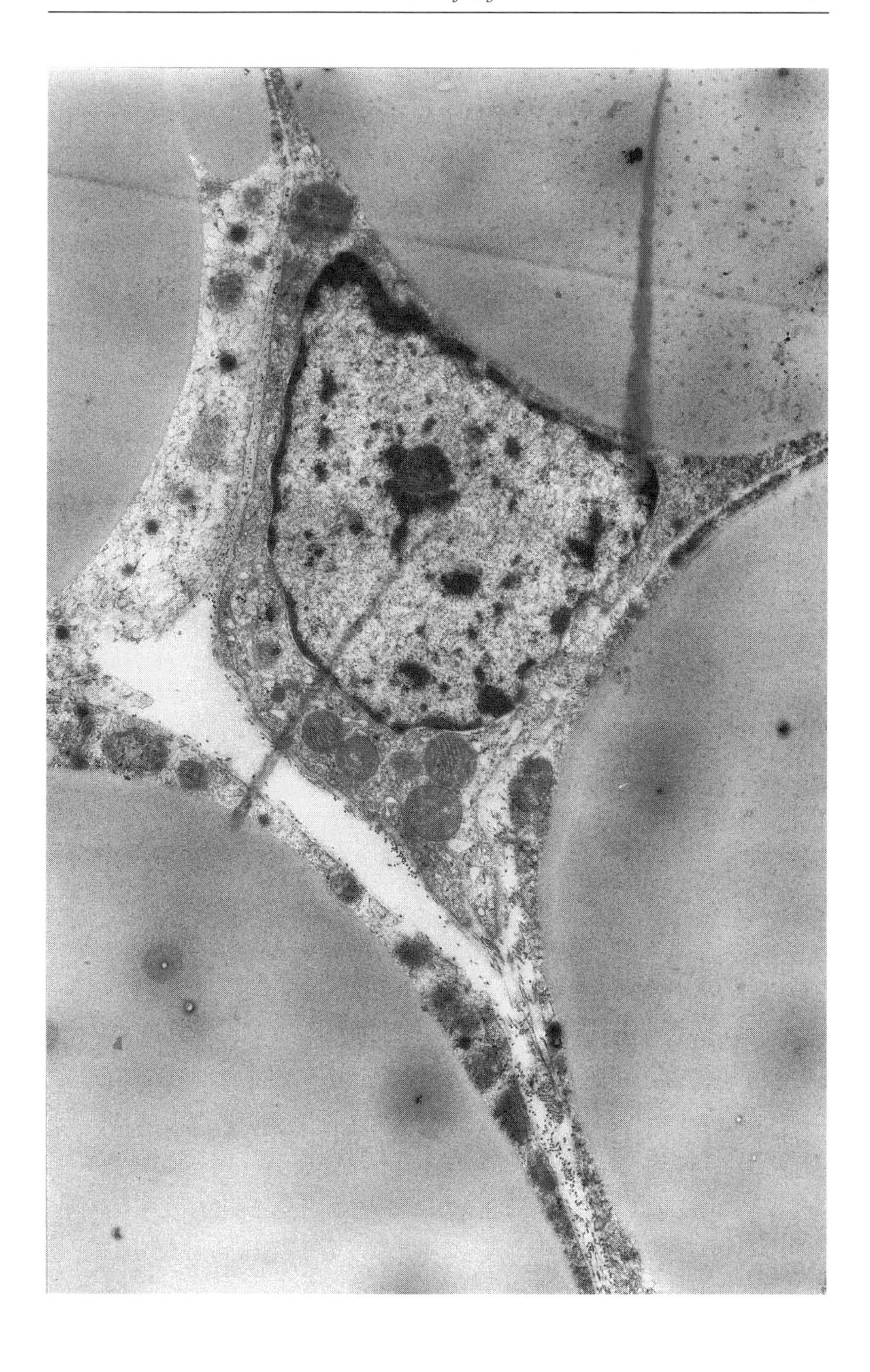

adipocytes are only concerned with storage and metabolism of lipids. Those of mammals (and possibly of other vertebrates) have the enzymes to take up and release glutamine, an amino acid that is only a minor component of proteins, and seems to be more important as a fuel, and as a vehicle for transferring amino (—NH_2) groups of atoms between different kinds of cells. Glutamine also seems to be stored in adipocytes, presumably in the aqueous phase of the cytoplasm.

Most kinds of cells can undergo reversible changes in volume, usually by taking up or releasing water, salts, or, especially in the cases of the liver and skeletal muscles, glucose from the blood. It is relatively simple to synthesise more outer membrane as the cell cytoplasm expands, and to reabsorb the excess as it shrinks. But volume changes of more than 50–100% are usually painful or harmful, often both: we all know how much a swollen boil or sprained ankle hurts![6] Red blood cells burst, and nerves may be irreversibly damaged if they swell too much by being exposed to diluted body fluids, as, for example, in drowning. But adipocytes can change in size by at least tenfold, probably more in severe starvation or pathological obesity – and you don't even notice until you try to zip up your skirt or trousers!

Pure triacylglycerols are colourless when melted and white when solid, so although adipose tissue from young pigs and sheep, as seen in butchers' meat is usually almost white, when alive in a warm body its colour is much duller. Small quantities of lipid-soluble pigments (mostly derived from the plants in the animals' food) may accumulate in the tissue, conferring on adipose tissue a colour ranging from pale cream to deep yellow. These substances remain associated with the triacylglycerols when they are released from the adipose tissue and reassembled in milk. So adipose tissue of pasture-fed cattle, and that of people who regularly eat the meat, milk, butter and cheese obtained from them, is often deep yellow. The colour seems to make no difference to the tissue's physiology.

By themselves, adipocytes are soft and fragile, and perfectly spherical, but intact adipose tissue has a rubbery consistency at normal temperature because the adipocytes, blood vessels and nerves are held in place by an encasing mesh of fine fibres, producing the tight packing shown in Figure 4. The fibres are composed of collagen, an ubiquitous structural protein in vertebrates that is produced by special connective tissue cells. Like other proteins, collagen has a fair amount of water associated with it, so the proportion of lipid in adipose tissue as a whole is always less than that of isolated adipocytes. Collagen shrinks on heating, bursting the adipocytes it encloses. So when bacon and other fatty meats are grilled or roasted, most of the triacylglycerols drip out as a liquid, leaving behind a crisp, net-like mass of collagen.

The abundance and arrangement of this material give the tissue many of its mechanical properties. The adipose tissue on the outer wall of the human belly contains little collagen and is soft and rubbery when warm, but the collagen-rich blubber of whales and seals is firm, tough and fibrous. Depletion of lipid, engorgement with blood, or accumulation of excess collagen all modify the mechanical properties of adipose tissue and hence its 'feel'. Adipose tissue of small children and young women is not only usually more abundant just under the skin, it feels softer and more pliable than that of men and elderly people.

Since the 1970s, much effort has been devoted to studying how adipocytes develop before and shortly after birth, with a view to finding ways of curtailing their formation or maturation, as a means of preventing obesity later in life in both people and farm animals. Mature, lipid-filled adipocytes are much too big and clumsy to be able to divide. Proliferation occurs in pre-adipocytes, which are tiny compared with what they could grow into. They have no large lipid droplet and their appearance under the microscope is indistinguishable from that of other kinds of immature connective tissue cells. Adipose tissue from young animals often includes numerous small, inconspicuous pre-adipocytes interspersed among the large, replete adipocytes.

It is much more convenient to study detailed biochemical processes in tissue culture than it is in the whole animal. Rapidly dividing cells such as pre-adipocytes taken from embryos or foetuses are usually easier to maintain in tissue culture than mature, non-dividing cells from adult animals. Exposure to even a very low concentration of fatty acids prompts cultured pre-adipocytes to mature into adipocytes: they stop dividing and start to synthesise and activate enzymes and other biochemical machinery that enable them to take up lipids and form a central droplet, thereby becoming larger but less dense.[7]

More lipid in the tissue culture medium prompts more pre-adipocytes to mature into adipocytes, and they enlarge faster, but there is still controversy over whether any connective tissue cell could become an adipocyte under appropriate conditions, or only certain cells are destined to do so. Pre-adipocytes may divide and lie low, masquerading as connective tissue cells interspersed among the huge mature adipocytes. They could remain in this state for years, maturing into functional adipocytes when prompted by a change of diet, exercise habits or hormone status, such as menopause.

Command and control

The main role of adipose tissue is provisioning other tissues, but as we shall see, it provides more than just lipids or lipid derivatives, and exercises fine control over when and how much of such materials are provided to which other tissues. Indeed, the need for accurate and rapidly adjustable control over uptake and release of storage materials may be one of the main reasons why, in terrestrial vertebrates, triacylglycerols are in specialised adipocytes, rather than just 'loose' in muscle or liver. The importance of such a role, especially in extreme situations like severe illness or starvation, should not be underestimated. The success of a military campaign depends as much upon an army's quartermasters as upon its generals, though traditionally accolades go only to the latter. So it is with adipose tissue: by eking out supplies and establishing priorities between competing tissues, adipose tissue enables orders from the brain to be carried out as efficiently as possible. And the brain gets all the credit.

Communication with other tissues is thus essential to adipose tissue's role in the body. Like other living tissues, adipose tissue is supplied by blood vessels, which can be seen as tiny threads or specks of red in quiescent, uninjured adipose tissue (see Figure 4, page 29). Many muscles in birds and mammals (and a few fish, notably tuna) contain special molecules that store oxygen, and contribute to its bright red colour: adipose tissue never has such equipment, and, especially when seen beside red meat (muscle) or kidney, appears pale and almost bloodless.

It is important to remember that this impression is based upon the appearance of the meat we buy in shops that comes from healthy animals that were never required to run far or fast, and were killed humanely, with the minimum of stress or excitement. In such animals, many of the blood vessels are constricted, and blood flow through the adipose tissue is quite low, though by no means negligible. If the vessels are dilated, which greatly increases perfusion, or if they are broken by bruises or cuts, the tissue appears red. Adipose tissue in animals that have been running fast, for example a deer or a badger killed running across a road, or suffering from a high fever, may look quite different.

Despite its appearance, adipose tissue contains numerous fine blood vessels, so many that scientists who are developing new ways of culturing materials for surgical transplantation to replace diseased or wounded tissues choose adipose tissue as a convenient source of the special kind of cells that lines small blood vessels.[8] Unlike the brain, liver, kidney or most muscles, few adipose depots are supplied from a single major artery: they are perfused from numerous smaller vessels arising from adjoining tissues. The shared blood supply is one of the main reasons why adipose tissue is often so firmly attached to muscles, skin or lymph nodes. This diffuse arrangement makes it much more difficult for surgeons to transplant living adipose tissue from one site to another.

As the Hoggans first described, adipocytes often form and their lipid droplets enlarge while they are near small blood vessels. The blood brings the materials from which triacylglycerols are built, removes the products of lipolysis (see Figure 2, page 12), and carries several kinds of molecular signals that control adipose tissue metabolism. Hormones, including one of the most widely studied of all, insulin, bind to receptors in the cell membrane of adipocytes (as well as to similar receptors in many other kinds of cells, notably muscle). Such blood-borne signal molecules are easy to study because they can be detected in a small sample of blood, but many others act on the same tissue in which they are produced. Only the residue or overspill enters the blood, usually in very small quantities. Adipocytes are now known to produce several locally acting messengers that probably help to coordinate the activities of adjacent cells.

The fine, delicate nerves of the sympathetic nervous system (which controls many aspects of blood pressure, movement and secretion in the gut, and other activities of which we are usually unaware) release noradrenalin, a small molecule that binds to other kinds of receptors on the adipocytes. One of the most thoroughly studied actions of noradrenalin on adipocytes is to increase lipolysis and the release of the fatty acids and glycerol so formed into the blood. The sympathetic nervous system plays a major role in exercise, exposure to cold, stress and many other circumstances, but its activity is not under direct voluntary control. We cannot 'will' stimulation of our adipocytes, only engage in activities, such as exercise, that entail activation of the sympathetic nervous system and hence stimulate the adipocytes to release fatty acids.

The binding of signal molecules to their receptors in the adipocyte membrane triggers a cascade of intracellular events that may lead to activation or suppression of particular genes. The adipocytes produce their own receptors, thus determining their capacity to respond to circulating hormones and other signal molecules. Receptor production not only changes with time, according to whether the animal is getting fatter or thinner, it also differs between adipocytes in different depots, and in different regions of a single depot, generating distinctive site-specific properties.

Because the receptors that mediate the action of these neural and blood-borne signals are located on the adipocyte membrane, it is often possible to simulate their action in intact animals by applying them to adipocytes that have been isolated from their framework of collagen (and the nerves and blood vessels in it) and are kept alive in an artificial solution. Such techniques are convenient for laboratory studies on how adipocytes respond to signal molecules in various combinations and at various concentrations and, as will be described in Chapter 4, have been used extensively.

But there is a snag: culture methods are usually most economical with around a thousand adipocytes, certainly no more than ten times that number – the reagents and materials are expensive and nobody wants to use more than necessary. So how should a representative sample be chosen from the tens of millions of adipocytes in a rat, billions in a human? Until recently, such questions were ignored, and tissue samples were chosen at random or from sites that offered easy access to plenty of material. Within a few years, the tradition of sampling tissue from certain sites became established, and everyone forgot, or never knew, why those cells were regarded as 'representative' of adipose tissue as a whole.

There were few obvious problems with such assumptions: improvements in microscopy, such as the electron microscope, revealed clear differences between muscles from different parts of the body and between neurons in regions of the brain, but the internal structure of adipocytes from different depots looked much alike. The most obvious difference was in size, which was a fiddle to measure accurately and might not matter anyway for metabolism. However, studies[9] in which samples from different adipose depots of the same animal were compared revealed that, in spite of their similarity in microscopic appearance, adipocytes differ in many metabolic properties, including their capacity to take up and release substances and in what nervous and blood-borne signals they can respond to. So, in assessing the physiological implications of measurements from small fragments of adipose tissue, it is important to consider how far the samples examined were really representative of the adipose mass as a whole.

As well as receiving chemical messages from other tissues, adipose tissue synthesises some of its own and releases them into the blood, thus meeting the most important criterion in the definition of a gland. In the late 1980s, the 'buzzword' was adipsin,[10] a protein produced by certain adipocytes and certain neurons. It turned out to have a curious combination of properties, under some conditions functioning as an enzyme but also having features in common with certain signal proteins that act on, and are usually produced by, cells of the immune system. Although at first hailed as the key to controlling obesity, its role in adipose tissue physiology was never clearly established. Enthusiasm for adipsin burnt itself out by the early 1990s, and biochemists' attention turned to a substance initially called *Ob* protein[11] (after the obese strain of mice in which it was first described), then given the more optimistic and sophisticated name, leptin,[12] derived from the Greek adjective λεπτος, which means peeled (as of fruit), slender or delicate.

Leptin is a medium-sized protein that seems to carry messages from the adipose tissue to the hypothalamus, the region of the brain that controls appetite and feeding behaviour. Like all messenger molecules, leptin's concentration in the blood is always very low, of the order of 1–4 ng (10^{-9} g) per millilitre of blood, or around ten to twenty millionths of a gram distributed throughout the entire blood system of an adult man. Following isotopically 'labelled' leptin molecules suggests that they remain in the blood circulation for about 3 hours. Slim people have around 10^{10} adipocytes, so assuming (probably erroneously) that all adipocytes contribute equally, and they are the only source, each adipocyte produces 10^{-15} g of leptin, the equivalent of about one millionth of its total mass every 3 hours – a negligible drain on its resources.

Biochemical methods of measuring such very dilute proteins are now much more accurate than they were as recently as 10 years ago. Thus it is possible to establish that after 48 hours without food, the leptin concentration in a mouse's blood falls to about a third of the value immediately after a meal. Mice normally eat several times every night, so a fast of this duration makes them quite hungry. When leptin levels are continuously low or absent, animals eat to excess and become obese, but they revert to normal appetite, and eventually to normal body mass, when leptin is injected or its endogenous production restored. As discussed in Chapter 8, there is much excitement about the prospects for curing human obesity safely and permanently by boosting leptin production, or administering synthetic leptin.

Techniques to separate, identify and synthesise proteins and other biological molecules are rapidly becoming cheaper and more efficient, so, as in the examples just described, biochemists are constantly adding to the list of messenger molecules, carrier proteins, enzymes and receptors that certain adipocytes can produce. Although it is sometimes not clear which, if any, of the properties that can be demonstrated under artificial conditions are significant in intact, living animals, the study of such processes keeps many biochemists busy for many years. In future, it may be possible to augment or inhibit some of these intermediaries, thereby getting adipose tissue to do some things it otherwise does only weakly or not at all.

Adipose tissue organisation

Among the terrestrial vertebrates – amphibians, reptiles, birds and mammals – adipocytes are broadly similar in chemical composition, microscopic appearance (although differing in the range of sizes they normally assume), and, from the limited information available, in the majority of their metabolic capacities. The most conspicuous differences between groups of animals are not in the microscopic appearance or biochemical properties of the adipocytes, but in the anatomical arrangement of the adipose tissue on the body.

In most amphibians, snakes and lizards, adipose tissue is located in the most sensible place for a tissue whose main role is to undergo large changes in size: it forms fat bodies, sometimes paired, that are located in the abdomen, very close to the centre of gravity of the body, and only loosely attached to other organs. In this situation, the adipose tissue can expand or shrink with the minimum of pulling or pressing on any adjoining tissues, and changes in its mass do not alter the body's balance during swimming or jumping. The tortoise's shell limits expansion of the abdomen, so as might be expected, their arrangement is a little different: adipose tissue forms in various sites around the limbs and in odd corners under the shell.[13] Some lizards also have adipose tissue in the lower part of the tail, where it surrounds the vertebrae, making the tail thicker, and, of course, heavier. But in reptiles, it never forms continuous layers associated with the skin, muscles, heart or guts.

This simple, apparently practical arrangement is never found in mammals and birds. Their adipose tissue is always fragmented into several distinct depots, scattered around the body. Some is located in patches between the superficial muscles and the skin, so it is often called subcutaneous, a term that implies, erroneously, that it is firmly attached to the skin. Some is associated with certain thoracic and abdominal organs, including the heart, intestines, spleen, kidneys and reproductive organs, and some, usually a small fraction, is embedded within and between skeletal muscles. Roast leg of pork, for example, includes some subcutaneous fat between the skin (which cooks to form 'crackling') and the meat (muscle), and some intermuscular adipose tissue between the muscles.

The categories subcutaneous, intra-abdominal or intermuscular, are useful to butchers and farmers because they distinguish between economically useful and wasted fat, and to scientists, because many methods of quantifying adipose tissue depend mostly upon measuring subcutaneous or intra-abdominal depots. So they continue to be used. But, as will become clear, these terms refer only to the tissues' anatomical positions: depots within the same category are not necessarily similar in size, and do not share more common physiological properties than those of different categories.

Defining depots

The first hint that in mammals there may be important differences between depots classified together in this way came from comparative studies of adipocyte volume.[14] The simple shape and clear outline of adipocytes (Figure 4, page 29) greatly facilitate the task of measuring and counting them. The volume of a sphere can be easily calculated from a single dimension, the diameter, which for adipocytes can be measured accurately with a simple light microscope. Furthermore, the collagen mesh that surrounds the adipocytes remains visible for days after an animal's death, even if the adipocytes themselves are dead.

If the average volume of about 50 adipocytes is measured from several different depots of the same specimen, a pattern of distribution of larger and smaller cells emerges as shown for squirrels in Figure 6(A). At first sight, there does not seem to be much rhyme or reason to the pattern. Larger adipocytes do not always make larger depots: on the contrary, some of the smallest depots, such as the popliteal behind the knee, consist of relatively large cells. Nor do adjacent depots, for example those either side of the forelimb, necessarily consist of cells of similar size.

If similar observations are made on adipose tissue from anatomically comparable depots in different species, some interesting similarities and contrasts emerge. In the adult camel from which the data on Figure 7A were obtained, over a third of all adipose tissue was in the humps, compared with 4–8% in the corresponding site of hares (Figure 6B), up to 4% in squirrels and a maximum of 1% in stoats (Figure 6C). But less than 1% of the camel's adipose tissue was in the inguinal depot, on the front of the thighs and the sides of the abdomen. This depot contained over a third of the total adipose tissue in stoats and squirrels, more than a fifth of that of the lioness (Figure 7C), and an eighth of that of horses (Figure 7B). There were also contrasts in the proportions of adipose tissue located on the dorsal wall of the abdomen, from nearly half in squirrels and hares (Figure 6A, B), to an eighth or less in the large animals (Figure 7). The ranges of the measurements in Figures 6 and 7B show that there are also substantial differences between specimens of the same species, perhaps related to age, fatness or just individual quirks.

The mean volumes of adipocytes also differed greatly between species, with those of the lioness ranging from 0.63 to 1.71 nl, while none of the stoats' depots contained adipocytes larger than 0.32 nl. It is easy to abandon all hope of finding any kind of consistency or order in such data but although the range of sizes of adipocytes differs greatly between species and between specimens, partly because some specimens were fatter than others, the anatomical *pattern* of relative sizes of adipocytes is similar in all terrestrial mammals. What differs between species is the relative abundance of adipocytes in each depot: the relative size of their adipocytes forms the same general pattern.

A great deal is now known about the biochemical steps by which mature adipocytes are formed, but since almost all of the information comes from the study of cells in culture, we cannot explain in detail why adipose tissue forms where it does, or what determines the species differences in the relative sizes of the depots. The gross anatomy must have implications for the animals' habits and capabilities because, as well as differing in relative size, the adipocytes in these depots differ in many physiological properties. Some seem to be equipped mainly for taking up excess lipid from the blood after a meal, others for quick responses to the onset of strenuous exercise (see Chapter 4).

Relatively minor depots may be almost undetectable in small animals, especially if the specimens are lean, so perhaps it is not surprising that larger animals (Figure 7) should appear to have more 'extra' adipose depots for which there are no exact equivalents in the smaller species (Figure 6). Some are just too big to be thus explained away: up to a third of all adipose tissue of horses (Figure 7B) and donkeys forms a sheet, sometimes several centimetres thick, on the inner ventral wall of the abdomen. There is a small

Figure 6. (opposite) The organisation of adipose tissue in various wild mammals. (A) The grey squirrel (*Sciurus carolinensis*), figures are the means of four specimens, body mass 0.38–0.67 kg, 3.1–14.4% dissectible adipose tissue. (B) European hare (*Lepus timidus scotius*), figures are the means of five specimens, body mass 2.4–3.5 kg, 0.3–3.2% fat. (C) Stoat (*Mustela erminea*), figures are the means of five specimens, body mass 0.19–0.4 kg, 3.2–6.3%. Upper sets of figures are the mean volume of adipocytes in nl (10^{-9} litres); lower figures are the proportions of the total dissectible adipose tissue to be found in the depot. The intermuscular depots are shaded.

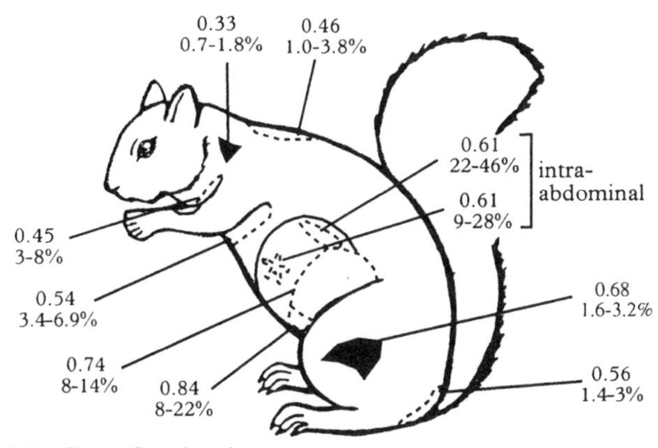

(A) Grey Squirrel

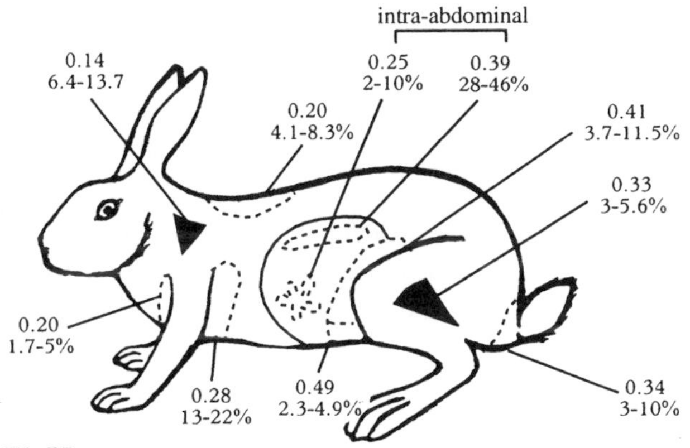

(B) Hare

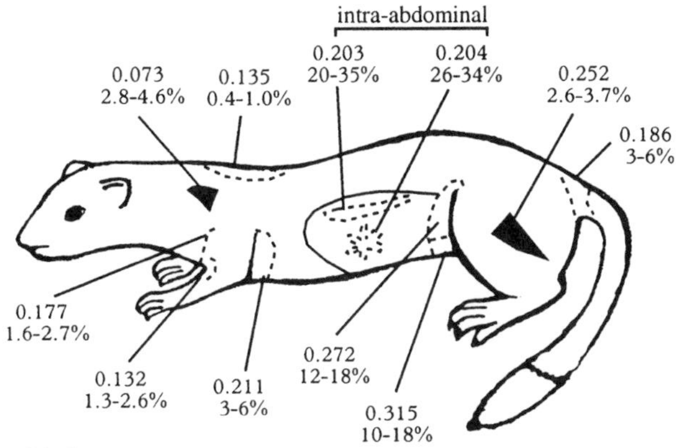

(C) Stoat

quantity of adipose tissue in the corresponding depot of large carnivores and camels (Figure 7A, C), but it is absent in most other species.

In primates, the most conspicuous such depot is the 'paunch' that arises from the mid-line on the outer wall of the abdomen and expands laterally as it thickens, sometimes becoming very thick at the mid-line, and extending as far as the chest and the hip bones. In ourselves, the paunch adipose tissue is centred around the navel, usually somewhat thicker above it in men, and below it in women. The paunch depots seems to accumulate fat selectively in primates: it may be minimal in lean (but not emaciated) specimens, but often becomes very massive in humans, and in monkeys and apes that become obese in captivity, while the other depots expand more slowly with increasing fatness, or, in the case of the intermuscular depots, hardly at all. An abdominal paunch with these properties seems to be a unique and distinctive feature of this group, that appeared early in their evolutionary history. It is as conspicuous in lemurs, primitive primates that occur only on the island of Madagascar (Figure 8), as it is in apes and ourselves. In spite of having this 'extra' depot, with the exception of modern humans, primates are generally no fatter than other mammals when living under natural conditions (though some do become obese in zoos).

There is no trace of this depot in hares, squirrels or any related small mammal, but Carnivora have a 'paunch' depot on the outer ventral wall of the abdomen. The depot can become thick in cats and their relatives – you may have seen it on fat cats – but it is relatively small in members of the dog, bear, and weasel and badger families. It does not expand disproportionately any more than any other superficial depot, even in very obese wild carnivores such as arctic foxes. The depot's surgical accessibility makes it a favourite site for taking biopsies and measuring skinfold thickness in people and monkeys. But we know very little about the physiological basis for its special properties because it is often technically impossible, as well as ethically unacceptable to conduct experiments on humans and other primates, and the paunch depot is completely absent in rodents, including rats, mice and guinea-pigs. If a tissue or organ is not there, it can't be studied, so the

Figure 7. (opposite) The organisation of adipose tissue in some large zoo-bred or domesticated mammals. Data are presented in the same way as in Figure 6. (A) Two-humped camel (*Camelus bactrianus*), a pregnant female, body mass 650 kg, 6.1% fat. (B) Domesticated horse (*Equus caballus*); figures are means of measurements from two geldings, body mass, 444 and 570 kg, 5.2% and 4.0% fat. (C) A pregnant female lion (*Panthera leo*), body mass 167 kg, 13.3% fat.

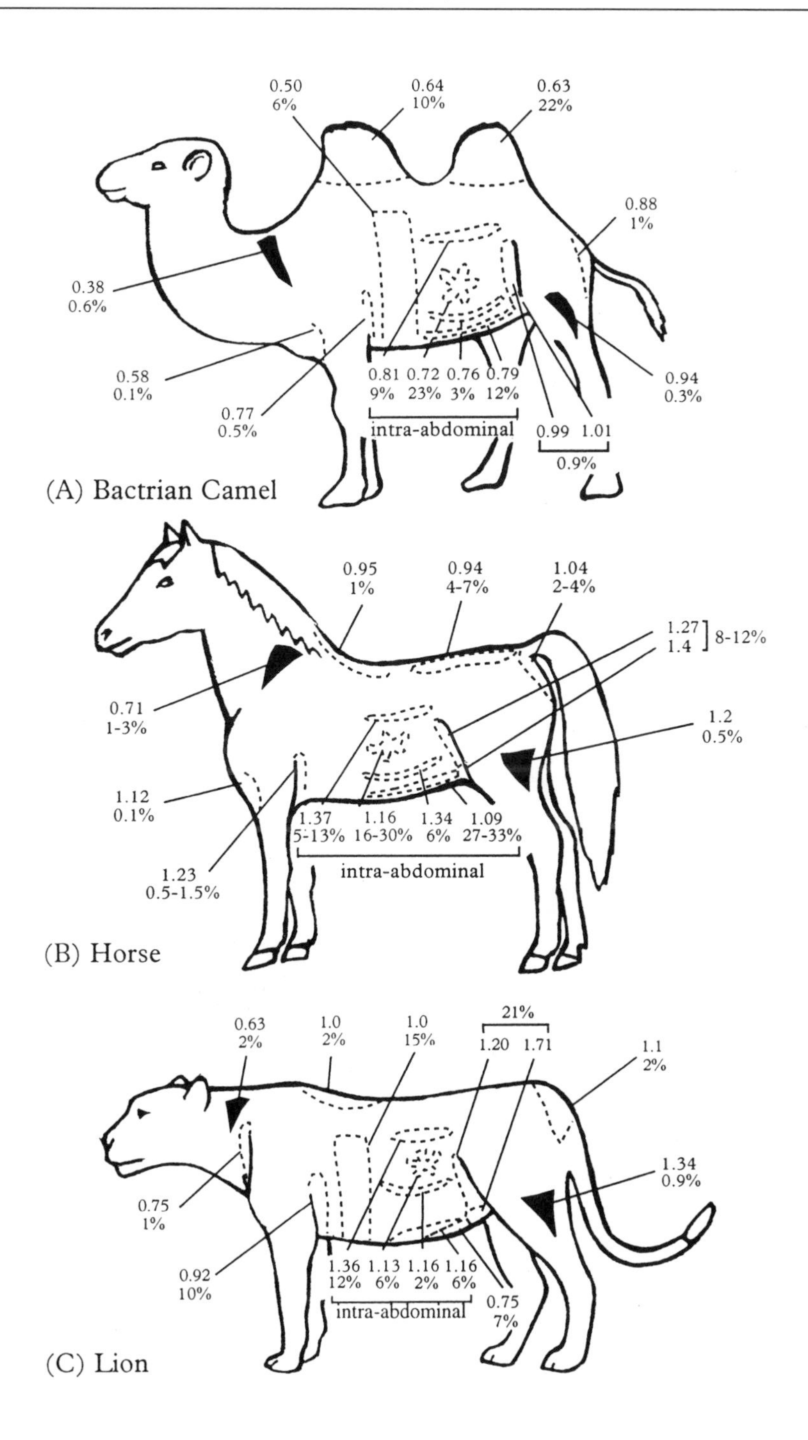
0.50
6%
0.64
10%
0.63
22%
0.88
1%
0.38
0.6%
0.58
0.1%
0.77
0.5%
0.81 0.72 0.76 0.79
9% 23% 3% 12%
intra-abdominal
0.99 1.01
0.9%
0.94
0.3%
(A) Bactrian Camel
0.95
1%
0.94
4-7%
1.04
2-4%
1.27
1.4
8-12%
0.71
1-3%
1.2
0.5%
1.12
0.1%
1.37 1.16 1.34 1.09
5-13% 16-30% 6% 27-33%
intra-abdominal
1.23
0.5-1.5%
(B) Horse
0.63
2%
1.0
2%
1.0
15%
21%
1.20 1.71
1.1
2%
1.34
0.9%
0.75
1%
1.36 1.13 1.16 1.16
12% 6% 2% 6%
intra-abdominal
0.92
10%
0.75
7%
(C) Lion

Figure 8. The paunch depot in various primates: ringed-tail lemur (*Lemur catta*), a primitive prosimian primate, body mass 2–4 kg (left); mandrill baboon (*Papio sphinx*), adult male body mass 30–54 kg (centre), the largest of all monkeys; and two kinds of ape, the orang-utan (*Pongo pygmaeus*), body mass 50–70 kg and human (*Homo sapiens*), adult male body mass 70 kg or more.

metabolic role of this familiar and distinctively primate adipose depot remains speculative.

Why depots matter

The relative size and exact position of each depot is of pressing concern to farmers. The appearance and cooking quality of meat depend greatly upon the abundance and arrangement of adipose tissue around the muscles of pigs, sheep and cattle. Meat animals could be raised much more efficiently if animals had less adipose tissue in total, but a greater proportion of it accumulated in sites around the muscles where it would be eaten (or at least left on the meat during marketing and cooking) rather than around the viscera, where it is simply removed, and either discarded entirely or used for low value products such as chicken feed.

Where adipose tissue forms on the body also matters to people. Although female slimmers aim for a narrow waist, flat tummy and slender legs, they often wish to maintain or even enlarge adipose tissue on the cheeks and breasts. As discussed in Chapter 7, these structures play a central role in sexual display and courtship. Women's breasts (though not the corresponding organs of most other female mammals) consist mainly of adipose tissue, which is sometimes artificially supplemented with breast implants made from materials that, at body temperature, have mechanical properties similar to those of living adipose tissue.

The relative sizes of adipose depots differ between individuals, even those of the same species, and change within a single individual from time to time. It's the job of adipose tissue to undergo large changes in mass, so it is difficult, if not impossible, to quantify its 'proper' mass. No other tissue is adapted to expand and shrink to such an extent: muscles, liver, even guts atrophy with disuse or during starvation but always remain quite recognisable for what they are, while the long-standing favourites of all anatomists, the brain and peripheral nervous systems, hardly change at all, even under severe starvation.

It is important to emphasise that the anatomical location of few other tissues varies in such a capricious manner within a single taxonomic group such as vertebrates. One of the major triumphs of biology between the late eighteenth century and mid-twentieth century was to demonstrate that the arrangement of major organs and tissues in each group of animals and plants follows a consistent body plan. Indeed, adherence to the plan was an essential qualification for admission of a particular species to a taxonomic group. Comparative anatomists could rely upon finding the brain, liver, uterus or major limb bones in the same region of the body, with the same arrangement of connecting nerves and blood vessels, in nearly all kinds of amphibians, reptiles, birds or mammals. In cases such as snakes and whales, where one or both pairs of limbs have disappeared in the adult, remnants could be seen in the developing embryo, and sometimes throughout life, as tiny 'vestigial' structures.

But adipose depots seem to appear and disappear without such formalities. While bones, nerves, major blood vessels, reproductive organs, guts and their associated glands follow the rules diligently, the appearance of adipose tissue not only varies somewhat between specimens of the same species (due mainly to differences in fatness) but also differs substantially between species with quite similar diets and habits, horses and pigs for example. Species that are regarded on other criteria as being closely related (i.e. as sharing a common ancestor quite recently in evolutionary time) usually have similar distributions of adipose tissue, e.g. goats, sheep, antelopes and cattle. But in only a few cases is it possible to suggest functional explanations for the major differences between less closely related species. The big picture of the arrangement of adipose tissue in a variety of wild mammals does not reveal any association between obesity (or the impossibility thereof) and certain distributions of adipose tissue.

It might have been these problematic features of adipose tissue that deterred nineteenth century anatomists from studying its arrangement in detail, and from proposing names and anatomical definitions for the major depots. By failing to fit neatly into the comparative anatomists' concept of a basic body plan, adipose tissue missed out on the descriptive and naming phases of the investigation, which are the foundation for comparative studies of other tissues. The consequences of this omission are still with us: many scientists, especially, but not exclusively, biochemists and physicians, do not believe that adipose tissue 'has any gross anatomy'.

The fact that some depots may be relatively massive in certain groups of animals, but so small as to be negligible in others has not been considered in the physiologists' and biochemists' choice of a representative depot for detailed study. Thus the adipose tissue associated with the epididymis (narrow tubes conveying sperm), which is among the largest and most conspicuous depots in male murid rodents (rats, mice and hamsters), is negligible (less than 1% of the adipose tissue) in many other mammals, including primates and carnivores. This depot is, of course, also absent in all female mammals. Nonetheless, the epididymal depot of the white rat is the biochemists' favourite source of adipose tissue. It can be extracted quickly from a freshly killed animal with minimal surgical skill and anatomical knowledge. It is large enough for most kinds of chemical assays, but small enough to survive for days when maintained artificially in tissue culture. Far more is known about the composition, development and metabolic abilities of the epididymal depot of rats and mice than about any other kind of adipose tissue.

Why adipose depots form in such unlikely places, in association with such a variety of other tissues, is an important question, which, as just mentioned, is of interest to livestock farmers and fashion models as well as biologists. Why, for example, is so much adipose tissue associated with the limbs, where its presence must increase the inertia of the legs, thereby adding to the energy required for locomotion? These and other aspects of the relationship between the anatomical arrangement of adipose tissue and its site-specific properties are discussed in greater detail in Chapters 4, 6 and 8.

The number of adipocytes

Adipocytes expand as animals fatten and shrink again as the triacylglycerols in them are withdrawn. In adult laboratory rats and mice, and in those few wild species in which the matter has been thoroughly studied, all such short-term changes in adipose tissue mass can be explained as due to the expansion or shrinkage of a constant population of adipocytes: the number of adipocytes does not change. In laboratory rats, quite severe manipulations such as undernourishment during gestation or from birth to adulthood, or feeding large quantities of exceptionally rich food over periods of several weeks, can double or halve the ultimate number of adipocytes. At least in rats and mice, the number of adipocytes is particularly sensitive to such manipulation during the suckling period, when adipose tissue (and most other tissues) are growing rapidly, but more modest increases can be induced by overfeeding adult animals.

The mean volumes of the lion's and camel's adipocytes (Figure 7) are apparently larger than those of the hares or squirrels (Figure 6). The rat experiments suggest a simple conclusion: the lion and camel were fatter, but in fact complete dissections show that this assumption is false, the lion was about as fat as the squirrels, and the camel as the fattest of the stoats. Clearly, species differ in the relationship between average adipocyte volume and fatness. Figure 9 shows some data on the numbers of adipocytes in various mammals that offer a predictive theory, if not a functional explanation, for this curious anomaly.[15]

Mature adipocytes are not necessarily the most numerous cells in adipose tissue – sometimes cells associated with the fibrous components, or with the blood and blood vessels, and immature adipocytes not yet able to take up lipid are abundant – but they are always by far the largest. These other kinds of cells are so tiny relative to the huge adipocytes that they occupy only a small fraction of the volume of the whole tissue, and their presence can be ignored in calculating the number of adipocytes present in an adipose depot. These circumstances make it simple to calculate fairly accurately how many adipocytes are present in an adipose depot from measurements of the average volume of its adipocytes, such as those shown in Figures 6 and 7.

To compile Figure 9, over 250 dead animals were collected at random, many of them road-kills, animals culled as vermin or redundant laboratory and farm animals. They ranged in size from mice and shrews to camels, horses and whales. All the adipose tissue was dissected from each specimen to establish its total fatness. In this sample, the larger species were not fatter than the smaller ones, in the sense that their adipose tissue accounted for a greater proportion of their body mass, nor was there any significant difference between the average fatness of the herbivores and carnivores. The mean volume of adipocytes in each depot was then measured, and from it and the depot's mass, the number of adipocytes in it were calculated. The values from all the depots were added up to obtain an accurate estimate of each specimen's total adipocyte complement.

All the measurements from each species are encircled, which means, of course, that the more specimens from any one species that are examined, the

Figure 9. (opposite) The numbers of adipocytes in various mammals of body mass ranging from less than 10 to more than 10 million (10^7) grams. The largest species studied, the fin whale (*Balaenoptera physalus*) weighs around 5×10^7 grams, or 50 tonnes. Both axes are on logarithmic scales. Points from all specimens of each species are enclosed by a ring. Non-ruminant herbivores are represented by the shaded rings and the solid line that shows that the number of adipocytes is proportional to (Body Mass)$^{0.74}$. Carnivorous mammals (including bats, insectivores and the marine mammals as well as Carnivora) are represented by the unshaded rings and broken line that shows that the number of adipocytes is proportional to (Body Mass)$^{0.78}$.

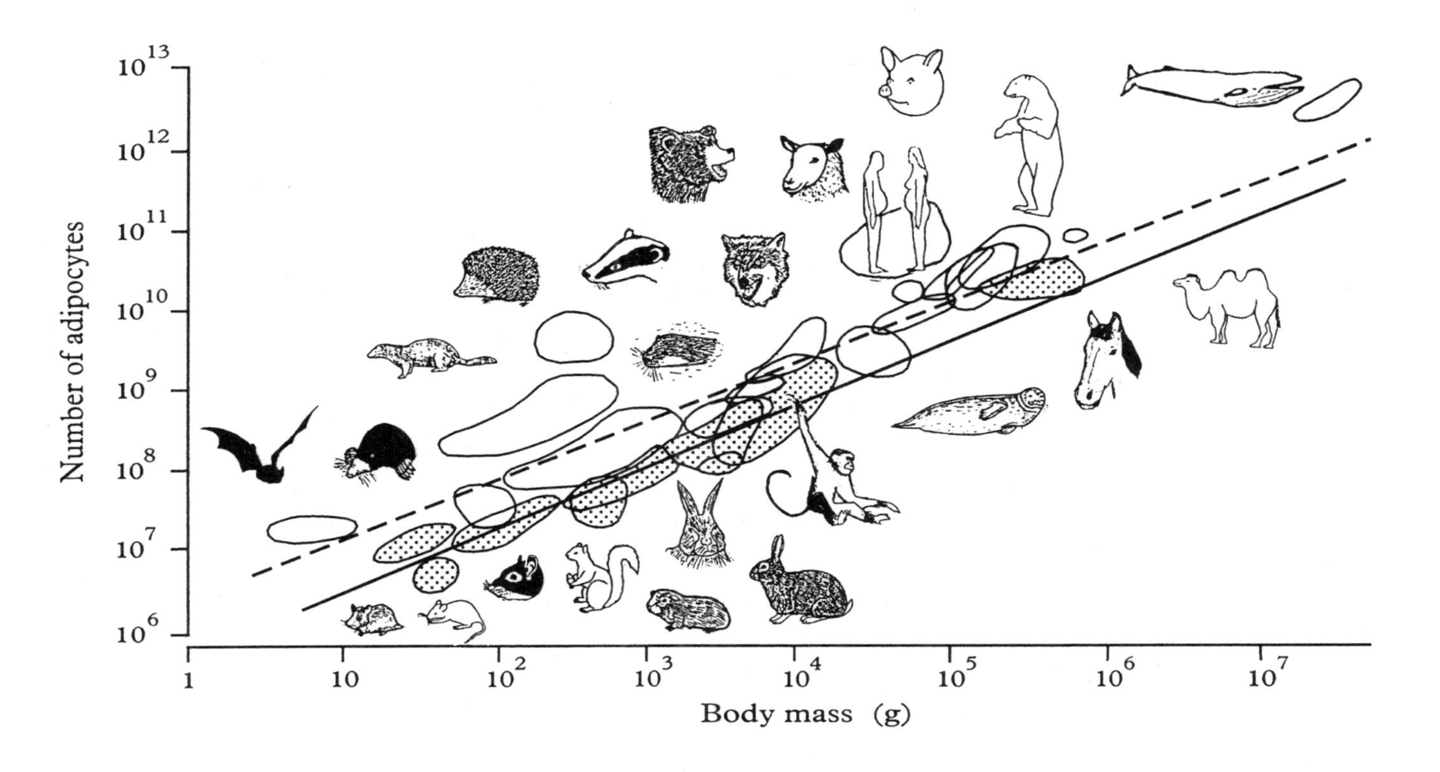
Number of adipocytes
10^13
10^12
10^11
10^10
10^9
10^8
10^7
10^6
1
10
10^2
10^3
10^4
10^5
10^6
10^7
Body mass (g)

larger the circle is likely to be, but even a very few specimens can give us some idea of the range of values. Some of the variability is simply due to differences in body size: larger mink, brown bears and horses simply have more adipocytes than smaller specimens of the same species. Apart from the fact that some of the specimens were not quite fully grown, many kinds of monkey and mink differ in size because adult males can be more than twice as big as mature females. Size differences in horses arise from human intervention: artificial selection has produced Shetland ponies and Shire carthorses. Even in species where adult lean body mass is more or less constant – moles, squirrels, arctic foxes, reindeer, etc. – total number of adipocytes can differ greatly, in the case of arctic foxes and monkeys through almost a factor of ten.

In spite of the variation, the measurements can be fitted to two parallel lines, indicating that, as one would expect, larger animals have more adipocytes than smaller ones. However, the slopes of these lines (0.74 for carnivores and 0.78 for non-ruminant herbivores) show that the larger species actually have fewer adipocytes in proportion to their body mass than smaller species. Since the larger animals were not proportionately leaner than the smaller ones, their adipocytes must be larger in relation to fatness. So we arrive at the conclusion that, in general, large animals have larger adipocytes, even if they are moderately lean. There is surprisingly little information about whether the same applies to other kinds of cells because it is much more difficult to measure accurately the volume of those that are not so conveniently spherical. Large birds such as swans also have proportionately few, larger adipocytes than tits, starlings and other small species.[16] It may be a general feature of all vertebrate adipose tissue.

The line fitted to the data from non-ruminant herbivores lies below that of the carnivorous mammals, indicating that, in general, the carnivores have about four times as many adipocytes, each about a quarter of the size, as non-ruminant herbivores of similar body mass. Although they had more adipocytes, the carnivores were not, on average, fatter. The cellular structure of adipose tissue is simply different in animals that have a diet that is high in protein and fat but low in carbohydrate, compared with those that obtain most of their energy from sugars and starches. So far as we can tell, this contrast makes little difference to the way wild animals fatten or how they control their appetite, but it may have implications for how and why people get fat, as explained in Chapter 7.

The equations fitted to these data provide a means of predicting how many adipocytes, and what range of sizes, we would expect to find in a mammal of a certain body mass. Such information is very useful for deciding whether a species we are not sure about – ourselves for example – fits into the general pattern. Figure 9 also includes some values for estimates of the adipocyte complements of modern western people.[17] The measurements upon which calculations of adipocyte number in humans are based are necessarily less rigorous than those that can be used for animals that can be dissected completely, and so the final figures are less accurate.

As with some of the wild animals, there is much variation between individuals, with some people having ten times as many adipocytes as others of similar body mass. But almost all the values lie far above the lines fitted to the data from wild animals, pointing to the conclusion that, compared with animals, nearly everyone has at least ten times as many adipocytes as would be expected from their body mass. Many people have proportionately more adipocytes than even notorious fatties like hedgehogs, bears, seals and pigs. Further studies show that, while grossly obese people always have more adipocytes than average, the relationship between fatness and adipocyte number is far from close: some apparently lean people of normal body mass turn out to have a surprisingly large number of adipocytes, while those of some moderately obese individuals are greatly enlarged, but are not more numerous than those of slim people.

Although often excessively numerous, human adipocytes are not very large, except in some (but by no means all) very obese people. In fact, they are so much smaller than expected from the comparison with wild animals that microscopists, including the Hoggans, and biochemists familiar only with rat and human adipocytes, have concluded that all adipocytes are the about same size, which, as Figure 9 shows, cannot be true for mammals of widely different body mass.

The death of adipocytes is almost as difficult to observe as their birth by division of tiny pre-adipocytes. The Hoggans concluded from their observations on the adipose tissue of a young man who had died from cancer, and that of a starving mouse, that severely depleted adipocytes disintegrate, releasing their remaining contents. But for much of the twentieth century, biologists believed that once fully formed and functional, adipocytes, like certain neurons, did not die during the lifetime of their owner. The inevitable conclusion from this notion is that the only way of getting rid of them is by brute force, such as surgical excision so research should concentrate on ways to curtail their formation. Much of our understanding of the formation and maturation of pre-adipocytes has been motivated by this aim.

During the past decade,[18] the Hoggans' concept has been revived with the finding that adipocytes and pre-adipocytes, like most other kinds of cells, can disappear by apoptosis, an orderly, often energy-consuming process in which the cells synthesise enzymes that destroy their fabric. Apoptosis of adipocytes is rarely observed because such large cells do not die very often, and death is swift, probably only a few hours from the first signs of decline to total disintegration. Other adipocytes and cells of the immune system remove the resulting debris so quickly and thoroughly that no visible trace of the cells' demise remains.

Watching adipocytes in the hope of catching a few in the act of suicide is obviously too boring to contemplate, but fortunately biochemical methods are now available that identify the activation of the genes that produce the enzymes needed for apoptosis. Such techniques show that loss of mature adipocytes is very slow indeed in healthy adults, but speeds up in prolonged starvation, diabetes and many forms of cancer. One of the main agents that promotes faster destruction of adipocytes, and curtails the rate of formation and maturation of pre-adipocytes is a messenger protein called tumour necrosis factor that adipocytes can produce themselves. The name recalls its first identified role: it has since been found to have many other actions on normal tissues as well as tumours, including a role in some of the complications of obesity in humans, discussed in Chapter 8.

Brown adipose tissue

Heat is always released when the chemical energy in food is converted into forms that can be used for movement, digestion, growth and tissue replacement. Nearly all of the heat that makes us 'warm-blooded' is generated in this way, as a by-product of general metabolic processes. We (and many other animals, including some insects) can increase such heat production by shivering: the muscles perform mechanically useless movements that generate heat warming themselves and the blood in them. The warm blood carries heat to the rest of the body. Most newly born mammals, including human babies, and the adults of some hibernating species such as hedgehogs and bats, contain brown adipose tissue, a uniquely mammalian tissue that is specially adapted to generate heat, supplementing that released as a by-product of other metabolic processes. Because brown adipose tissue produces heat without any unnecessary movements, its activities are sometimes called non-shivering thermogenesis.

Brown adipose tissue is most abundant and conspicuous in neonates and in adult hibernators. In fact, the depots in the neck and around the shoulder blades in hedgehogs and dormice used to be called the 'hibernating gland', distinguished from 'common fat' by its firmer texture and darker colour. Its rich blood supply, and the presence of numerous mitochondria, the tiny cell enclosures in which fatty acids and glucose are broken down and their components combined with oxygen, make it pinkish brown in life, darkening to dull beige after death.

Brown adipose tissue can make its own energy-providing molecules in the same way as other tissues, but when it is required to generate extra heat, it activates its unique capacity to switch to breaking down lipids or glucose in a 'deliberately' inefficient way, so that over 90% of the chemical energy it consumes is released as heat instead of being used to produce ATP.[19] The 'uncoupling' of fuel utilisation from the formation of ATP in the mitochondria is implemented by a special protein named uncoupling protein, which seems to be unique to mammals. When fully activated, for example during rapid warming of the body at emergence from hibernation, brown adipose tissue uses fuels and oxygen at up to ten times the rate that muscles do, and can produce enough heat to raise the body temperature by more than 2 °C per hour.

So much of the volume of brown adipocytes is occupied by mitochondria that, as Figure 10 shows, there is not much space for storing lipid, which is present as numerous tiny droplets. Active brown adipose tissue is only about 10% by weight lipid. Like white adipose tissue, brown adipose tissue is richly innervated by the sympathetic nervous system, and in both cases, noradrenalin stimulates lipolysis. In brown adipose tissue, the fatty acids so produced remain in the cell and are consumed for heat production, while those of white adipocytes are released into the blood, where other tissues can take them up and use them. In the newly born and in awakening hibernators, much of the brown adipose tissue's fuel comes from the white adipose tissue. The latter's stores therefore determine the animal's ability to maintain its temperature until it can find food: mother's milk or a rich, digestible meal.

There is no doubt that this tissue makes an essential contribution to rewarming the body in hibernators and cold-stressed neonates. It is much less certain what role, if any, brown adipose tissue plays in non-hibernating, adult mammals. Suggestions include 'burning off' excess glucose and/or lipids circulating in the blood, before they are stored away as fat in white adipose tissue. There was much excitement in the 1970s about this possibility

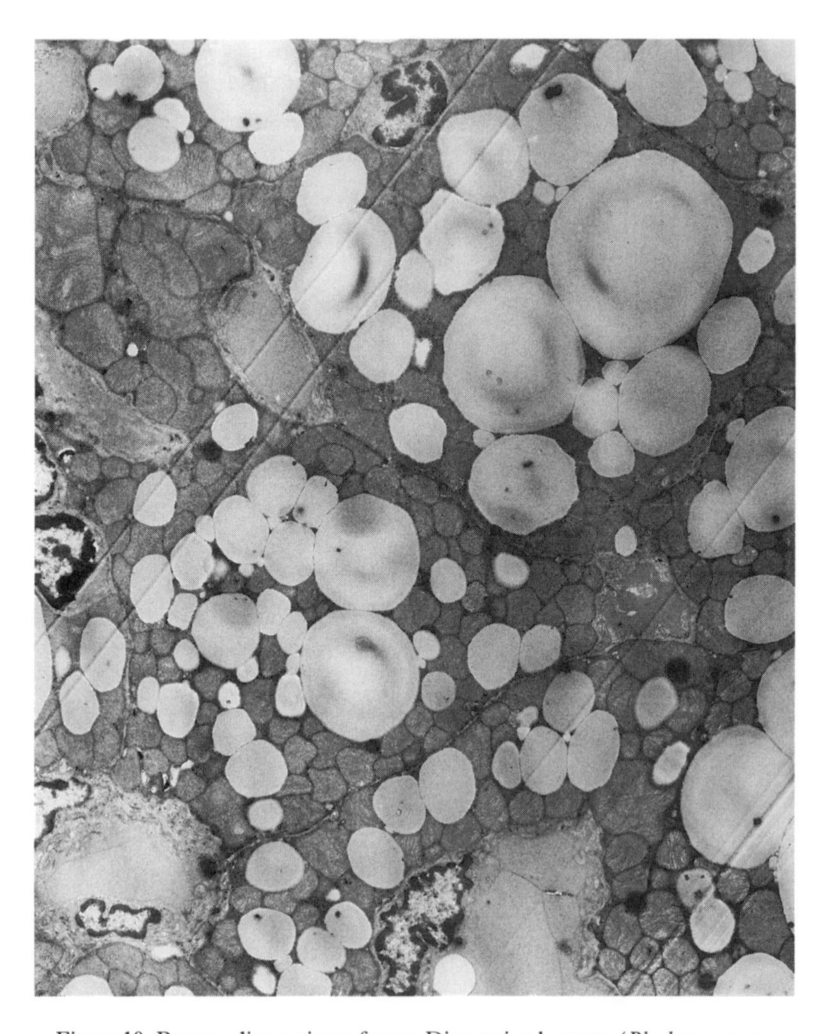

Figure 10. Brown adipose tissue from a Djungarian hamster (*Phodopus campbelli*) that had been kept for 3 months during the winter at about 4 °C. This electron micrograph is fixed and stained in the same way as Figure 5 and magnified 3600 times, so the field of view is about 35 μm (0.035 mm) across. Brown adipocytes are much smaller than white adipocytes and consist of a densely stained nucleus (centre, top) and numerous mitochondria tightly packed together, plus clear lipid droplets of various sizes up to 10 μm across. Cross-sections of several blood vessels are also visible, including a thick-walled arteriole on the lower left, and a thin-walled venule, centre left. (Courtesy of Heather Davies.)

because, if it were true, activating brown adipose tissue might increase the body's capacity for disposing of unwanted lipid. It certainly seems to do so in mice. Recent experiments show that leptin plays a part in determining the neural commands sent to mouse brown adipose tissue. By stimulating heat production, leptin may increase energy expenditure at the same time as it reduces energy intake by suppressing appetite. Mice given drugs that stimulate brown adipose tissue, and have a mutation that removes normal suppression of its activities, remain lean even when fed on a very high-fat diet for several months, regardless of how much leptin they produce.

If things worked that way in humans, it should be possible to develop a drug that would enable people to eat as much as they liked without the risk of becoming obese. Unfortunately, it turned out that adult humans have so little brown adipose tissue, and stimulating it artificially with most known drugs carries serious risks, especially for the heart and for the body's capacity to regulate its temperature, that research to develop such a means of treating, or preventing, human obesity has now largely been abandoned as unworkable. More detailed study of the biochemistry of brown adipose tissue may lead to a drug that stimulates it selectively, thereby avoiding these side-effects.

Brown adipose tissue is best known in eutherian (placental) mammals but uncoupling protein has also been found in a small nocturnal marsupial in Australia. Many of the sites in which it actively generates heat in newborn mammals contain normal-looking white adipose tissue in the adult. There is much argument about how, if at all, brown adipocytes can develop into white adipocytes. In small mammals that live in cold climates, such as dwarf hamsters, at least some depots consist of a mixture of brown and white adipocytes throughout life. The proportion of space occupied by the brown ones increases when the animal is placed continuously in a cold environment, especially if the light/dark cycle is adjusted to resemble the long, dark nights of winter. If subjected to severe enough biochemical treatments, proteins characteristic of active brown adipose tissue can develop in adipocytes in sites that never normally contain such cells.

Fatness

Because adipose tissue in birds and mammals is often associated with the edible muscles of the limbs and back, and is thus familiar as a component of meat, butchers and biologists make a distinction between 'fat' and 'lean' tissues. 'Fatness' is the mass of 'fat' as a percentage of total body mass. That much seems obvious, but quantifying fatness reliably, and deciding what constitutes being too fat or too thin, have proved to be much more difficult.

Fatness could be calculated from measurements of the mass of storage lipids. However, most simple chemical means of isolating lipids extract phospholipids as well as triacylglycerols, so such methods only work well for fairly fat specimens in which the phospholipids represent only a small fraction of the total, and the adipose tissue is replete with triacylglycerols. Fatness measured as the mass of adipose tissue always produces a higher value because only a fraction, in wild animals often less than half, of whole adipose tissue is lipid. Dissecting out the fat bodies of dead specimens of frogs or snakes is easy enough, but because adipose tissue is partitioned into so many depots in birds and mammals, measuring their fatness by direct dissection is tedious and time-consuming.

Measuring fatness in people

Recording the fatness of people after they have died is at best of academic interest only, but measurements in living people can be a useful indicator of general health. The simplest measure of fatness is just body mass. Comparisons between one person's body mass and another's can be made more accurate using the equation: (body mass in kg)/((standing height in metres)^2). This formula is now called the body mass index (BMI) but when first used, it was known as the Quetelet index, after its inventor. Lambert Adolphe Jacques Quetelet (1796–1874) was a Belgian physicist, astronomer and meteorologist who found time between directing the Royal Observatory in Brussels and writing numerous textbooks on physics to examine human growth and intellectual development. He compiled long Tables[20] of detailed measurements of the human body with the same diligence as he had catalogued properties of the heavenly bodies. Such radical shifts in research project met with less disapproval in the nineteenth century than they do now.

Calculated in this way, average BMI for slight teenagers is between 18 and 22, and around 23–26 for mature adults. In famine situations, the BMI of starving women may be as low as 9 (the minimum for men is 14.5–15), but such people are obviously emaciated and very weak. BMI over 27 is regarded as 'overweight' and people whose BMI exceeds 30 are classified as obese, but physicians differ somewhat in their definitions of these terms.

Except under ideal conditions, measurement of BMI is never very accurate. Most people are about 3–5% heavier in the evening than before breakfast, but most of the extra mass is water, salts and glycogen, not increased fat stores.

Heavier people have more adipose tissue, but total body mass also depends upon the density and relative mass of the other tissues. A bulky person who has a lot of fat, wasted muscles and bones depleted of their mineral content may well weigh less than a more slender person who has lots of muscle and very dense bones. Many young athletes have such a physique: they may weigh more than average for their height, but no-one would describe them as obese. The minerals that increase the hardness and density of bone are often greatly depleted in elderly people, especially women, making the skeleton smaller and lighter. The mass of the musculature also differs substantially between people of the same age, and may change during a person's lifetime.

Other ways of measuring fatness in living people including underwater weighing, which depends upon the fact that lipid has a lower density than aqueous tissues. Accurate underwater weighing is technically difficult for living, air-breathing animals such as people, as well as being unpleasant for the subject. People can alter the overall density of the body by expelling air from the lungs, as whales, seals, ducks and other specialist divers do routinely before submerging. Such modifications are more than enough to confound estimates of body density based upon underwater weighing.

Another approach is to measure the water content of the body, which comprises 73% of the body's fat-free mass, and to calculate fat content as the difference between lean mass and total mass. The person drinks a glass of water with a known proportion of a stable isotope of hydrogen or oxygen.[21] The ingested water is distributed throughout the body surprisingly quickly, in less than an hour. The total volume of body water can be calculated from the final concentration of isotopes in a blood sample. There are also various forms of X-ray systems that can distinguish between fat, lean soft tissues and bone, and produce two-dimensional pictures, from which tissue volumes can be estimated.

Other methods of estimating fatness depend upon assessing the thickness of particular adipose depots. For a mammal of its size, a human's skin is exceptionally thin and pliable, as well as being almost hairless. It is thus fairly easy to measure the amount of subcutaneous adipose tissue in a fold of skin using special callipers, like giant calibrated tweezers. The easiest sites from which to measure skinfold thickness are the back of the upper arm below the shoulder blade, or on the hips, although they are not necessarily the most representative of the adipose mass as a whole.

By making various assumptions, a formula for the relationship between the thickness measured and total fatness is derived. Skinfold thickness measurements are never very reproducible, even in expert hands. They work best on children and young adults, because the proportions of adipose tissue in their superficial and internal depots are fairly constant. The method does not quantify intra-abdominal adipose tissue, so if such depots expand relative to the superficial sites, as often happens in middle-aged men and older people of both sexes, skinfold thickness becomes less and less representative of the whole adipose mass. We have all seen people with huge 'pot-bellies' and lean, spindly limbs.

Measuring fatness in animals

If estimating fatness from body mass is complicated for humans, it is even more tricky for other mammals and birds. The equivalent of body mass index is not easy to measure accurately for a quadrupedal animal. Body conformation differs between species, so the standards must be worked out for each one. In lean specimens, the food in the stomach may be heavy compared with the mass of adipose tissue and so makes a large, but variable contribution to total body mass.

Lipids are much poorer conductors of electricity than the watery components of cells and body fluids. This contrast has been exploited in the development of apparatus that measures whole-body conductance of electric current as an indicator of the proportion of lipid present. More fat produces lower whole-body conductance. The apparatus has to be used carefully, because hair, feathers and the pockets of air they enclose also have high resistance to electric currents, but it is a useful technique, especially for small birds and mammals, and works equally well on living or freshly dead specimens.

The skin of most mammals is too thick and tough, and the underlying adipose tissue too thin and patchy, for skinfold thickness measurements to be of much use. Validation of methods based upon measurements of only part of the adipose mass clearly requires a knowledge of the depots' properties and under what circumstances they are representative of the total mass of the tissue. However, people continue to try using the method for lack of anything better. Especially for zoo and laboratory animals, which are generally fatter than those living wild, it can provide a crude estimate of the mass of superficial adipose tissue.

A high-tech version of skinfold thickness measurements is the ultrasound scanner, which can measure fat thickness fairly accurately in large mammals, especially if, as in pigs, the hair is sparse. The method depends upon the fact that sound vibrations travel at different speeds through chemically different materials, and are reflected, producing an echo, at the interfaces between the two tissues. A pulse of high frequency sound is directed into the body from a source placed on the skin, and the timing and intensity of the echo is recorded. With a resolution of, at best, only about 1 mm, ultrasound scanners are not accurate for animals smaller than dogs.

These new techniques are impractical for most large animals – their skin is too thick and their superficial adipose tissue too thin for accurate measurement – so dissection is still widely used. For more than 30 years, assessment of lipid reserves in wild vertebrates *in vivo* and *post mortem* [22] has depended mainly upon empirically derived 'indices' which are based upon the dimensions of the whole body, or that of the thickness (measured by skinfold callipers) or mass (measured by *post mortem* dissection) of one or a few adipose depots. The sample sites are chosen at random or for some technical reason such as simplicity of dissection. The enormous diversity of sample sites and indices is cumbersome and makes comparisons between studies of the same species very difficult and imprecise, and those between species almost impossible.

One of the favourites is the kidney fat index, the mass of the perirenal adipose tissue as a percentage of that of the kidneys. This index works fairly well for lean deer, antelope and similar ungulates, unless the mass of the kidneys changes seasonally, which can happen in some drought-adapted species. It is less accurate in animals such as carnivores (see Figures 6 and 7) in which the perirenal depot represents only a small fraction of the total adipose tissue, and is indistinguishable from that on the rest of the dorsal wall of the abdomen. Wildlife biologists also try to estimate total fatness of mammals from the lipid content of the marrow in the long bones. This rough and ready measurement eliminates the need for a complete dissection, which saves time in the field and avoids spoiling the eating quality of the meat. Some physiological reasons why such 'marrow fat indices' work well on certain species but are completely useless for others are discussed in Chapter 8.

A further problem with using the mass of one or a few depots as an 'index' of fatness is that the partitioning of adipose tissue between depots seems to be inherently variable. In all species for which there are sufficient data, substantial inter-individual variation in the relative masses of depots that cannot be attributed to age, sex or lean body mass is always found, even in genetically homogeneous wild populations and in animals bred and maintained under carefully controlled laboratory conditions. Such 'fat patterning' has been intensively studied in humans and captive monkeys, but attempts to relate the data to normal physiological activities or to specific diseases have, on the whole, been disappointing. Very little is known about how these differences in partitioning of adipose tissue affect the workings of the whole animal, because it has so far proved impossible to manipulate substantially the relative sizes of the major depots by any means less drastic than surgical ablation.

As wild mammals get fatter, superficial adipose tissue always expands more rapidly than that inside the abdomen. The depots shown in Figures 6 and 7 (pages 41 and 43) become thicker and expand laterally to cover a greater area. In very fat specimens, including many humans, adjacent superficial depots overlap and merge so that they become difficult to distinguish, but measurements of adipocyte volume can reveal their identity. Since expanding depots spread out as well as thicken, thickness alone is only a crude measure of the total adipose tissue, even if the site from which the measurement is taken is kept strictly constant.

Of the intra-abdominal depots, the dorsal wall of abdomen depot (that includes the perirenal) undergoes the largest changes, especially in deer, sheep, cattle, antelopes and other ruminant animals: it expands greatly in extreme obesity and shrinks to almost nothing after prolonged starvation. Its anatomical position means that, like the fat bodies of lower vertebrates, such large, and sometimes rapid, changes in size have minimal impact on the shape or size of any adjacent organs. As discussed in Chapters 4 and 8, the main, probably sole function of the whole of this intra-abdominal depot is as an energy store for the entire body. The intermuscular depots undergo the least change, so they are relatively massive (compared with those elsewhere in the body) in lean specimens, and relatively small in obese ones.

Prolonged captivity or other grossly abnormal ecological conditions may affect this differential expansion and so mislead those who use measurements on captive specimens as a guide to methods to be applied to animals in the wild. In certain humans, the intra-abdominal depots enlarge enormously, while the superficial depots remain constant or even shrink. There is no *a priori* reason why such anomalous fattening could not happen in other species of mammals: there is slight evidence for accumulation of more intra-abdominal adipose tissue (particularly in the depot on the inner ventral wall of the abdomen) in bears[23] that have been in captivity for a long time, and the intra-abdominal depots can become massive in some caged monkeys and lemurs.[24]

How fat is too fat or too thin?

Such questions are surprisingly difficult to answer. Optimum fatness certainly varies from species to species, as described later in this chapter. Evolutionary and ecological theory suggest some reasons why the capacity for fattening and optimum fatness may alter readily as an adaptation to changing conditions or habits. However, it is impossible to be sure that the average fatness, or the most frequently recorded range of fatness, for animals in a particular population corresponds to the optimum for that species under those conditions. For example, 'average' body mass almost certainly does not equal 'optimum' for modern humans living in industrialised countries, or for that of their cats or dogs. Most wild animals and birds, if they survive at all in captivity, become fatter once they have adapted to confinement, and may live as long or longer in that state than they would have done in the wild.

Healthy animals living wild usually have around 4–8% dissectible adipose tissue under ideal nutritional conditions. Adipose tissue may be proportionately much more massive for part of the year, just before some special situation such as breeding, migration or hibernation. Many animals that are below 4% dissectible adipose tissue are probably having trouble finding enough food, though not yet in danger of starvation, but for others, having much less adipose tissue seems to be normal. Some common wild animals, such as rabbits, hares, moles and foxes, seem to live and breed satisfactorily with less than 1% of the body mass being adipose tissue.

This estimate of typical fatness in wild animals is far below what is considered 'normal' for modern humans. Normal 'lean' young men have around 15% dissectible adipose tissue, young women 17–25% (10–15 kg adipose tissue), but over 30% is by no means exceptional for either sex. The fatness of young recruits to the United States Army fell from about 15% of the body mass to 4–6% (2–4 kg of adipose tissue) after 8 weeks on an endurance training course.[25] At least a quarter of this amount must have been structural adipose tissue which is never depleted even in severe starvation (see Chapter 6). Below this level of fatness, the young men started to lose protein and become weak. Thus the minimum fatness for healthy humans is around the average value for many wild mammals.

Laboratory animals are generally fatter than their wild relations, but usually do not become as fat as humans unless given drugs that impair the regulation of appetite. Few mammals, and still fewer birds, fatten indefinitely even if large quantities of highly appetising food are provided. Adult guinea-pigs[26] allowed to eat as much as they like are 4–14% adipose tissue if they run around freely out of doors, 10–20% if confined in small indoor cages. Those allowed access to food for only part of the day can be slimmed to 2–10% adipose tissue, though those whose fatness falls below 5% are clearly hungry. Like humans, guinea-pigs kept on similar diets and conditions can differ greatly in fatness, and, again like humans, the males become slimmer just with regular moderate exercise, but females do not lose fat unless their food is restricted as well.

'Average' or 'modal' values of BMI in contemporary western people should not be mistaken for the 'natural' or 'physiologically ideal' situation. Very roughly, estimates using advanced X-ray and computer technology indicate that a healthy woman in her fifties with a BMI of 24 (i.e. well within the 'slim' category) is about 38% by weight adipose tissue, an enormous proportion compared with that of wild animals. Adipose tissue may be by far the most abundant single tissue in the body of obese people. Government health warnings and the popular press are quick to point out that most obese people eventually develop metabolic disorders such as diabetes and heart disease, and can expect to die younger than average. But perhaps we should be more surprised that very fat people are able to function as well as they do: disproportionate growth of other tissues (the skin, skeleton, heart etc.) causes far more severe problems and discomfort.

Obese people seem to be healthier than many 'genetically obese' animals, that are bred in the laboratory for research into the mechanisms of obesity. Mutations in at least four different mouse genes, and three rat genes, are known to enable animals to become enormously obese on a diet of standard laboratory chow. The most widely studied are the Zucker rat, and the so-called *ob/ob* mouse, which when fully grown can turn the scales at 100 g, compared with a body mass of around 20–25 g (less than an ounce) for a normal wild mouse. Many such mutants are also diabetic, or incapable of any exercise, or intolerant of even brief exposure to slightly lower temperature, and homozygotes[27] are often infertile, so breeding them is difficult.

Each abnormal gene gives rise to different biochemical defects that cause obesity in different ways, so they can be useful for certain kinds of investigations into the physiology of obesity, among them the role of leptin in the regulation of appetite and body mass. It turns out that the *ob* gene contains the instructions for making leptin itself. The mutant form produces an incomplete protein that fails to bind to its receptors in the brain, so neurons that control appetite do not receive the message from the adipose tissue. Another recently discovered genetically obese strain of mice, called *diabetes*, produces defective leptin receptors, so although normal leptin is secreted when required, the brain cells fail to receive the message it conveys.

Genetically obese rats and mice are usually born lean, so in many cases, neonates that have inherited one, two or none of the mutant genes are indistinguishable. But they usually fatten faster than usual during suckling, and are clearly obese at or shortly after weaning. The various disruptions to the controls on appetite and energy expenditure caused by the lack of leptin gradually combine to make mice that inherit two mutant genes become massively obese: *ob/ob* mice do not obviously eat more until they are a month old, by which time they already have about four times as much adipose tissue as their normal littermates, presumably because reduced energy expenditure has enabled the adipose tissue to expand without overeating. Increased appetite then perpetuates and enhances obesity, and the adults become so fat they can hardly walk. The adipocytes expand to as much as five times their normal size but the basic pattern of site-specific differences in adipocyte volume shown in Figures 6 and 7 persists.[28]

These mutant rodents are so obviously defective in so many ways that it is not surprising that none has been found in wild populations: if any were born, they would not survive long enough to breed, and the trait would not be passed on to the next generation. Although interesting for research into basic physiological mechanisms, we should not assume that human obesity or that of any wild animal is caused by similar mutations or involves similar defects.

Naturally fat

Organisms in the wild have to find their own food and avoid being eaten themselves, at least for long enough to breed successfully. These processes, collectively called natural selection, determine physiological features such as running speed, food-gathering techniques and digestion, and psychological attributes such as curiosity, alertness, and how easily they are scared into hiding or running away. These attributes set the balance between how much food an animal can obtain, and how much is expended in everyday activities, but in many cases, the situation is complicated by the fact that all days are not alike: seasonal activities such as migration, mating or breeding take priority over feeding.

Fasting may become essential because the food is simply not available at the time and in the place where the animals have to be to attend to these activities. For example, there is nothing for ground-feeding birds to eat while flying high, or for seal-eating polar bears when giving birth in a snow den many kilometres inland. Some animals fast in the midst of plenty of food: the animal does not eat because it is too busy doing other things – migrating, establishing territories, mating or tending offspring – or because there is not enough space in the abdomen to accommodate full guts as well as mature eggs or foetuses. Animals accumulate fat just before periods during which, for one reason or another, they place themselves in situations in which food is absent, unobtainable, or indigestible. Fatter animals can survive longer between meals, so they do not need to venture out to search for food during unfavourable weather or poor hunting conditions, and can devote themselves full-time to more important activities such as long-distance migration, courting mates, or caring for eggs or young.

There are costs as well as benefits to accumulating fat stores: the additional mass, or the way it is arranged on the body (or both), may hinder movement, making the animal easier prey but a less efficient predator. In the case of small birds that are subject to heavy predation and escape by quick take-off and powerful flight, being even slightly overweight (or underweight due to wasted muscles) may increase the risk of becoming a hawk's meal. Body mass is not quite so critical for swimmers and for many kinds of running animals, because being a little heavier only slightly affects agility or efficiency of movement. But for walking and running, the energetic cost of transport is related to body mass almost independently of the gait used, so heavier animals unavoidably use more energy for each kilometre travelled.

By definition, nestlings do not forage for themselves: their parents bring food. They also rarely have to escape from predators: the parents choose inaccessible nest sites (such as cliffs or trees) or breed only in places where predators are scarce. The main business of nestlings is rapid growth and maturation of lean tissues, but many young birds and mammals also fatten readily if their parents bring them enough food. Being fat has few disadvantages for nestling birds, and would give them a better chance if they were orphaned before fledging, and during periods when the parents' foraging is delayed by bad weather. Seabirds such as petrels, auks, penguins and gannets, whose parents forage over long distances and so may be absent for days at a time, are often fatter as chicks than at any other time in their lives. They usually slim down before attempting to fly: some species routinely fast for many days, even weeks, after their parents abandon them. They stay on or near the nest and do not leave until their flight feathers are fully formed and they have lost much of their fat.

Nestlings that are fed on a diet that is rich in energy but poor in protein and calcium are especially likely to become very obese. Many kinds of birds, including common garden species such as finches, eat mainly seeds and other plant material when adult, but parents bring insects and other highly nutritious animal food to their chicks. An exception is the Venezuelan oil bird, *Steatornis caripensis*, which lives entirely on fruit that contains only about 1% metabolisable nitrogen (i.e. proteins). The chicks are fed large quantities of such fruit and are obese at fledging, weighing up to 57% more than their parents.[29] As sedentary nestlings, they are unable to burn off the excess energy that they ingested to obtain enough protein to support the growth of their lean tissues. Fruit-eating bats (and other mammals) may be able to live on fruits with a lower nitrogen and mineral content than birds can because their brown adipose tissue can dispose of the excess energy, thereby avoiding obesity.

The concept of costs and benefits to accumulating fat stores has led to much theorising, but rather little experimentation, about the ecological conditions that might define optimum fatness, and how they change during the animal's life history. Ornithologists in particular have a taste for computer programs that predict exactly how much adipose tissue is appropriate for birds engaged in certain activities under certain conditions. Some such attempts at 'modelling' the relationship between body composition and energy utilisation simulate real situations quite accurately. But direct measurements of many species in the wild show that there is often quite a bit of variation in fatness among members of the same species at the same times of year, which cannot easily be linked to habits or the ecological situation in which they live.

Where the optimum fatness lies depends critically on circumstances, so average fatness can be very different in different localities (and probably also from year to year, though there is less information on this topic), even for wild animals of the same species. For example, the common red fox (*Vulpes vulpes*) is usually lean where food is available regularly but predators abound, even when, as now often happens, it frequents towns and eats mainly human refuse. Populations living in Scandinavia, where the food supply is very unpredictable, and predators and potential competitors are scarce, are substantially fatter than those elsewhere in Europe.

Birds that are fat for only part of the year just before migration may have habits that reduce the risks associated with being heavy and sluggish. They may live and feed in flocks or herds where the 'safety in numbers' principle applies, or spend time only in places where predators are few or absent.

Small birds such as great tits that feed only by daylight become measurably heavier towards evening, and lose weight during the hours of darkness, especially if the night is cold or prolonged. How much fat they carry from day to day correlates with the likelihood of obtaining food, which depends upon both the presence of suitable prey, and being senior enough in the 'pecking' order to get sufficient access to it, and upon the prevalence of predators. Great tits in Britain became significantly heavier when their principal predator, the sparrow-hawk, disappeared in the late 1950s and 1960s, as a result of poisoning from agricultural insecticides.[30] When improved farming practices allowed sparrow-hawks to re-establish themselves in western Britain in the 1970s, the mean body mass of great tits declined, though that of wrens, a species rarely caught by hawks, remained unchanged.

High latitude, high fatness

The climate at high latitudes and in deserts is not only severe for large parts of the year, it also changes rapidly and varies irregularly from year to year and from place to place. The conditions that promote seasonal or transient fattening in animals living in more predictable, temperate climates prevail almost continuously in such habitats, so many of the most spectacular examples of natural obesity are found among arctic or desert animals.

Temperate-zone bears, including the brown bear (*Ursus arctos*) and black bear (*Ursus americanus*) have a mixed diet of plant matter, especially fruit and seeds, plus fish, small prey and carrion. They fatten rapidly in autumn and spend much of the winter sleeping[31] in caves or dens, living off the lipid stored in their adipose tissue. The maximum fatness and the proportion of the year spent being fat depend upon local conditions. In general, those living in colder climates sleep for longer and so become fatter before retiring, but the fattest bears of all are polar bears (*Ursus maritimus*), whose unique habits and physiological adaptations show that it IS possible to be fat, fit and active.

Unlike other species of the genus *Ursus*, polar bears are top carnivores: they eat seals plus the occasional young walrus and, when really hungry, they may resort to a bit of carrion. Their favourite prey is the smallest arctic marine mammal, the ringed seal (*Phoca hispida*) which weighs about as much as a man when fully grown. Bears swim by paddling like a dog, and in open water they have no hope of catching adult seals, which swim faster and with greater agility, and can dive more deeply. Seals have to surface briefly to breathe every few minutes, and their pups are born and suckled on the ice, so therein lie the bears' best chances.

In the summer and autumn, natural gaps and cracks in the ice provide plenty of places at which to surface, but as the sea freezes over in winter, the seals have to make their own breathing holes, by gnawing or scraping against the underside of the ice with their teeth and the claws of their short, strong forelimbs. They seem to have a 'circuit' of up to a dozen such holes which they visit in turn, tearing them open again if they have frozen over. Polar bears may stalk a seal while it is resting out on the ice, or wait quietly beside a breathing hole, sometimes for many hours. When a seal comes up to breathe, they leap onto it, grabbing it by the nose and hauling it out onto the ice, where it cannot move fast or put up much resistance to its predator. Seal pups make easier prey: until weaning, they shelter from the severe weather in holes that their mothers dig in snow drifts, but the bears' acute sense of smell and powerful clawed forelegs enable them to locate and excavate a concealed pup through several metres of snow.

Seals have specialised adipose tissue called blubber which forms an almost continuous layer over the massive musculature that makes them such powerful, agile swimmers (there is more about blubber in Chapters 5 and 6). Seal carcasses provide ample meat and fat, but most bears eat only the blubber, leaving most of the meat. Only lactating mothers and hitherto starving bears eat the entire seal. This somewhat surprising habit is discussed further in Chapter 7.

A diet of seal blubber facilitates the process of fattening but is not the main reason why bears are almost continuously fat. Polar bears are fat because access to seals varies unpredictably from place to place and time to time. If the ice is too thick, the seals cannot make breathing holes and go elsewhere. As soon as the ice breaks up in the spring (by currents and storms as well as from melting), the seals can breathe anywhere so the bears cannot catch them. Summer is a lean time for polar bears. Where the ice has broken up, they hang about on coasts and islands, eating little except a few berries and a bit of seaweed and carrion if they find it. If weather and currents are unfavourable, hunting may be little better in winter, but the bears are far from idle. Tracking by signals emitted from radios attached to free-living bears shows that they travel huge distances, sometimes hundreds of kilometres, over land or across frozen sea to feeding grounds. How they know where they are going remains a mystery, but they make long journeys over snow and ice at a brisk walk.

Moderate levels of obesity seem to be essential to such habits. Every year a few individuals, adults as well as newly weaned juveniles, are found in a severely emaciated condition, sometimes attributable to an injury, but sometimes apparently just following a run of bad luck. Their superficial adipose tissue, which over the rump might be more than 10 cm thick in a successful hunter, has shrunk to a flimsy sheet of watery tissue. Their muscles are wasted, so the hips are narrow and the legs thin, and at first glance, they look more like a dog than a bear.

Finding food early in the winter as soon as the sea freezes is especially important for pregnant females. They fatten rapidly while still hunting near the coast, but around November those living around Hudson Bay and the islands off the north coast of Canada travel inland, sometimes more than a hundred kilometres, and dig a snow den where they give birth to their cubs in mid-winter. They remain in the den for up to 4 months, suckling the cubs, before returning to the coast in April. The mothers do not feed during this period – there is no suitable food around – so the final stages of pregnancy and the first few months of lactation are fuelled entirely from reserves. No wonder they need to fatten so much early in pregnancy.

Being fat has many advantages, and incurs few natural risks for animals with their lifestyle. Although they can gallop quite fast over short distances, the bears' hunting technique does not call for speed, and they do not normally have to run away from danger, because they have no predators other than humans. However, it has become clear to scientists who try to catch bears (to mark them and take samples) that, especially in the summer, large polar bears are at serious risk of overheating and may collapse from exhaustion if forced to gallop continuously for longer than a few minutes.

Polar bears share much of their range with arctic foxes (*Alopex lagopus*) which also readily becomes fat when they find abundant food. However, the average fatness of those on Svalbard, an archipelago half-way between the north coast of Norway and the North Pole, does not differ consistently during the year, in spite of the very large seasonal changes in food availability and environmental temperature at such high latitude. Unlike most other arctic animals, these foxes do not undergo seasonal fasting or migration: they are opportunists that gorge themselves when food comes their way, but may search for weeks and find nothing. Their food supply is so irregular and unpredictable that they have to be continuously fat enough to sustain themselves through the lean times.

The food supply of grazing animals is much more predictable in time, and fattening in most arctic herbivores is closely tied to the seasons. Two different lineages of cud-chewing, hoofed mammals have become specialised to living in the treeless arctic tundra: they are reindeer (*Rangifer tarandus*), a kind of deer in which females as well as males grow antlers, and muskoxen (*Ovibos moschatus*), which, as their generic name suggests, are somewhere between sheep (*Ovis*) and cattle (*Bos*). Reindeer occur over much of northern Europe, Greenland, Canada, Alaska and Russia, with some local variation in size, shape and colour. Muskoxen probably had a similar distribution but due to hunting during the last few thousand years, they are now restricted to the New World Arctic. Like all ruminants, they do not hibernate, but continue feeding throughout the winter, when necessary digging through the snow to reach the dead vegetation underneath.

The reindeer that occur on Svalbard are the smallest of the living subspecies of *Rangifer tarandus*, with about the same body mass as ourselves (50–75 kg). Reindeer have probably been living there since the ice-sheets began to retreat at the end of the last glaciation (about 10000 years ago) but their usual predators, wolves and lynx, never colonised the area. In their absence, Svalbard reindeer have evolved to become short, stout, slow – and fat, much fatter than any of the other subspecies of reindeer that live on the mainland. At the beginning of winter, adipose tissue can exceed 25% of the total body mass, and the subcutaneous depot over the rump can be more than 8 cm thick. This quantity of fat, combined with sedentary habits and thick fur, enables the reindeer to survive for weeks with little or no food, when the snow becomes too deep or too hard to dig through. Svalbard reindeer have not become fat because their food is exceptionally rich or abundant. On the contrary, their forage is notorious for its indigestibility and poor nutritional quality: the other subspecies of reindeer would starve on such a diet.

Muskoxen on Victoria Island off the north coast of Canada sometimes become as fat as Svalbard reindeer. They cannot run as fast as the leaner subspecies of reindeer that share their habitat, but both sexes have horns and the adult males (some weighing over 500 kg) mount a formidably effective defence against wolves. There is no evidence that any of these naturally obese wild animals are exceptionally sensitive to cold (as many genetically obese rats and mice are), or suffer from any of the metabolic complications observed in severely obese people.

Fat and fit

Continuous severe restriction of food intake, especially of lipids, greatly prolongs life expectancy in laboratory rats and mice, experimental monkeys and possibly also humans. Such a diet would also produce a very lean body. So do naturally obese wild animals have shorter lifespans than naturally lean members of the same species? This question is difficult to answer because so many wild animals die from predation, rather than from disease as most humans and laboratory animals do, and there are many other differences between obese and lean populations. But we can say there is little evidence that obesity in wild animals *shortens* longevity.

The usual cause of death in Svalbard reindeer is starvation, often hastened by worn teeth or exceptionally severe winter weather. On average, these seasonally obese reindeer live longer than the lean reindeer on the mainland, mainly because the latter are subject to predation from wolves and other carnivores. The maximum longevity of arctic foxes on Svalbard also seems to be longer than that recorded in Canada and other places where they live surrounded by more predators, although the average lifespan is probably not much different, because so many of them die from starvation in their first or second year. There is no evidence that migratory birds or mammalian hibernators have shorter lives than similar species without these habits. If anything, the opposite is true: for example, temperate-zone bats, which fatten during the late summer and hibernate for months each winter, live a long time for a mammal of their size.

Many animal populations are probably heterogeneous for the tendency to fatten. Such genetic variety means that shifts in the average fatness of populations as a whole can evolve rapidly, when the species' habits and habitat favour fatter individuals. Becoming obese seems to be a relatively simple evolutionary change, and so to have appeared independently in many species. Genetic analysis is now fashionable in biology and medicine, and much effort has been devoted to identifying a genetic basis for obesity. The tendency to become obese, and probably also the ability to fatten in a way that combines fatness with fitness, are at least partially inherited. As for any multifactorial process, some aspects of lipid metabolism, appetite control and food choice are found to be inherited. But genes for obesity cannot be identified in the same way as those for blue or brown eyes: development of the condition depends upon the circumstances in which the animal is living, as much as upon its genetic constitution. Animals and people cannot become obese unless they find a lot to eat.

Many small, short-lived animals that breed continuously put any extra energy they obtain towards breeding earlier in life and having larger litters more frequently. If provided with more food, wild rats produce more offspring, with the result that populations sometimes reach plague proportions. But there are always a few individuals that would put the extra energy into becoming fat if natural selection, in the form of faster or more agile predators and competitors, did not eliminate them. In captivity, of course, food is more easily obtained and animals are protected from predation. Farmers' choices of which animals are to breed (instead of being harvested before maturity) may be independent of, or even positively correlated with, fatness. In other words, those aspects of natural selection that discriminate against higher fatness are greatly reduced in domesticated animals.

Fatter livestock are often more docile and easier to handle or, putting it the other way round, low energy reserves tend to make animals prone to wander (looking for food) or behave aggressively. Lean and hungry individuals escape, or are harvested when they become a nuisance, leaving the fat and contented to produce the next generation and pass on their genes for the tendency towards obesity and docility. Faster growth, early maturity and high fecundity are also associated with greater fatness, so it is not surprising that obesity frequently evolves under domestication, and domesticated strains fatten much more readily than their wild relations. Until recently, adipose tissue was not only eaten in greater quantities than is now fashionable, but it was also valuable as a raw material for making candles, soap and lubricants.

How many adipocytes?

On the face of it, numbers of adipocytes would seem to be a possible factor that sets limits on how obese, or how lean, an animal or person could become. But do animals that are habitually obese, or are capable of becoming very fat under certain circumstances, have more adipocytes than those of similar size that do not fatten to any extent? Genetically obese rats and mice have a few more adipocytes than predicted from their size, but the increase is not as great as might be expected. The *ob/ob* mouse has only about 50% more adipocytes than its lean littermates, although it has about six times as much adipose tissue: its adipocytes are just greatly enlarged. The question of whether more adipocytes are essential to efficient fattening in wild animals cannot be answered by studying laboratory rodents with obviously maladaptive genetic mutations. The naturally obese arctic mammals just described provide an ideal opportunity to determine the contribution of adipocyte number to the capacity to fatten.

Thorough studies of a number of specimens of the same species in the wild always reveal much inter-individual variation in the total number of adipocytes in the body as a whole, that cannot be attributed to age, sex or any obvious feature of dietary history. For example, it turns out that most Svalbard reindeer have only 2–3 times more adipocytes than would be expected in temperate zone and tropical mammals of similar size (see Figure 9). All specimens examined for this study[32] were collected in December when the reindeer were close to the peak of their annual cycle of fatness, since they could still reach vegetation from the previous summer by digging through the snow.

The number of adipocytes (relative to total body mass) proved to be even more variable in carnivores. Some arctic foxes and wolverines (*Gulo gulo*)[33] have up to five times as many adipocytes as expected from their overall size, but a significant minority actually proved to have fewer than the expected number. At the time the specimens were collected, there was no consistent relationship between the total number of adipocytes and fatness, measured as the relative mass of adipose tissue in the body. We have to conclude that the number of adipocytes in the adipose tissue is not a major determinant of the capacity for fattening. If fewer adipocytes are present, each must expand more to accommodate as much storage lipid as specimens that, for whatever reason, have proportionately more adipocytes.

Such inter-individual differences in numbers of adipocytes obscure attempts to use the mean volume of adipocytes as a measure of fatness, much to the disappointment of biologists, who have long nursed hopes of estimating fatness from small (less than 0.1 g) biopsies of adipose tissue. The data summarised in Figure 11 were obtained as part of an attempt to do just that: 370 wild polar bears in the Canadian Arctic were sedated for a few minutes by injecting drugs with a dart gun from a helicopter, and biopsies of superficial adipose tissue taken.[34] After measuring their length and girth, the bears were rolled onto a stretcher and weighed, suspended from a tripod. The scientists then placed the bear in a position from which it could easily get to its feet, and retreated with their instruments and records.

The data are divided into three groups: solitary males, solitary females and mothers, and cubs that were still with their mother, and presumably suckling from her and/or eating the food she caught. Ninety per cent of all the measurements in each category fall within the shaded areas around the

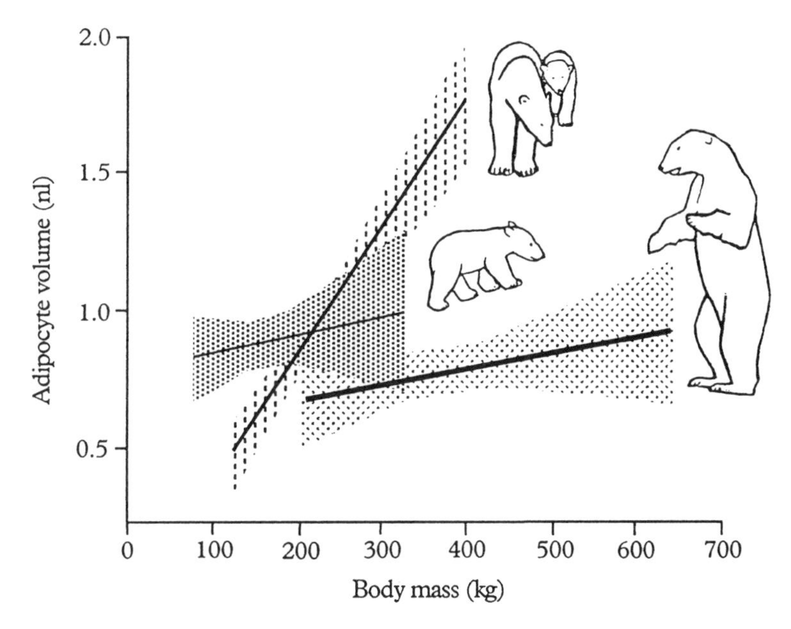

Figure 11. Relationship between body mass and the average volume of adipocytes in a sample of adipose tissue taken from the buttock area of wild polar bears in northern Canada. The lines represent the mean values for juveniles (of both sexes) accompanied by their mothers, solitary males from weanlings to large adults, and weaned females, including pregnant and lactating mothers. The shaded areas around the lines enclose 95% of the values for each category.

lines, which show the mean values. The shaded areas overlap on the left side of the graph because some solitary cubs, probably recently abandoned by their mothers, were similar in size to those still living in family groups. As you might expect, the cubs that were able to take advantage of their mothers' hard work and skill in catching seals were generally heavier than those that were obliged to hunt for themselves.

The correlation between body mass and adipocyte volume was strong for adult female bears, in which skeletal growth stops when they reach sexual maturity at about 4 years old. Thereafter, they fatten during each pregnancy in preparation for suckling their cubs, while their lean body mass remains more or less constant. But for the 92 solitary male bears old enough to be living independently of their mothers (about 2.5 years old), and for a similar number of cubs from the same area, adipocyte volume is almost useless as a measure of fatness. It correlates very weakly with body mass, although there was every reason to believe that the specimens differed substantially in fatness.

Male polar bears do not reach sexual maturity until they are about 8 years old, by which time they are twice as big as mature females, and they may continue growing at least slowly for several years after that. Much of the increase in body mass is lean tissue, not fat. Bigger seems to mean better for male polar bears: heavier animals have the advantage in male–male displays of strength, and so far as it is known, females prefer large males as mates. The mass of the adipose tissue probably increases as the animals grow by the addition of more adipocytes, as well as by enlargement of those already there (as in almost every other tissue except the brain), so their mean size remains about constant over a wide range of body masses. The same seems to be generally true for other large mammals, including humans, that grow relatively slowly over a long period, during which food availability may be erratic. In spite of their obvious vagaries and fallacies, biopsies are so easy to collect and measure that some biologists are still tempted to use them as a means of estimating fatness.

In humans, and laboratory animals kept under highly artificial regimes of diet and exercise, the number of adipocytes increases with age after sexual maturity, but all these effects are slight, generating at most a doubling of the adipocyte population. They cannot be induced in wild caught, naturally obese mammals such as dormice (*Glis glis*) and there is almost no evidence for an increase in adipocyte number after maximum adult size is reached in any of the wild mammals so far investigated. Because the natural range of numbers of adipocytes in wild mammals (and humans) seems to be much greater than that which can be created by manipulating laboratory animals, the metabolic or functional implications of inter-individual differences in how many adipocytes are present, if any, are not known. Except in humans, there seems to be little association between inter-individual differences in adipose tissue distribution and those of adipocyte number, although the matter has been investigated in very few species.

The overall impression from such comparisons is that the relative masses of adipose depots are approximately constant whether the tissue consists of a large number of small adipocytes or of fewer, larger adipocytes. In having an indeterminate number of cells, adipose tissue is probably not different from any other tissue that consists of long-lived cells that turn over very slowly: there are probably also individual differences in numbers of cells in the liver, gut, nervous system and muscles. The irregular shapes of these cells make them much more difficult to quantify than spherical adipocytes.

Final words

Viewed under the microscope, adipocytes are among the dullest of all cells. In spite of their large size, they have very little visible internal structure. Most of their diversity is biochemical, so its revelation has had to await the development of sophisticated methods for keeping cells alive outside the body and measuring their activities. Comparative anatomy had long been out of fashion by the time such techniques became available, so adipose tissue missed its opportunity for recognition as an anatomically organised tissue.

Its variability is another reason why comparative anatomists have ignored adipose tissue for so long. Animals can fatten over a few weeks, then use up the stored energy in a matter of hours. Those in one area may fatten readily, while other populations of the same species hardly fatten at all. Species differ in the anatomical arrangement of their adipose tissue and in its cellular structure, but the implications of such contrasts for energy storage and other physiological processes remain to be established. Although most wild animals are lean, some are obese during certain periods of their lives. A minority are obese almost continuously, but they do not develop the metabolic disorders that attend the condition in humans. Obesity is thus unusual among human illnesses in that apparently similar situations occur naturally in certain wild animals without adverse consequences for other body functions.

The 'optional' and 'indeterminate' features of the biology of adipose tissue suggest to some scientists that they may not be physiologically important. To others, they exemplify the essence of biology: the relationships between structure and function, and between inheritance and environmental interactions. The next chapter is about how all these factors combine to produce such a variety of fatty acids, and the circumstances in which the different kinds are synthesised and retained in plants and animals.

3

Diverse fatty acids

Until less than a century ago, the education of scientists was not specialised as it is now. Laboratory chemists and physicists would have a working knowledge of botany and zoology, and readily took an interest in characterising wild plants and animals, especially those of economic importance. Using the crude and laborious methods of extraction and purification pioneered by Chevreul, nineteenth-century chemists succeeded in isolating and elucidating the basic chemical structures of most of the common fatty acids, thereby revealing them to be the most varied components of natural lipids. The legacy of the close association between biology and chemistry lives on in the names of many of the fatty acids that are abundant in wild and cultivated plants, or, less often, animals, even though nowadays most lipid research is carried out by industrial chemists and biochemists.

The chemists' point of view

Separating and identifying biological materials on the basis of differences in their chemical reactivity, boiling points, freezing points and the kinds of crystals they form was slow and often so inefficient that large quantities of starting material were needed to obtain a tiny amount of a purified product. Fatty acids presented a particular problem because many kinds tend to decompose or combine with oxygen when heated. The invention of gas-liquid chromatography in the late 1950s changed all that: it is now possible to separate fatty acids that differ only very slightly in chemical structure and measure their concentrations to below 0.1% of a mixture. Fatty acids are detached by hydrolysis from glycerol, or whatever else they were attached to in life, and re-esterified to a smaller molecule (see Figure 2, page 12). The resulting fatty acid esters are chemically stable but evaporate more readily than whole triacylglycerols. They are then vaporised in a gas such as hydrogen, helium or nitrogen with oxygen strictly excluded, and passed through a very long (50 m or more), very narrow (0.25–0.5 mm) coiled glass tube, lined with a thin layer of a material that reversibly absorbs them. The fatty acid esters take from a few minutes to over an hour to reach the far end, where their abundance is measured. Smaller molecules (i.e. those with fewer carbon atoms), and those of simpler structure emerge sooner than larger or more complex molecules.

This apparatus has not only revealed the existence of far more kinds of fatty acids than earlier scientists anticipated – 199 different ones have been detected in human breast milk alone[1] – but has also made possible the discovery that apparently minor chemical differences between them are sometimes biologically very important. This chapter is about the biology of some of the most abundant, and from the human point of view, most important fatty acids. It is convenient to think of fatty acids individually, although we should remember that in living tissues, they are usually incorporated into larger molecules such as triacylglycerols and phospholipids, upon which they confer many specific and important properties.

Differences in the way that the carbon atoms are joined together, as well as in the number of carbon atoms in the chain, determine the chemical properties, and hence the biological roles, of fatty acids. Adjacent carbon atoms may be joined to each other and to two hydrogen atoms by single bonds, as shown in Figure 1 (page 10), or they may be linked to each other by a double bond and hence attached to only one hydrogen atom each. The presence of each double bond means the absence of two hydrogen atoms, which increases the ratio of carbon atoms to hydrogen atoms in the whole molecule, making it less 'saturated' with hydrogen. Hence fatty acids with only single carbon-carbon bonds are called 'saturated', and those that contain double carbon-carbon bonds are called 'unsaturated'. Monounsaturated[2] fatty acids have one pair of carbons linked by a double bond, polyunsaturated fatty acids have two or more such bonds.

It is easy to suppose that a 'double' bond would be stronger than a single bond, just as two ropes are stronger than one. In fact, as we shall see, the opposite is the case: double bonds between carbon atoms participate more readily in most chemical reactions with other molecules than neighbouring single bonds. Double bonds also alter the way molecules pack together to form crystals: they bend chains of carbon atoms into kinks, and they limit the rotation of one carbon atom with respect to its neighbour.

Saturated fatty acids

The saturated fatty acid that has 14 carbon atoms (C14:0, the ':0' means it has no double bonds) arranged with their attendant hydrogen atoms in an unbranched chain is called myristic acid because it was first identified in 1841 in the oil extracted from the seeds of an evergreen tree, *Myristica fragans*, in which it is exceptionally abundant. Commonly known as the nutmeg tree, *Myristica* is native to the Molucca Islands, which are now part of Indonesia, but the species has long been cultivated in warm, lowland tropical areas throughout the region of South-East Asia that was for centuries known in Europe as the 'spice islands'. As well as nutmeg oil, the tree yields the spices nutmeg and mace that were, and still are, highly valued imports, amply justifying its epithet 'fragans' (sweet-smelling).

Similar to myristic acid, but with 16 carbon atoms, is palmitic acid (C16:0), first isolated in 1840 from the oil in the fruits and seeds of palms, notably those of the oil palm (*Elaeis guineensis*) and the coconut palm (*Cocos nucifer*). Coconuts are dispersed by marine currents and grow naturally only on or near tropical beaches, where storms and high tides deposit their huge seeds. They can be planted inland, as long as the soil is fairly sandy, but, like nutmeg and chocolate-bean trees, they cannot tolerate even brief periods of low temperatures and can thus only reliably be cultivated on tropical islands and lowlands. Palmitic acid is also the commonest saturated fatty acid in mammalian triacylglycerols. It is especially abundant in the adipose tissue of ruminant animals such as sheep and cattle, which was the source of the lipids that Chevreul and the Royal Institution researchers studied. Figure 1 (page 10) shows palmitic acid.

Lauric acid, C12:0, was first found in the seeds of laurels and other Lauraceae, but is also present in substantial quantities in coconut and palm oils. Baytrees and laurels are native to the Mediterranean, but they can survive when planted in sheltered places in the cooler climate of southern Britain and central Europe. Most other members of the plant family Lauraceae, including the tree *Cinnamomum zeylanicum* from which the spice cinnamon is obtained, are native to central and southern India and Sri Lanka. A few species, notably the avocado pear (*Persea americana*) originate from tropical America.

The cacao beans from which chocolate is prepared are large seeds that are rich in lipids. When extracted from the bean, these lipids are very aptly called 'cocoa butter' and are chemically quite similar to real butter, containing large quantities of palmitic acid (C16:0) and the 18-carbon saturated fatty acid, stearic acid (C18:0). The name 'stearic' comes from στεαρ, the Greek word for 'hard fat' i.e. tallow or beef suet,[3] in which it is the most abundant saturated fatty acid. Around 60% of the fatty acids in cocoa butter are saturated (and hardly any are polyunsaturated), a greater proportion than in any other commercially important edible fat and among the highest of any plant lipids yet studied.

The chocolate that we eat is a mixture of sugar, cocoa butter and relatively small quantities of the unique chocolate essence that is generated from non-lipid components by fermenting the beans for about a week with the yeasts that naturally infect them. The flavour and cooking properties depend critically upon how much cocoa butter is included in the mixture. Good quality chocolate may contain as much saturated fatty acids as a similar mouthful of beef suet or mutton tallow. In an ordinary freezer, chocolate becomes hard and brittle – and tasteless because the taste is greatly influenced by the melting of the fats in the mouth. So the best advice for eating very cold chocolate is to bite off only tiny pieces.

The cacao, or chocolate tree (*Theobroma cacao*, the name means 'god's food', though there is no evidence that the South American people who first used its products regarded it in that light), is native to the warm, damp, equable climate of the Amazon rainforest, and outside that region it grows well only in lowland areas of Central America, tropical Africa, Sri Lanka, the Philippines and Indonesia. The fatty acid composition of the lipids in the seeds seems to be a major factor limiting the species to this range. The tree is so sensitive to temperature that specimens cannot easily be grown anywhere in Europe or North America, even in heated greenhouses. The demand for chocolate is so great that confectionery companies would be delighted to have cheaper, more reliable supplies nearer to home. Agriculturists have tried to cultivate the species in cooler climates, but so far *Theobroma* has resisted attempts to displace it from its warm, wet habitat.

Unsaturated fatty acids

The enzymes in mammalian cells (and probably those of most other animals) that desaturate fatty acids (i.e. convert them from saturated to unsaturated), always insert the first double bond between the ninth and tenth carbon atoms, counting from the —COOH group, which forms the ester bond, as shown in Figure 2 (page 12). The commonest monounsaturated fatty acids are oleic, with 18 carbon atoms, written C18:1, and palmitoleic, with 16, written C16:1. The position of the double bond along the chain of carbon atoms is important to the identity, and often to the physiological role, of fatty acids. Confusingly, older systems numbered carbon atoms starting from the acid end, but modern nomenclature of fatty acids identifies bond positions counting from the methyl end. In the case of palmitoleic acid, the double bond is between the seventh and eight carbons counting from the methyl end of the molecule, and is thus correctly written C16:1*n*–7, while that of the most abundant of all, oleic acid (C18:1*n*–9), has a double bond at the ninth carbon atom counting in either direction.[4]

All kinds of animals in which the matter has been investigated are found to be capable of synthesising these monounsaturated fatty acids from the saturated equivalents (i.e. C18:0 and C16:0) with specific desaturase enzymes, but very little desaturation takes place if monounsaturates are plentiful in the diet. Plants can make monounsaturated fatty acids too: both palmitoleic and oleic acids take their names from the presence in large quantities (up to 70% of the total fatty acids) of oleic acid in olive oil, from which it was first isolated.

Olives (*Olea europaea*) are the most ancient perennial plants to have been cultivated in Europe for extraction of oil: the fruit may be eaten raw, or it and the hard seed ('stone') pressed to release the oil which is purified and used for various purposes, including, but by no means exclusively, cookery. The Oleaceae are really a tropical family, and most of the olive's living relatives occur in tropical Asia and Africa, where they are of little economic importance. The olive's closest wild relatives in northern Europe are ash and privet, neither of which has ever been a significant source of human food.

The presence of the ancestors of the domesticated olive in southern Europe is believed to be a relic of the flora that evolved around the Mediterranean in the mid-Tertiary period, about 40 million years ago. Most such species disappeared during the Pleistocene Ice Ages, but olives hung on, expanding their range during the warm interludes between glaciations. Even now, olives grow well only in the warmer, coastal areas around the Mediterranean: Spain, southern France, lowland Italy, Greece and Cyprus. Olive cultivation has a long and venerable history, going back at least 5000 years, but the trees mature slowly, and yield only around 200 kg of oil per hectare at peak productivity (compared with 1000 kg per hectare from commercial plantations of oil palms), so after millennia as a major crop, their importance is now declining fast.

Lipids containing a high proportion of saturated fatty acids are often called 'hard' fats, because they are solid at room temperature (around 20–25 °C). They readily solidify because their unbranched, regular chains of carbon and hydrogen atoms pack together neatly, but this tidy arrangement is disrupted by the presence of unsaturated fatty acids. The double bonds bend the fatty acid molecules, like a kinked straw, so they no longer pack so neatly side by side in triacylglycerols or phospholipids, and the melting point of the whole molecule is lowered. Olive oil thus remains liquid at cooler temperatures than mutton tallow or cocoa butter. Although the majority of the triacylglycerol fatty acids are oleic acid, olive oil is a mixture of triacylglycerols of several different compositions, so it forms a grease when stored in the fridge, rather than the sharp crystals characteristic of a pure, uniform substance.

Fatty acids readily crystallise once they have been hydrolysed from the glycerol and are free to assemble with their own kind. Molecules of similar shape crystallise together, each kind forming distinctive crystals. Crystallisation enabled Chevreul to purify oleic acid from mutton fat, but he could not separate palmitic from stearic acid by that means. The two saturated fatty acids are so similar, they crystallise together. Believing it to be a pure substance, he named the lustrous, off-white mixture of C16:0 and C18:0 that he isolated from mutton fat 'margaric acid', from the Greek word, μαργαριτης (margarites), a pearl.[5] Many years later, improved techniques of purification and separation revealed its true composition.

Polyunsaturated fatty acids

Fatty acids with more than one double bond are collectively known as polyunsaturates. The positions of the double bonds become more critical in polyunsaturated fatty acids, because once a double bond is in place, it cannot be moved (except by hydrogenating it and desaturating it again under the same rules). The position of the second double bond is determined by that of the first, and by the quirks of the enzymes involved. It is at this point that animals (with the exception of certain insects and single-celled organisms) and plants part company. Animal desaturases can only add double bonds between an existing double bond and the —COOH group, and there must be at least one single carbon to carbon bond in between. So the animal enzymes can add double bonds to oleic acid (which has its first double bond on the ninth carbon) at the twelfth, and/or the fifteenth carbon, with or without elongating the carbon chain by two carbon atoms at a time. Unfortunately, the fatty acids so produced are not nearly as useful to mammals, in terms of the variety of metabolic processes to which they can contribute, as those in which double bonds form nearer to the methyl end of the molecule.

Desaturase enzymes in plants have different properties: they add double bonds between an existing such bond and the methyl end. So they can produce fatty acids with double bonds involving the third, and/or the sixth carbon atom, as well as the others starting at the ninth carbon atom. Green plants, including algae, can synthesise, starting from oleic acid (or from its saturated equivalent, stearic acid (C18:0)), linoleic acid, C18:2*n*–6, with two double bonds and α-linolenic acid, C18:3*n*–3, with three double bonds where the *n* refers to the position of the double bond nearest to the methyl end of the molecule.[6] These 18-carbon fatty acids with two or three double bonds are usually the most abundant polyunsaturates in terrestrial green plants but those of algae can have up to five double bonds. Animals have the most unsaturated fatty acids: those of membrane phospholipids in the mammalian nervous and immune systems, and storage triacylglycerols in some cold-blood aquatic animals have up to six double bonds.

Linoleic and linolenic acid, whose importance in mammalian physiology was only recognised long after chemists first identified them, have confusingly similar names because both were first isolated from the oil extracted from the seeds of the flax plant (*Linum usitatissimum*). The diverse uses of this plant, a tall meadow herb that grows as a weed in some areas, are reflected in its scientific name, which means 'linen, most useful'. Flax was among the very first plants to be domesticated in Europe and western Asia, and its cultivation may be almost as ancient as that of wheat, grapes or peas. The fibrous stems are the raw material for the manufacture of linen, and crushing or grinding the tiny seeds yields linseed oil, used in paints and preservatives as well as in food. Selective breeding has produced distinct modern strains of the plant, some used to produce the best quality linen, and others grown for their high yields of oil, but originally both materials were obtained from the same crop.

The ancient Egyptians wore linen clothes and sandals, and used the fibres to manufacture a wide range of equipment from ropes for building pyramids to wicks for lamps and candles. They also used both woven linen bandages and linseed oil for embalming dead people and animals, many of which have lasted in the mummified state for more than 5000 years. Within living memory, linen was widely used for clothes, 'bed-linen' and furnishings, but these days, its use as a cloth is almost eclipsed by cotton, and synthetic fibres such as polyester.

Linseed, tung and various other plant oils are often called 'drying' oils, although nothing, certainly not water or the oil itself, actually dries (though volatile solvents added to improve spreading qualities may evaporate): the polyunsaturated fatty acids in the oil combine with oxygen in the air and they link together ('polymerise'). The process is fairly slow, but it produces a 'dry', firm, non-greasy, cohesive coating. Putty is a mixture of finely powdered chalk and linseed oil, that has long been used to fix glass into window-frames. It takes many hours to oxidise[7] ('dry'), even on a warm, breezy day, but once hardened, putty sticks very well, and remains effective as a sealant even after years of exposure to severe weather.

The role of linseed oil in embalming corpses and in paints and varnishes depends upon similar chemical processes: if allowed to 'dry' slowly in air, it forms a firm, impermeable surface that binds fine particles of pigment and resists bacterial and fungal decay. Its susceptibility to oxidation required the precious oil to be stored and transported in tightly sealed vessels. The manufacture of such containers might seem straightforward to us now, but the need stimulated those supplying Bronze Age embalmers and artists to further developments in the skills of manufacturing pottery and, later, glass.

Linseed oil is a major ingredient of artists' oil paints, and is still widely used to condition wood, notably cricket bats. The best quality linseed oil, containing triacylglycerols with as much as 60% linolenic acid, is an essential ingredient of oil paints. Only flax grown in cool, temperate climates yields oil of this composition. The availability of suitable oils in the Netherlands contributed to the development of oil painting there early in the fifteenth century. Until the middle of the twentieth century, flax grown in the cool, wet climate of Ireland was the basis for the manufacture there of shirts, sheets and other linen products, as well as linseed oil.

Linseed oil is nutritious because it contains both linoleic acid, C18:2*n*–6, and linolenic acid, C18:3*n*–3, both of which, as explained below, are essential components of the human diet. Linseed oil is no longer eaten in developed countries, though in a purified form, it certainly could be, and people in Ethiopia and the Near East, where the plant is still cultivated extensively in cool highland regions, use it as a flavouring. The implications of the 'drying' properties of these and other polyunsaturated fatty acids for their metabolism in the body are discussed in Chapters 4 and 8.

In most green plants, polyunsaturated fatty acids of the *n*–6 family, linoleic acid and its derivatives, far outnumber those of the *n*–3 family. But comparative studies on wild and semi-domesticated plants are highlighting some exceptions to this rule. One of the best sources of linolenic acid, C18:3*n*–3, is purslane (*Portulaca oleracea*), a fast-growing, drought-tolerant plant that grows as a weed in open, semi-arid habitats throughout the world.[8] Its fleshy, fairly tasty leaves contain 0.3–0.4g per 100 g fresh-weight linolenic acid, five times more than a similar quantity of lettuce or spinach. In some rural areas, it has been used as a salad plant and as an anti-inflammatory poultice or medicine. There are moves to improve its palatability to humans, and make it more widely accepted as a food.

The plants' point of view

Lipids may be most concentrated in seeds and fruits, but the highest proportion of polyunsaturated fatty acids is usually found in the leaves and other green tissues. Leaves are rarely more than 1% by weight lipid, almost all of it as a waxy coating on the outer surfaces, or as phospholipids in the internal membranes, including those of chloroplasts and mitochondria (see Figure 3, page 21), and the outer cell membrane. Membrane lipids in both plants and animals contain a high proportion of polyunsaturated fatty acids though, as in case of seed triacylglycerols, the exact composition differs between species, depending upon climate and habits. The membranes may become leaky, or break down entirely if they are abruptly exposed to temperatures to which the fatty acid composition of their phospholipids is not suited.

The importance of appropriate matching between membrane composition and temperature is most clearly illustrated by some familiar properties of bananas. Although now cultivated mainly in Central America and Africa, bananas (*Musa*) are native to the warm, humid climate of lowland tropical South-East Asia. The storage material in the fruit is almost entirely starch, but the internal cell membranes consist of phospholipids as usual. We prefer to eat bananas after the skin has turned from green to yellow, but before they are completely ripe, and would, in life, fall off the tree. As bananas ripen at room temperature, enzymes are released that soften the starch and other complex carbohydrates, turning the stiff fibres in the skin and edible pulp into a soft mush. Other enzymes facilitate reactions with oxygen in the air to produce a black pigment chemically similar to melanin, the substance that darkens our own skin after exposure to bright light.

Both processes are greatly accelerated if the fruit is cooled below about 10 °C, by, for example, being placed in a refrigerator for as little as an hour or two, but the effect depends not upon the enzymes themselves, but upon their lipid guardians. At about 10 °C (the exact temperature differs with the variety of banana, and the conditions under which it was grown), the lipids in the cell membranes 'freeze' and cease to pose an effective barrier to the diffusion of enzymes and their substrates out of the compartments in which they are normally segregated, so the normal ripening processes begin prematurely. Bringing the bananas back to room temperature after they have been cooled just makes matters worse: although membrane function may be partially restored, the escaped enzymes cannot be recaptured and in fact work faster at warmer temperatures. Apples, pears and other fruits native to cooler climates are not adversely affected by being in a refrigerator because their cell membranes maintain their integrity at such temperatures, though most become mushy by similar mechanisms when thawed after being frozen hard in a deep freeze.

Many annual plants avoid such problems because their seeds do not germinate until they have been exposed to higher temperatures for several consecutive days, a reliable indication that spring has arrived. Gardeners often sow seeds in artificially warmed soil, forcing them to germinate prematurely. The resulting seedlings then have to be 'hardened off' gradually before they can be transplanted outside without risk of damage from cold. The physiological mechanisms include adjustments to the fatty acid composition of the membrane phospholipids as well as modifications to the enzymes and other proteins.

Seeds

In many temperate-zone and arctic plants, the roots and seeds, where the storage materials are concentrated, are the only parts that survive the winter: the rest dies back when cold or lack of light make photosynthesis inefficient. To take advantage of a short growing season, the seeds germinate early in the spring while the soil is still cold. The chief function of energy stores in seeds is to supply the germinating seedling until it grows its own leaves and can photosynthesise for itself. The storage materials of the seeds of plants native to cool or cold climates, such as oats, rye, acorns, beech-nuts and hazelnuts, usually contain starches, with little or no lipids. But as already mentioned, the seeds of many tropical and temperate-zone plants contain lipids in sufficient quantities for people to be able to harvest them efficiently.

Complete breakdown of unsaturated fatty acids releases slightly less metabolic energy (about 1–2% less for each double bond) than the saturated equivalents with the same numbers of carbon atoms. Triacylglycerols containing saturated fatty acids also pack more neatly (i.e. they have higher melting points), than those containing mono- and polyunsaturates, and homogeneous assemblages solidify at higher temperatures than mixtures. So in general, more energy can be stored in a small space by means of saturated triacylglycerols, the more homogeneous, the better. The seeds of coconuts, oil-palms and nutmegs are not only large, but the lipids in them comprise an especially concentrated store of energy. Their seedlings are thus well provisioned with supplies that last them weeks or months.

The temperature at which major physiological processes take place seems to be the most important determinant of the composition of plant lipids. Most biochemical reactions take place in solution so the reactant molecules must be dissolved in liquid for metabolic processes to proceed. Unsaturated fatty acids make phospholipid membranes more fluid, so other molecules can move within and across them more readily, and lower the freezing temperature of triacylglycerols. Lipids containing mixtures of different kinds of fatty acids also remain liquid at lower temperatures than those in which all the fatty acids are similar. The lipids of nearly all temperate-zone and polar species are mixtures of many different fatty acids.

This statement must strike you as absurd: it is obvious that almonds and peanuts are not liquid. However, lipids are unusual in that at the molecular level they have many of the essential features of the liquid state, such as greater movement within and between molecules, at about 50 °C below the temperature at which there is a visible transition from liquid to solid. Such 'melting' lipids start to take up latent heat of fusion (the extra heat required to turn a solid into a liquid or a liquid into a gas, over and above that required to raise it to the appropriate temperature) at such temperatures. These properties contribute to the unique texture of intimate mixtures of sugars, starches and lipids in nuts, and in synthetic products such as chocolate, pastry and biscuits, that we associate strongly with 'real food'. Few other foods generate quite the same taste and 'chewy' sensation.

The correlation between the fatty acid composition of lipids in seeds and fruits, reflected in the common names for the fatty acids, and the geographical distribution and ecological habits of the plants tells us something about their metabolic role. All the plants in which saturated fatty acids are abundant as storage lipids are native to the warm, equable climates of tropical lowlands. Plants such as coconut, oil-palm, nutmeg and cacao, that live only in hot climates don't need polyunsaturated fatty acids to keep their storage lipids fluid: they have the luxury of being able to maximise the energy obtainable from the smallest volume of storage organ, by having triacylglycerols consisting mostly of saturated fatty acids. Those of the seeds of the cinnamon tree, for example, are up to 95% lauric acid (C12:0), although the fatty acids in the lipids of the fruits of the same species, and of its relative, the avocado pear, are around 50% oleic acid and 25% palmitic acid, with very few medium-chain saturates, a typical composition for an oily, rather than a fatty, fruit.

Olive trees, with fruits and seeds containing mostly monounsaturated fatty acids, grow in somewhat cooler climates but are still not worth cultivating for their oil outside areas that have long, hot summers. Plant oils rich in polyunsaturated fatty acids are produced from crops grown in temperate climates in which winters are cold. It is possible, though inefficient, to extract the oil from walnuts or, in principle, most other kinds of tree-borne nuts. Walnuts, almonds and chestnuts are among the most concentrated sources of linoleic acid in the human diet. They have long been cultivated in the parts of central Europe and western Asia, where summers are long and hot but it snows in winter. Most of the world's pistachio nuts come from Iran and Turkey, where they are native, but they, like the other Old World nut crops, are now also planted in southern USA, where the soil and climate are similar.

Most nut trees mature slowly, and although the flavour of olive and nut oil is regarded as superior, the yield cannot compete with that from the annual crops, such as cultivated varieties of sesame, safflower, sunflower, oil-seed rape, maize, soybean and peanuts, as well as flax (linseed). These plants grow and set seed during the warm summer, missing the cold winters entirely. Like the nuts, the storage tissues of these annuals contain starches and proteins as well as lipids, and in many cases, including linseed, soya, maize and sunflower, the residue of crushed seeds from which the oil has been pressed can be fed to livestock as 'cattle cake', or further refined as human food.

The oldest oil-bearing crops, sesame, safflower and flax, grow best in the cooler regions of Europe, northern India and China, and do not flourish in tropical climates. Sunflowers, peanuts and maize are native to the New World and have become economically important much more recently, but prefer climates similar to those of the Old World oil-producing plants. Warm summers are important: the best harvests are obtained from the mid-west of America, and central or southern Europe and Asia; there are no oil-producing annual crops in northern Scotland or Iceland.

The relative abundance of the major saturated, monounsaturated and polyunsaturated fatty acids in seed lipids differs somewhat between species: several possible mixtures of fatty acids in triacylglycerols may have similar physical properties and each plant may be adapted to germinate under slightly different conditions. Some species can adjust the composition of their seed lipids according to the climate under which they are growing. The fatty acids of the seeds of the sunflower (*Helianthus annuus*) are 44–72% linoleic acid (C18:2*n*–6), the remainder being mostly oleic acid (C18:1). Those growing in the coolest climates contain the most linoleic acid, but under such conditions, the total yield is smaller, and the crop takes longer to mature.

The tiny seeds of gooseberries, blackberries and certain other temperate climate plants are rich in both linoleic acid and other fatty acids of the same family (*n*–6) with more double bonds. One of the richest, and most conveniently harvested sources of such fatty acids is the seeds of the evening primrose (*Oenothera biennis*) which is not botanically related to ordinary primroses, or even similar in general appearance, being a tall annual or biennial weed, but its flowers are primrose yellow in colour. For reasons to be described in Chapter 8, the longer-chain, more unsaturated *n*–6 fatty acids have medicinal properties that became widely known in the 1970s, stimulating demand for oils that contain them. Commercial cultivation of the evening primrose for its seed oil is now expanding rapidly in the USA and northern Europe.

Spirulina, a tiny, primitive kind of single-celled organism that lives in salty, alkaline lakes in sunny tropical regions is an even more abundant source of fatty acids of the *n*–6 family than the seeds of these higher plants. It grows well in Lake Chad, on the edge of the Sahara Desert and in ecologically similar lakes in northern Mexico. Local people traditionally eat them dried, because they are rich in protein as well as rare fatty acids, but there are plans to cultivate them on a much larger scale for export as a health food.

The seeds of the jojoba bush (*Simmondsia chinensis*) are almost unique in that the storage material is a wax, not a triacylglycerol. Jojoba oil is liquid at normal temperatures (in contrast to bees' wax, which is used by bees as a solid) because both the fatty acid and the alcohol components of the wax are long-chain molecules. The fatty acids are mostly (around 74%) C20:1, with smaller quantities of C22:1, and oleic acid (C18:1), and the alcohols also contain 20 or 22 carbons. Although the proportion of lipid in the seeds varies widely between strains and according to the conditions under which the plant is grown, the fatty acid composition of the wax, in contrast to that of triacylglycerols, is remarkably constant. Jojoba oil, like other waxes, is indigestible to most animals, including humans, but the plant is now cultivated widely in California and other warm, semi-desert areas of the USA. The extracts are used as lubricants and in cosmetics. Jojoba oil would also be suitable as fuel, and has been proposed as a renewable substitute for petrol (gasoline).

The animals' point of view

The compositions of storage and membrane lipids in cold-blooded animals are governed by many of the principles that determine the composition of plant lipids. Both animal and plant cells can synthesise the common saturated and monounsaturated fatty acids and (except in the case of ruminant animals), they pass unaltered from plant to herbivore. Palmitic, stearic and myristic acid are the commonest saturated fatty acids, and oleic the most abundant monounsaturate, in the adipose tissue triacylglycerols of most terrestrial mammals and birds as well as in plant seed lipids. The proportions of these fatty acids, and their arrangement in the triacylglycerol molecules of most animal fats (e.g. lard, tallow[9]) are such that they are usually solid at room temperature, though they are fluid at the higher body temperature of birds and mammals.

The lipids of cold-blooded animals generally contain a higher proportion of unsaturated fatty acids than those of warm-blooded animals, with species native to colder regions having the most polyunsaturates. The compositions of membrane and storage lipids are controlled partly by inheritance: different species have slightly different combinations of genes that can program the synthesis of enzymes and other proteins involved in lipid metabolism that may promote the selective uptake, synthesis or retention of certain lipids or preferential oxidation of others. But like most other aspects of lipid biology, circumstances and opportunity also play a major part.

Tiliqua rugosa, known variously as the stump-tailed skink, bobtail, blue-tongued skink or shingle-backed lizard, names that reflect some of its distinctive characteristics, is quite common in the deserts of Western Australia.[10] Its natural diet is mainly plants (at least when adult) but also includes carrion, large insects and the occasional nestling bird or mammal, so in captivity, it readily eats artificial food, in which the lipid composition can be controlled. After as little as a fortnight on experimental diets, those given food containing sunflower oil (rich in polyunsaturated fatty acids) chose to spend their days (and nights) in places that were up to 5 °C cooler, so their bodies were also cooler, than those fed on mutton fat (containing mostly saturated fatty acids). These experiments show that what the lizards found to eat during the preceding weeks can affect the times of day at which they are active, and the places they choose to frequent. Such habits, of course, determine what they are likely to find for their next meal, who could be their mate, how long their energy reserves can last, and what predators are likely to eat them.

Similar experiments on other, distantly related species of lizard pointed to the same conclusion. Alteration of the fatty acid composition of the diet brings about small but physiologically significant changes in the composition of the phospholipids in the muscle and liver cell membranes, and in the temperature at which the lizards choose to spend most of their time. The diet of such animals may determine their habits and habitats to a much greater extent than has hitherto been understood. If your pet tortoise, lizard or snake is lethargic and won't eat, try giving it a sun lamp or heater where it can warm itself up, and offer it worms, snails or other cold-blooded animals rather than dog or cat food (which is often made from scraps of mutton or beef, both rich in saturated fatty acids).

Local adaptation

Even in tropical mammals, the skin surface, and often whole appendages such as ears, tails, legs, hooves and paws, are cooler than the core body temperature while the animal is at rest. In cold climates, these appendages may be as cool as 5 °C (about the same as a refrigerator), while the brain and viscera are at 38 °C. The mixtures of triacylglycerols of the composition usually found in internal adipose depots would solidify at the temperatures of the limbs. For the reasons to be explained in Chapter 4, ruminant triacylglycerols necessarily contain a high proportion of saturated fatty acids, but the composition of fatty acids in lipids located in cool appendages is nonetheless adapted to the temperatures at which they are maintained.

Site-specific differences in the composition of triacylglycerol fatty acids in adipose tissue were studied in Svalbard reindeer during the winter,[11] when the average air temperature was around –20 °C. These small reindeer have thicker fur than other deer, and it extends further than usual over their legs and hooves, making them look like a cross between a sheep and a shire horse. A greater proportion of unsaturated fatty acids was found in triacylglycerols in the adipose tissue near the skin than in samples from the inner side of the same depot. There was a similar gradient in composition of triacylglycerol fatty acids in the fatty bone marrow from hip or shoulder to hooves that makes it clear that the feet are normally cool, much cooler than the internal organs. The unsaturated fatty acids help to keep the mixtures of triacylglycerols fluid, and hence metabolically usable, at lower temperatures. How such selective accumulation (or, conversely, selective release) of certain fatty acids could be achieved is described in Chapter 4.

The distribution of triacylglycerols of different fatty acid composition follows a similar pattern in other hoofed animals that live in cool climates. The special flavour and properties of neat's foot oil ('neat' is the Anglo-Saxon word for cattle) have been known and exploited since the Middle Ages, and probably earlier. This material, extracted by boiling the hooves of slaughtered cattle or horses, is liquid at room temperature while suet and other beef fats are solids, because it contains much more oleic acid and fewer saturated fatty acids, especially stearic and myristic acids, than triacylglycerols in adipose tissue around the kidneys, between the muscles or in other warm parts of the body.

Neat's foot oil is certainly edible, but was chiefly valued as a lubricant for guns, cartwheels and other machinery. The use of such animal fats for greasing bullets and keeping gunpowder packets dry was one of the issues that provoked the mutiny of Indian troops against British officers in 1857. Each bullet and the correct quantity of gunpowder were loaded by hand from small paper packets that were torn open by holding the folded end between the teeth. Close contact with grease derived from cattle tissues offended the religious beliefs of Hindus, and that from pigs was unacceptable to

Muslims. The British military authorities tried to meet the soldiers' objections by ordering that clarified butter (acceptable because its production did not involve killing cattle) be used instead, but the change did not quell the discontent and the mutiny spread through much of northern India. Until mineral oils became widely available towards the end of the nineteenth century, the only other lubricants of similar quality were prepared from extracts of the blubber of whales or seals, raw materials not readily obtainable in central India. Nowadays, only small quantities of neat's foot oil are extracted from a few carcasses for making high-grade leather goods and other specialist applications.

Marine lipids

The sea is always cold compared with the warm bodies of mammals and birds. Currents, upwellings and seasonal changes in warming from the sun produce substantial local and seasonal differences in water temperature, but the changes are not as large or as abrupt as in terrestrial habitats. Marine algae tend to have more highly unsaturated fatty acids than terrestrial plants, and the animals that eat them both incorporate the algal fatty acids unchanged into their own tissues, and elongate and/or desaturate them further. Thus most plants cannot build fatty acids with more than 18 carbons, but many marine animals can elongate oleic acid to form gadoleic acid (C20:1*n*–9) and various 22- and 24-carbon fatty acids. Gadoleic acid gets is name from the Latin word for cod (*Gadus*) and is common in the lipids of many kinds of fish, especially those living at high latitudes.

Other fatty acids are also readily elongated, but those of the *n*–6 series derived from linoleic acid are even more unstable (in the sense that they readily participate in chemical reactions, especially oxidation) than those of the *n*–3 series, and so are usually less abundant in marine lipids. Marine algae are the basis of nearly all food chains in the sea, so any animal that eats them, either directly or by eating smaller animals that eat them, takes in far more *n*–3 polyunsaturates than animals that feed in food chains based upon terrestrial plants.

Most marine fish readily convert α-linolenic acid (C18:3*n*–3) into the long-chain polyunsaturates, eicosapentaenoic acid, C20:5*n*–3 (the name just means '20' (carbon atoms) and '5' (double bonds)) and docosahexaenoic acid, C22:6*n*–3 (the name means '22' (carbon atoms) and '6' (double bonds)). These processes are much less efficient in the few species of terrestrial vertebrates in which they have been investigated thoroughly (really only rats, humans, dogs, cats and a few kinds of farm animals), and if such animals can obtain the long-chain *n*–3 polyunsaturated fatty acids they need from the diet, they synthesise very little of them. The enzymes that generate these polyunsaturates from α-linolenic acid can also desaturate and elongate linoleic acid (C18:2*n*–6) to form, among other products, arachidonic acid, C20:4*n*–6. Its name reflects its first identification in peanuts, the seeds of the annual leguminous herb *Arachis hypogaea*, but recently it has been found to have several physiologically important roles in vertebrates, especially in the brain.

The storage and structural lipids of almost all animals that feed in or from the sea – ragworms, mussels, oysters, shrimps, crabs, fish, seals, whales, polar bears, gannets, albatrosses and penguins, to name but a few – are rich in *n*–3 polyunsaturated fatty acids. Medium-size to large fish are the seafoods most frequently eaten in Europe and the USA, so such mixtures of polyunsaturates are often referred to as 'fish oils', though they are not unique to fish, and the basic fatty acids from which the others are derived are not even produced by fish.

All fish tissues contain small quantities of membrane phospholipids, but they provide few fatty acids compared with the triacylglycerols. Most aquatic predators swallow their prey whole, but how much of the nutritious lipids reach our guts depends upon where the oil is in the fish. As mentioned in Chapter 1, the livers of gadoid fish such as cod and halibut are rich in oil, but their muscles (the part we eat) appears pale and watery because they contain almost no storage lipid. Conversely, the lipids of fast-swimming fish such as mackerel, herring, sprats, capelin, sardines and pilchards are stored in the muscles, close to where they are to be used. Their firm, dark flesh can exceed 20% by weight lipid. 'Fish oils' also concentrate in seals and whales that feed on fish and smaller aquatic animals. These days, whales and other marine mammals, if they are hunted at all, are valued for their rich, tender meat, but until the mid 1950s, the main product was oil. Except for the Japanese and Icelanders, for whom whale meat is a traditional delicacy, the flesh was eaten at sea by the whalers or turned into animal feed.

Cool-water fish such as herring, sprats, capelin, mackerel and halibut tend to store more lipid than their tropical relatives, because at high latitudes, the supply of most kinds of food changes with the seasons, just as it does on land, and many species routinely live on their fat reserves for weeks or months. But the fatty acids in the fishes' triacylglycerols include more monounsaturates than polyunsaturates because they come from waxes that were the storage (or flotation) materials of the planktonic animals they eat. Fish such as sardines, pilchards and anchovies live in warmer waters where the small swimming invertebrates rarely contain waxes. Their triacylglycerol fatty acids are derived from the membrane phospholipids of their prey, and so have a greater proportion of the highly nutritious 'fish oils', *n*–3 polyunsaturates.

Some of the food of freshwater animals is derived from aquatic algae, but detritus from terrestrial plants and the animals that feed on them also make a substantial contribution, so fish and insects living in lakes and rivers usually have more fatty acids of the *n*-6 series than oceanic animals. The *n*–3 polyunsaturates are generally still in the majority, however, and all aquatic insects that have been studied can elongate and desaturate polyunsaturated fatty acids at least as efficiently as mammals can. As for other animals, the fatty acid composition of the predatory fish follows that of their diet: 'farmed' salmon held in artificial enclosures and fed on chow made from scraps of slaughtered livestock and other materials originating from the land acquire a correspondingly 'terrestrial' fatty acid composition. This feature impairs their flavour and nutritional value to human consumers, and may also undermine the fishes' health, so synthetic feed must include a substantial proportion of fish meal made from oily fish. Wild-caught salmon command a much higher price, mainly because of the superior flavour conferred by the lipids, rather than any difference in the muscle proteins.

Animals can, and sometimes do, break down polyunsaturates as fuels, but most highly unsaturated fatty acids are preferentially used for alternative roles. Polyunsaturated fatty acids are major components of membrane phospholipids: the complex shape created by the double bonds enables them to maintain the structure's fluidity and support membrane proteins that act as receptors for specific molecules, or form channels that selectively allow certain small ions to pass into or out of cells. Some serve as precursors for the synthesis of important messengers between cells of the immune system, and thus coordinate and regulate the body's response to disease, as described in Chapters 4 and 8. In general, the majority of polyunsaturated fatty acids in the membranes of energy-using and protein-producing tissues, such as the muscles, liver, kidney and adipose tissue, are of the *n*–6 series, while *n*–3 fatty acids predominate in the nervous system and in the light-sensitive cells of the eye.

Flies, butterflies and other advanced insects have excellent colour vision and some species have remarkably complex behaviour. The principal fatty acid in the phospholipids of their nervous systems and the neural components of their compound eyes is eicosapentaenoic acid, C20:5*n*–3. The vertebrate brain and eye retina are rich in docosahexaenoic acid, C22:6*n*–3. As in the case of sterols, these two fundamentally different groups of animals seem to have 'got started' on nerve membrane phospholipids in which slightly different combinations of fatty acids predominate, and built their neural complexity with all its attendant receptors and ion channels in parallel but contrasting ways.

In cold-blooded animals, the composition of structural lipids has to be adapted to the ambient temperature as much as that of storage lipids. The phospholipids in the eyes of crayfish (*Procambarus clarkii*) contain unsaturated and saturated fatty acids in the ratio 2.2 : 1 when they are maintained at 4 °C, but the proportion falls to 1.5 : 1 in those kept at 25 °C. Electron micrographs similar to Figure 3 (page 21) reveal that extensive disruption to the internal membranes of the eye appears within 3 hours of the crayfish being abruptly transferred from one regime to another.[12] It is difficult to believe that their vision was not also impaired, at least for some hours. Avoiding such problems must be among the advantages of living in rivers deep enough for temperature changes to be quite slow.

Essential fatty acids

Once animals have the raw materials, linoleic acid, C18:2*n*–6, and α-linolenic acid, C18:3*n*–3, their own desaturases can add further double bonds to both these fatty acids, and their elongase enzymes can elongate the chain of carbon atoms, as they can to oleic acid. But vertebrates (and probably many other kinds of animals) cannot make these 18-carbon polyunsaturates for themselves. They are called 'essential' fatty acids because they are indispensable to vertebrates, including ourselves, but are obtainable only from the diet. Although all essential fatty acids are polyunsaturates, not all polyunsaturates are essential: some have no known role as precursors for synthesis, and so are useful only as fuels and as components of some structural phospholipids.

Until very recently, only plants and certain micro-organisms were believed to be able to synthesise the basic 18-carbon polyunsaturated *n*–6 and *n*–3 fatty acids so it was thought that they, and only they, are the ultimate source of such lipids for the animals. But improved techniques for breeding and raising small invertebrates allow their metabolism to be studied in the laboratory. Prowess in biochemical synthesis often turns out to be far superior in insects than in ourselves or other vertebrates, and recent research suggests that some primitive insects such as cockroaches, termites and crickets may be able to make linoleic and linolenic acid in small quantities by desaturating oleic acid.

The brine shrimp, *Artemia*, a common crustacean in salty lakes and pools throughout the world, seems to be able to convert *n*–6 to *n*–3 fatty acids,[13] a transformation hitherto believed to be impossible. The possibility that this achievement is due to symbiotic bacteria or other micro-organisms living in their gut was excluded by sterilising the water in which they were raised from eggs. No animal or plant is known to be able to convert fatty acids of the *n*–9 family into those of the *n*–6 or *n*–3 families and *vice versa*. Although there are good reasons for expecting that such transformations could not take place, we should heed Darwin's advice, and never say 'never', or 'always' about biological processes.

This dependence upon fatty acids produced by plants is not as impractical as it seems, since most mammals are primarily herbivores: nearly all rodents, rabbits, elephants, primates, horses and rhinos and their relatives, pigs and hippos eat leaves, roots, seeds or fruit, as do the ruminants (deer, antelopes, cattle, sheep and goats), although there are important differences in the way they digest plants. Their diet of raw plant material includes plenty of *n*–6 and *n*–3 polyunsaturated lipids, and the carnivorous animals – snakes, eagles, lions, etc. – eat these herbivores, lipids and all. It might seem that eating leaves or seeds of green plants, or marine organisms (and most animals have access to one or the other) would furnish plenty of polyunsaturates for all, but there are complications that arise from some universal properties of polyunsaturated fatty acids, and how animals digest and metabolise them.

Cis and trans

There are two kinds of double bonds, *cis* and *trans*, that bend the molecules in different ways, radically changing their overall shape and thereby altering their physical properties such as melting temperature, and their capacity to bind to large biological molecules such as enzymes and receptors. The *cis* and *trans* forms of fatty acids that are otherwise identical in structure and atomic composition may have quite different biological properties and contrasting physiological roles, so the terms *cis* and *trans*[14] may be included in the full scientific name of a particular molecule.

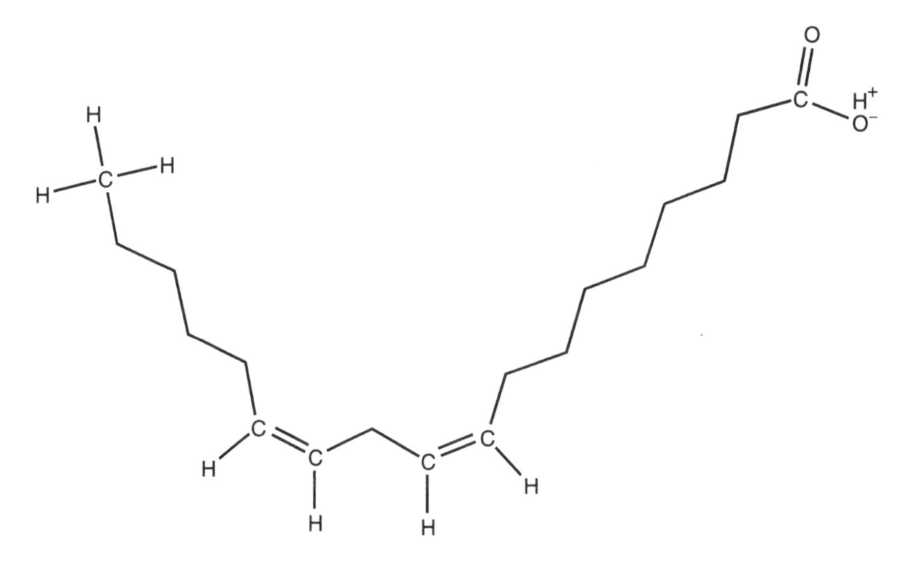

Figure 12. Linoleic acid (C18:2*n*–6) with *cis* double bonds at the ninth and twelfth carbon atoms. The *cis* bonds bend the chain of carbon atoms into a curve.

Cis double bonds bend the chain of carbon atoms to form an angle of about 135°. Fatty acids with several such bonds, for example, linoleic acid shown in Figure 12, are bent into a U-shape. As explained in Chapter 4, this configuration can facilitate the formation of ring structures and the addition of further groups of atoms, that convert them into lipid-based messenger molecules. The *trans* configuration is straighter, so even molecules containing several such double bonds are only slightly kinked, and pack more neatly side by side. Consequently, the melting and boiling points of *trans* unsaturated lipids are nearly always more similar to those of their saturated counterparts than the corresponding *cis* molecules. For example, the melting point of pure stearic acid (C18:0) is 70 °C, that of the common unsaturated fatty acids with *cis* double bonds are 16 °C for oleic acid (C18:1*c*) and –5 °C for linoleic acid (C18:2*c*), but that of the much less common elaidic acid, C18:1*t* with a *trans* double bond, is 43 °C.

Of course, in living tissues, fatty acids do not normally occur in a pure, non-esterified form, and most triacylglycerols contain at least two different kinds. But these figures give an idea of how much the configuration of a single bond in a fatty acid could affect the physical properties of a complex lipid molecule. As we have seen, lipids containing a high proportion of natural (i.e. mostly *cis*) polyunsaturated fatty acids are fluid at room temperature.

Under certain conditions, the bonds can switch between the two configurations. Colour vision in ourselves and probably other vertebrates that can see colour, such as lizards, birds and certain fish, depends upon alternation between *cis* and *trans* bonds in a fatty acid-like molecule. Certain cells in the retina at the back of the eye, called Müller cells after their discoverer, bind the all-*trans* form of retinol, a molecule very similar to and derived from vitamin A. The binding converts the bonds from *trans* to *cis*, in which form retinol becomes available to the cones, crucial receptor cells in colour vision, where it undergoes further chemical changes while absorbing light. Their job done, the retinol molecules revert to the all-*trans* form and back to the Müller cells, and the cycle continues.

The lower melting points of fatty acids with *cis* double bonds and their capacity to form more complex derivatives may be major reasons why so many organisms have plumped for enzymes that desaturate in this way. All unsaturated fatty acids built or modified by animals have *cis* bonds, but the desaturase enzymes of plants and bacteria are less specific, and can form either *cis* or *trans* bonds in fatty acids, so almost all plants have a few structural lipids with *trans* double bonds. The majority of seed oils, including all those used as human foods, consist of triacylglycerols with *cis* fatty acids, but a few contain high proportions of those with *trans* double bonds. The principal fatty acid of tung oil, eleostearic acid (C18:3), has one *cis* and two *trans* double bonds. This drying oil is extracted from the seeds of the tropical shrub, *Aleurites montana*, that is now cultivated mainly in Malaysia and Malawi. It is used in the manufacture of paints, varnishes and oilcloth, and is much prized as a waterproof, semi-glossy finish for wood in furniture and carvings.

Trans fatty acids in food

The principal biological sources of *trans* fatty acids in the modern human diet are ruminants. Most of the enzymes that perform the initial stages of breakdown of plants in the large rumen (forestomach) are not produced by the animal itself: after the sheep, goat or cow has chewed its food into small fragments, tiny micro-organisms coat the particles and produce enzymes that digest them. In the strictly anaerobic conditions that prevail inside this huge, warm 'compost heap' of fermenting plant material, some unsaturated fatty acids are saturated, and other double bonds are transformed from the *cis* to the *trans* configuration. These modified fatty acids are absorbed along with other nutrients, when the digested material passes into the animal's intestine. Thence they may be broken down for energy production, or incorporated into storage triacylglycerols and membrane phospholipids, or secreted in milk.

Trans fatty acids are so strongly associated with micro-organisms that the Finnish biologists[15] who found significant quantities of them in the adipose tissue triacylglycerols of free-living beavers (*Castor canadensis*) suggested that these large, tree-felling rodents must have ruminant-like digestion. Perhaps symbiotic micro-organisms help to maximise the nutrients obtainable from a diet of leaves, roots, bark and wood-chips.

When other animals eat plants or micro-organisms, or the meat, adipose tissue or milk of ruminants, they take in small quantities of lipids with *trans* double bonds. Many such fatty acids are oxidised to release energy, but a few find their way into storage and structural lipids, where they stay without causing any obvious harm, interspersed with the *cis* majority. About 5% of the fatty acids in triacylglycerols in adipose tissue and milk of pasture-fed cattle contain double bonds of the *trans* configuration, the most common being elaidic acid (C18:1*t*).[16] Humans are the only animals to eat significant quantities of the milk of another species, and this habit only arose within the last 9000 years, when goats, sheep and later cattle were domesticated. So in that sense, eating butter, cheese and full-cream milk, the main sources of naturally synthesised *trans* fatty acids in our modern diet, is no more 'natural' to humans than eating artificially hydrogenated fats, or indeed than riding a bicycle.

These days, a large proportion of the *trans* fatty acids in processed foods are synthesised artificially in an industrial process called hydrogenation. Oil, these days usually vegetable oils extracted from maize, sunflower or oil-seed rape but in the past, seal or whale oil, is heated to about 180 °C in the presence of hydrogen and a finely powdered catalyst such as nickel. Various transformations take place, including the addition of hydrogen atoms in place of many of the carbon to carbon double bonds, thereby making the fatty acids more saturated, and conversion of some of the *cis* bonds to *trans* bonds. By adjusting the temperature and time for which such treatment is applied to the composition of the original oil, mixtures of triacylglycerols with the desired combinations of saturated, monounsaturated and polyunsaturated fatty acids can be produced. Up to 60% of the double bonds in hydrogenated fats can be of the *trans* configuration, including the conversion of up to 90% of the oleic acid (18:1*c*) into elaidic acid (18:1*t*).

The different physical properties and flavours of such artificially modified triacylglycerols suit them for use as spreading margarine or as frying oil, etc. Many of the double bonds in the fatty acids of margarines that contain a large proportion of polyunsaturates and are solid, or at least buttery, at room temperature, must be of the *trans* configuration, otherwise the mixture would be liquid, like the natural seed oils that are rich in *cis* polyunsaturates. Pies, pastries, biscuits (cookies), cakes and crisps (potato chips) made with hydrogenated cooking fats are less susceptible to spoilage by exposure to air or sunlight than those made with unaltered lipids like butter. Such products keep better, and so are easier to market cheaply as convenience foods.

Trans double bonds synthesised by hydrogenation are identical to natural ones (though they may form on different parts of the molecule): the industrial processes simply make them more abundant in our food. Does it matter whether the lipids we eat have *cis* or *trans* double bonds? This issue is hotly debated. It is true that most lipases, desaturases and other enzymes involved in lipid metabolism work best with *cis* fatty acids, which is hardly surprising, since they are always in the majority in animal fats. But most such enzymes can deal with fatty acids that have *trans* bonds, albeit more slowly than those with *cis* bonds. Ruminants certainly don't seem to come to any harm, even though a large proportion of the small quantities of unsaturated fatty acids that they absorb (most ruminant lipids are saturated) are of the *trans* configuration.

The fiercest arguments centre on whether, in humans, *trans* polyunsaturates in phospholipids impair the structure and properties of cell membranes, especially those in crucial tissues such as the immune system and the brain. Only about 0.5% of the phospholipid fatty acids in the brain are *trans* polyunsaturates, compared with up to 14% in adipose tissue triacylglycerols (usually much less). At present, people in the USA and western Europe eat between 2 and 12 g of *trans* fatty acids per day, about 7% of their total fat intake. Any effects of dietary *trans* lipids cannot be immediate or spectacular, otherwise margarine, potato crisps and many kinds of pies and pastries would be obviously poisonous even if the diet provided plenty of the *cis* equivalents to meet metabolic demands. Some authorities[17] believe that experiments claiming to show minor impairment of physiological function are flawed, and that *trans* fatty acids are harmless. But their action could be more insidious, perhaps affecting the growth of young children.

There is a suspicion that the availability of long-chain polyunsaturated fatty acids, most of which are formed by enzymatic desaturation and elongation of linoleic or linolenic acid, may not meet the demand of tissues such as the brain and the sense organs (which contain large quantities of long-chain polyunsaturates) during periods of rapid growth and development, such as during gestation and suckling. In brief, too many *trans* fatty acids may impair the maturation of important non-regenerating organs like the nervous system and the eye, or the capacity of tissues to respond effectively to injury or infection. Cows' milk (used to make powdered baby milk) contains more *trans* fatty acids than women's milk. As well as eating large quantities of ice-cream, butter, hamburgers and other foods rich in fats of ruminant origin, many youngsters these days consume yet more *trans* fatty acids in the artificially hydrogenated fats of such children's favourites as potato crisps, biscuits and pastries.

Plant metabolism is best served by triacylglycerols being liquid at ordinary temperatures (5–25 °C), but cooks would prefer them to be solid, so they can be spread on bread or rubbed into flour. Artificial hydrogenation was invented to confer upon plant oils the physical properties of fats derived from warm-blooded animals, such as butter and lard. There is another solution to this conundrum that avoids the complications of rearranging the carbon to carbon bonds within single fatty acids. Instead, whole fatty acids are rearranged within triacylglycerol molecules (using specific enzymes) in a process called interesterification.

As explained in connection with plant seed lipids, melting temperature depends upon how well the molecules pack together, which depends upon the arrangement of the fatty acids within single triacylglycerol molecules, as well as upon their composition. Very few natural triacylglycerol molecules have three identical fatty acids (the middle one is almost always different from the two outer ones, and often all three are different) but artificial interesterification can rearrange the fatty acids so that a much greater proportion have three similar ones. Such reorganised lipids are more solid at room temperature and so are suitable for use as margarine, although the shift in melting temperature is not as great as can be achieved by converting unsaturated to saturated fatty acids.

The cooks' point of view

The subjective flavour of oils is very complex. Pure triacylglycerols have little taste, but lipid-soluble substances dissolved in them can impart a strong taste. The taste also depends greatly upon the physical state of the lipids in the food at the temperature at which it is eaten. As well as the natural flavours of the oils themselves, lipids are better solvents than water for many of our spices. Dissolving the key ingredients of flavourings is a major reason why adding cream, butter or oil to dishes of white fish, or low-fat meats such as veal, improves their taste, as well as making them seem more 'filling' as meals. The active ingredients of spices derived from tropical seeds, such as pepper, allspice and nutmeg, are especially likely to be lipid-soluble. Most animals cannot eat large quantities of the complex chemicals that we call flavours, so their presence in seeds and fruits protects the plant tissue from herbivory.

Shortening, as used in baking, is so called because it 'shortens' the strands of starch that form when wheat flour is cooked, so 'shortbread' biscuits and pastry are light and flaky, instead of tough and stringy as bread is. The fat must be rubbed into the starch particles; the texture, and to some extent the flavour, of pastry depend upon how thoroughly these ingredients were mixed before baking. Traditionally, animal fats such as butter or lard were used, but nowadays, synthetic shortening made from hydrogenated coconut or palm oil is preferred for the mass production of baked foods: consisting almost entirely of saturated lipids, it mixes more readily with flour, and does not melt or go rancid as easily as the natural fats.

As well as the common saturated fatty acids, coconut oil and palm kernel oil contain exceptionally large proportions of lauric acid, C12:0, which is rare in animal triacylglycerols and most common in the seed oils of tropical plants. Triacylglycerols containing lauric acid are just as digestible and nutritious as those with any other medium-chain or long-chain saturated fatty acid – cocoa butter and chocolate for example – and coconut oil is eaten locally where it is produced, but its flavour does not appeal to western taste, so imported coconut and palm oils are processed to make 'shortening' and other cooking fats or used in soaps, notably Palmolive soap.

Lipids in water

Both the colour and flavour of foods containing lipids are determined by how they are mixed with other the components, as well as by their chemical composition.[18] Butter and cream are chemically almost identical, but they differ in the physical state of the lipids and in the proportion of water present. Butter is a water-in-oil emulsion of around 10–20% by weight water and salts, suspended in non-globular lipids. Cream is an oil-in-water emulsion of fine lipid globules, the natural state in which lipids are suspended in fresh milk. The contrasts in taste and culinary properties between cream and butter are due almost entirely to this difference in physical state. Creamy milk is converted into butter and buttermilk by 'churning' it briskly until the globules coalesce together and separate from the watery components.

The size of the water droplets contributes to the colour. Beating butter (e.g. in making cakes or butter sauces) makes it paler, because as in cream, the water droplets in the emulsion become smaller, and scatter light more efficiently. Melting and recrystallisation under carefully controlled conditions make factory-produced butter darker, and, at the same temperature, firmer, than the home-made version. Ghee, or 'clarified' butter, has been boiled to remove the water; the remaining lipids form a clear mixture, tinted only with the pigments that are dissolved in them. Milk from pasture-fed cattle contains variable quantities of yellow pigments called carotenes, derived from the plants they have eaten, including, but by no means exclusively, buttercup flowers. Carotenes have various roles in animals, among them the formation of pigments in the light-sensitive cells of the eye. In mammals, they are transferred from mother to offspring in the milk, so they appear in butter and other dairy products.

Thin cream contains more water than thick cream, sometimes as much as 80% by weight. Cream cannot be whipped stiff unless the lipids are concentrated enough and they are crystalline, which enables the lipid globules to hold tiny pockets of air and prevents them from fusing together. All crystalline states are altered by temperature, and cream that has been frozen or boiled does not afterwards whip satisfactorily. Real ice-cream must be churned vigorously as it cools, so that the water freezes as numerous tiny crystals that, aided by the added sugar, cling to the lipid globules, instead of separating from them to form butter. Commercial ice-cream often contains little or no real cream (except brands that are positively identified as 'dairy' or 'real' ice-cream), but is made from various vegetable oils and large quantities of sugar held into a confection, of which over 50% of the volume is air, by artificial stabilising agents, some of which are made from extracts of seaweed.

Cheese can be almost any shade from orange and red to white, blue or brown depending upon the physical state of the lipids, proteins and water, the metabolic products of the bacteria and fungi that ferment the milk derivatives, and additives such as wine or herbs. The adverse effects on its flavour of heating or freezing cheese are due more to irreversible changes in physical state than to chemical alteration of the flavourings themselves. Such modifications may be unappetising but, in contrast to oxidation, or contamination with the wrong strains of micro-organisms, they do not necessarily impair the nutritional quality of cheese. The relationship, or the lack thereof, between the subjective experience of the palatability of foods and their physiologically accessible nutrient content, is believed to contribute to the undermining of our ability to match appetite to food requirements, as explained in Chapter 8.

Lipids in air

The practice of feeding cod liver oil[19] to infants and exhortations to eat more fish may strike you as curious: our ancestors lived for millennia on the African savannah without ever going anywhere near a cod liver. Nearly all children (and a great many adults) do not like the taste of fish oils and would not eat them voluntarily. So how come we need such exotic foods when our ancestors managed so well without them? The answer lies partly in the diet itself, and partly in our ways of cooking and storing plant lipids.

Double bonds between carbon atoms react much more readily with oxygen, and certain other molecules, than the single bonds in long-chain fatty acids. As we have seen, plant oils rich in polyunsaturates are 'drying' oils: the double bonds oxidise and polymerise within a few hours of exposure to air. These processes can be hastened by leaving bottles of oil in bright sunlight for days or weeks, or by heating it. Artists use both techniques to thicken 'drying' oils to a consistency appropriate to making paints that are easy to apply and do not drip or run. 'Raw' and 'boiled' linseed oils each have their own special uses as artists' materials. Ordinary oil paints take a week or more to dry completely, much longer if applied as thickly as Van Gogh often used them. Metallic catalysts such as cobalt, manganese or lead accelerate solidification of drying oils. They are sometimes sprinkled over pictures to speed up the drying process after the paint has been applied.

Unsaturated fatty acids in foods behave in exactly the same way: their double bonds take up oxygen from the air, forming substances whose harmful effects can spread beyond lipids to proteins, membrane lipids and even the genetic material. They make the food taste 'rancid' and impair its nutritional quality. Fatty acids with several double bonds oxidise faster than those with just one, and the higher the temperature, the faster such degradation proceeds. Some oxidation of lipids is inevitable in living tissues, and the process is believed to impose limits on the maximum proportion of polyunsaturates that biological lipids can contain.

Most foods containing polyunsaturated lipids do not keep fresh for long. Fish, especially oily fish such as mackerel and salmon, and nuts such as walnuts, are delicious when very fresh, but soon start to taste rancid. The process has nothing to do with spoilage by bacteria or fungi, which affects mainly the proteins and requires contamination with living organisms, although the two processes may occur simultaneously and combine to make the food taste 'rotten'. Modern 'vacuum packaging' for cheese, bacon, fish and similar foods not only excludes air-borne bacteria and other living agents of putrefaction, it also prevents oxidation of unsaturated lipids.

Oxidation of lipids is accelerated by exposure to light and higher temperatures, so rancidity in foods such as salad oil, butter, nuts and biscuits can be slowed by keeping them in opaque containers in a cool, dark place. Unbroken nutshells hidden in soil or crevices (e.g. by squirrels) protect their contents very well, but of course such arrangements are not feasible for nuts in cakes or confectionery. Where air-tight packaging is inconvenient or expensive, manufacturers of processed foods resort to using various synthetic 'anti-oxidants' which delay the deterioration of the lipids. Some such 'food additives' are suspected of being harmful to infants and people of weak constitution if eaten in large quantities.

Beef suet remains wholesome for months when stored at room temperature in a paper packet because it consists mainly of saturated fatty acids plus a few monounsaturates. Extracting cocoa butter from cacao beans is straightforward because nearly all the fatty acids are saturated, so they do not degrade even when boiled in open pans. In spite of its high lipid content, chocolate keeps for weeks in a paper box or packet. But to avoid oxidation of their polyunsaturates, most other vegetable oils must be extracted at low temperatures, in air-tight vessels. 'Cold press' is a term used to describe 'virgin' oils that retain more of their unsaturated lipids, but many are removed or inactivated during the purifying and bleaching procedures that involve heating. Brands of olive oil described as 'virgin' ('vierge' in French) are always more expensive because their taste is regarded as superior. The total yield of oil is increased by heating the fruit or seeds during pressing, so the cheaper grades of oil have been subjected to high temperatures and solvents that dissolve out the last few drops. The greater the proportion of polyunsaturates in the plant oil, the more susceptible it is to spoilage during extraction and storage.

Vegetables (e.g. potatoes) and meat cook faster by frying than when boiled in water, because triacylglycerols, especially animal fats that contain mostly saturated fatty acids, can be heated to a higher temperature than boiling water with only slight evaporation (but still enough to coat the kitchen ceiling over several months). However, there are few more efficient ways of degrading polyunsaturated fatty acids than cooking, especially frying in an open, bubbling pan, which both heats them enough to break the molecules into fragments, and maximises exposure to oxygen in the air. Dangerously high concentrations of toxins may be formed in cooking oil that is cooled and reused several times for frying.

There are, of course, good reasons for cooking food, quite apart from its effect on the taste: heat kills parasites that are often present in fish and meat (tapeworms in pork are among the most persistent) and can establish themselves in the consumers, causing chronic debility. People find many plant tissues that are rich in carbohydrates (e.g. potatoes, rice, wheat, maize, beans, turnips, chestnuts, acorns, cassava and many more) distasteful and indigestible when raw, but palatable and nutritious after baking or boiling. Such cooking loosens the tightly packed starches and promotes uptake of water, making the food softer and more accessible to water-based digestive enzymes. Our digestive processes are not alone in this respect. Those of most organisms, including fungi and bacteria, work more efficiently on hydrated starches: rice, wheat flour and beans rot very quickly after cooking or after soaking, but resist spoilage for months at normal temperatures if kept dry.

But on the whole, the taste and digestibility of fats and oils are impaired by cooking, and in many cases their nutritional quality is also compromised, especially by high-temperature processes such as frying. Early humans obtained all the fatty acids they needed from plants, because they ate much greater quantities of leaves and nuts, and, most important of all, they ate them raw. Modern people who cook much of their food are among the few mammals likely to experience a shortage of certain fatty acids in the diet. Typical values for adipose tissue triacylglycerols in people eating a modern diet are 35–75% saturated fatty acids, 20–30% monounsaturates and at most 20% polyunsaturates, often less than 5%. The abundance of the latter depends largely upon the person's diet during the months before sampling.

Green vegetables

Exposure to light hastens the degradation of the lipids in seeds and nuts, but because photosynthesis is fuelled by light, leaves and other green tissues of plants must be exposed to bright sunlight for long periods. Photosynthesis involves splitting water molecules apart, generating, among other products, powerful oxidising agents. Yet the phospholipids in the internal membranes of such tissues contain among the highest proportion of polyunsaturated fatty acids of any natural material. Their presence seems to be essential to photosynthesis and some polyunsaturated fatty acids are synthesised *de novo* only by green plants and photosynthetic algae. Green plants protect their vulnerable but indispensable fatty acids from oxidative damage with an impressive array of anti-oxidants, including vitamin C (ascorbic acid), beta carotene, alpha tocopherol (vitamin E) and glutathione.

Fresh, green leaves and stems are clearly a good source of anti-oxidants for herbivorous animals. The provision of anti-oxidants is one of the main reasons why fresh, uncooked plant foods are 'good for you'. The biological reason for their presence in such abundance and variety is the need to protect cell components, including polyunsaturated fatty acids in membrane lipids from oxidative damage in a chemical environment infested with potentially harmful molecules. Such molecular bodyguards are less necessary in seeds, of course, because they do not photosynthesise when mature. So although many seeds are rich in nutritious triacylglycerol fatty acids, they do not provide significant amounts of vitamin C and similar nutrients. The quantities of anti-oxidants seem to be readily adjusted to requirements, and wild plants often have more vitamin C than domesticated varieties of the same species.

Exposure to additional sources of highly reactive, potentially toxic substances such as tobacco smoke, vehicle exhaust fumes and reused cooking oil can be partially offset by eating large quantities of green vegetables, but the body's stores of anti-oxidants are easily overwhelmed. Experimental rats were fed on cooking oil that had been reused several times for frying and the composition of lipids in the membranes of various tissues were examined *post mortem*. Those of the brain, heart, kidney and testes were the most extensively altered after several weeks. It is easy to imagine how impairment of membrane function in any of these vital organs could lead to ill-health and premature death.

Strange fats

The lipids and starches concentrated in grains, nuts and seeds provide a rich source of food for animals including ourselves, but they are not there to make the plants' seeds more appetising and nutritious to animals. Quite the opposite: seeds such as walnuts, almonds, cacao (chocolate) beans and coconuts protect their nutritious contents with a hard shell (in many species, the nutshell is much harder than the tree's wood) that only the most powerful beaks or persistent gnawing teeth can open. Other nuts, such as peanuts, form underground, avoiding seed predators that cannot dig to reach them. Small seeds such as those of wild strains of flax, sesame and sunflower scatter as they fall, making it hard work for animals to collect them, but their coats are still very tough relative to their size. One of the differences between the domesticated strains of these plants and their wild ancestors is that the seed capsules do not break open and scatter their contents, making the crops easier to harvest, though still difficult to crush.

Many plants have evolved chemical as well as mechanical means of deterring predation on their seeds and the nutritious storage materials that are essential to their successful germination. Domesticating such plants for use as human food presents special problems. Oil-seed rape (*Brassica napus*) is closely related to turnips, swedes and cabbages. Like these crops, it grows well in areas of Europe too cold to support olives or any of the other oil-producing annuals, such as sesame. Rape has been cultivated on a small scale for its oil since at least the thirteenth century, but during the last quarter of the twentieth century, political incentives have enormously increased production in north-west Europe, mainly for the manufacture of cheap margarine. The crop's most conspicuous feature is its brilliant yellow flowers that are now a common sight over much of lowland Britain in early summer. As might be expected for a cool-climate plant, the seed triacylglycerols are rich in unsaturated fatty acids, including linoleic, linolenic and oleic acids, and contain only a little palmitic and stearic. But in wild strains of rape, the most plentiful of all, over 40% of the total, is erucic acid, C22:1 (the name refers to its presence in cabbage seeds), which occurs widely in other members of the family Brassicaceae (formerly called Cruciferae), including mustard.

Erucic acid is rare in animal fats generally and almost absent from those of mammals, and the usual mammalian enzymes that metabolise lipids cannot deal with it. The molecules are too large to be excreted, so when rapeseed oil is first fed to rats, triacylglycerols containing erucic acid accumulate, especially in the heart muscle, impairing its function.[20] After a week or so, the rats produce additional enzymes that can metabolise the unusual fatty acids, oxidising them to liberate the chemical energy they contain, or converting them into a common kind, and their symptoms disappear. However, the effect that eating erucic acid might have on people with weak hearts is hotly debated, and much effort has been put into breeding *B. napus* that has a lower proportion of the contested fatty acid in its seed lipids, and into selectively extracting that which is present without harming the others. The latter course is not easy: separation procedures involve heating, which as explained, easily degrades the polyunsaturated fatty acids that are the oil's main culinary and nutritional advantage.[21]

A major natural function of the high proportion of erucic acid is probably to render the seeds toxic to weevils and other seed-eating insects as well as to vertebrate seed predators such as voles, mice and finches, thus protecting them from herbivory. The caterpillars of the cabbage white butterfly feed exclusively on the leaves and flowers of cabbage and related brassicas, such as cauliflower, broccoli and oil-seed rape. Although concentrated in the seed oils, erucic acid occurs in small quantities in other parts of the plant. The caterpillars that eat them have not, however, acquired more of a taste for erucic acid than mammals have: even when present in quite high concentrations in the diet, the unusual fatty acid is absent from phospholipids and triacylglycerols in the insects' tissues. Either it is not absorbed through the gut or the enzymes involved in esterifying fatty acids into these lipids do not bind to it, or, less likely, it is all used as fuel at once before it has a chance to become incorporated into structural or storage lipids. Erucic acid even seems to be toxic to the brassicas themselves: highly specific enzymes direct it into seed triacylglycerols only. None is found in phospholipids or any other non-storage lipid, suggesting that its presence in such molecules would disrupt membrane function.

In western countries, scientists have made considerable progress in selectively breeding, and more recently genetically engineering, plants in which the seed triacylglycerols contain as little as 5% erucic acid. But in some developing countries, notably China, people still eat rape-seed oil that contains a substantial proportion of erucic acid. The refined oil from the improved plants is sold under the name canola oil. Insects as well as people find the artificially modified plants much more appetising than the wild forms, and large quantities of pesticides are needed to protect rape crops that yield oil low in erucic acid from herbivore damage.

Over 90% of the triacylglycerol fatty acids in the beans of the castor oil plant, *Ricinus communis*, are ricinoleic acid, a monounsaturated 18-carbon fatty acid, that differs from oleic acid (C18:1) only in that it has an —OH in place of an —H on one of the middle carbon atoms (some distance from the double bond). This fatty acid does not normally occur in animals,[22] and most cannot absorb or utilise it efficiently. Eaten in small quantities, castor oil acts as a lubricant and slight irritant to the guts, and by this means, relieves constipation in people and dogs and cats. As in the case of rape-seed oil, rats and probably other mammals can develop the means of metabolising ricinoleic acid sufficiently well to survive on synthetic diets in which most of the lipid is castor oil, but it deters some potential seed predators.

By incorporating ricinoleic acid into the storage lipids of its seeds (together with some even more toxic proteins), the castor oil plant protects its offspring from destruction by seed predators, including boring insects such as beetles, gnawing mammals such as mice and squirrels, and nut-cracking birds such as parrots. The germinating seeds can metabolise triacylglycerols containing ricinoleic acid as efficiently as any other organism can use its own storage fats. All that is needed is special enzymes that can cope with such fatty acids. The extra —OH also disrupts the packing of adjacent fatty acids so triacylglycerols containing ricinoleic acid remain more fluid at lower temperatures than triacylglycerols with three oleic acids.[23] Clearly, *Ricinus communis* has no need for a mixture of different kinds of fatty acids when a single special fatty acid protects the seed from the effects of cooling and helps to deter seed predators. Its purity and qualities as a 'drying' oil make castor oil useful in the manufacture of paints, polishes, crayons, lubricants and in protective coatings for fabrics, which are now far more important than its medicinal uses.

Final words

In this chapter, we have considered only a tiny fraction of the known fatty acids, but even such a brief acquaintance is sufficient to demonstrate their variety, and some of the main kinds of chemically minor but often biologically important differences. The diversity of fatty acids is in fact quite modest compared with that of proteins, where each species synthesises thousands of proteins, each of which is, at least in many cases, different enough to be distinguished by immunological methods.

Knowledge of natural lipids has accumulated gradually, and they and the fatty acids isolated from them were given names that reflected their origins or applications. Special properties of particular oils were usually known and exploited long before their biological functions were understood. Husbandry of the plants and animals that produce them and their extraction and purification made possible the manufacture of paints, lubricants and preservatives, as well as less enduring products such as fuels, cosmetics, perfumes and elaborately prepared food. These days we take such amenities for granted, but our ancestors devoted much ingenuity to improving the technology for extracting and adapting natural oils to different uses. Each new development enabled people to have brighter lights, or build more complex and powerful machinery, as well as contributing greatly to the variety and sophistication of their cultural lives.

Modern science has not only explained why certain oils have special properties, but also produced some general principles that explain why certain combinations of them occur in certain plants or animals. The main reason why most plant lipids contain a mixture of fatty acids of different chain length and unsaturation is maintenance of the necessary tissue fluidity at low temperature. But such general concepts cannot predict the exact composition of lipids in the seeds of any particular species or individual: there is nearly always more than one biochemical solution to any biological situation, and lipids are no exception. The nutritional benefits of such variety to animals and people are just a spin-off from the plants' needs.

Animals can make only some of the fatty acids they need for themselves. Others, notably the 18-carbon fatty acids of the n–3 and n–6 series, must be obtained from the diet and they can be quite rare in certain habitats. Thus the requirement for specific lipids pins animals to particular diets, habits and habitats to a much greater extent than the requirement for proteins or carbohydrates. The following chapter is about how organisms deal with these large and somewhat intractable molecules.

4

Lipids in action

Nearly all the best known biological processes – replication of genes, synthesis of proteins, transport and utilisation of oxygen – take place in the watery compartments of cells. Waxes and triacylglycerols are quite literally excluded from water-based metabolic processes unless their participation is facilitated by 'mediators', usually molecules of which part is water-soluble, and part lipid-soluble. One of the advantages of lipids is that they can be concentrated into secure energy stores such as seeds or adipose tissue, but because they are so condensed, such special biochemical mechanisms are needed to bring them into metabolic processes.

Animals obtain many of the lipids they need from their food, but they have to be digested, absorbed and transported to the appropriate destinations. Plants (except fungi) do not take up lipids from other organisms but make their own where they are to be stored and used. So plants are generally more adept at the intracellular synthesis and utilisation of lipids, and animals at their extracellular digestion and transport. This chapter is about the mechanisms by which lipids are digested, absorbed, transported, synthesised, deployed and eventually broken down.

Lipid digestion

The first step in lipid digestion is emulsification, the breaking up of pure lipid into tiny droplets suspended in an aqueous medium, as in milk. Emulsification is especially important to the digestion of oily fish livers or adipose tissue, which contain large quantities of densely packed lipid. In vertebrates, the first stage is mechanical, the powerful muscles in the wall of the stomach churn its contents, breaking up the lipids into tiny globules. The gut contents then leave the stomach and pass to the intestine, where bile, a chemical emulsifier, completes the process. Bile is synthesised in the liver and released into the intestine via the gall bladder and bile duct when required. The active ingredients of bile are not enzymes but complex salts of cholesterol that emulsify fats in much the same way as soap does, and thereby make them accessible to enzymes that operate best in a watery, rather than a fatty, environment.

Lipid-digesting enzymes are called lipases and they break the ester bonds linking the fatty acids to glycerol (see Figure 2, page 12). In all vertebrates that have been investigated, the main digestive lipase is released into the small intestine beyond the stomach from the pancreas, a gland with a wide range of important functions (that of cattle and pigs is sold by butchers as 'sweetbreads'). The stomachs of certain mammals, among them humans, rabbits and pigs, also produce a gastric lipase, so lipid digestion begins there. Normally only about 10% of fat breakdown takes place in the human stomach, but regular consumption of a high-fat diet can induce the production of more gastric lipase. Although more than 75% of the molecular structure of gastric lipase is identical to that of pancreatic lipase (from the same species of animal), the two enzymes are different enough to attack preferentially ester bonds at opposite ends of the glycerol molecule, thereby liberating slightly different mixtures of fatty acids.

Working in conjunction with co-lipase, a molecule that forms a link between the emulsified lipid droplets and the enzyme, lipase breaks up triacylglycerols by releasing the two outer fatty acids and leaving the middle ones attached, forming monoacylglycerols. A similar enzyme, phospholipase, removes one of the fatty acids from phospholipids. Non-esterified fatty acids and monoacylglycerols, but not whole triacylglycerols, can be absorbed through the lining of the gut and into the blood. Digestive lipases can hydrolyse the ester bonds of a wide range of fatty acids, though those of humans and rats act fastest on triacylglycerols containing short- and medium-chain fatty acids (e.g. milk fats), and slowest on those of very long chain polyunsaturates, such as those of fish oils. These properties mean that the essential fatty acids are less efficiently digested than those whose main function is energy production.

Whipped cream and certain cooked fats are deemed 'rich' foods, that some people find indigestible. In fact, as pointed out in Chapter 3, cream is no richer in lipids than butter (and may actually contain less fat by weight), but their physical state makes it more difficult for the lipases to hydrolyse them, so digestion is slower, and the sense of satiety persists for longer.

For humans and rats, digestion and absorption of lipids are relatively slow, much slower than those of all but the most complex carbohydrates. In a normal mixed meal, the glucose, sucrose and other simple sugars have usually been digested, absorbed and distributed to tissues such as muscles before significant quantities of lipids even get into the bloodstream. Adult humans may take more than 6 hours to complete the digestion and absorption of a meal that included large amounts of 'rich' fats, although small birds probably complete the processes much faster, in part by restricting themselves to foods such as nuts, in which the lipids occur as tiny droplets encased in membranes, an arrangement that makes them more digestible.

As might be expected in view of the unique chemical composition and physical state of milk, lipid digestion in mammalian neonates has some special features that disappear at weaning. The saliva of human infants, rat pups and calves (and probably other mammals) contains a lipase that hydrolyses triacylglycerols, so digestion begins in the mouth. Since milk is already a lipid-in-water emulsion, emulsification by stomach churning and bile secretion is much less important to its digestion. The milk of humans, gorillas (and possibly other primates) and carnivores also contains a lipase, which supplements the action of lipases that the neonates produce for themselves. It has been proposed that adding a lipase could make artificial infant food more digestible for premature or sick babies.

Fungi and many micro-organisms also produce lipases as part of their battery of enzymes that break down the tissues of dead and dying organisms. Those of certain fungi are extracted and used in the food industry for interesterification of edible fats (see Chapter 3) because they are much less specific than most vertebrate lipases: they deal with triacylglycerols containing many different kinds and arrangements of fatty acids equally efficiently.

Pharmacologists take an active interest in lipid-digesting enzymes to develop drugs that aid slimming. Their research has produced drugs[1] that inhibit pancreatic lipase sufficiently to prevent about 30% of triacylglycerols eaten from being digested thoroughly enough to be absorbed through the gut lining. The undigested, unabsorbed lipids pass right through the gut and are eliminated in the faeces. Many drugs work by inhibiting (or less often facilitating) an enzyme, but to be effective and safe, the dose must be carefully monitored. The body tends to respond to interference by producing more of the affected enzyme, often enough to overwhelm the inhibitor. Total suppression may block important metabolic processes, so the elimination of lipases, as happens in certain rare inherited disorders of lipid metabolism, would, over a long period, lead to serious deficiencies in the intake of essential lipids and associated nutrients.

Absorption

Until very recently, fatty acids and monoacylglycerols were believed to diffuse passively through the membranes of cells lining the gut and into the blood (in contrast to amino acids and glucose). It is now clear that the absorption of these products of lipid digestion is facilitated by various kinds of carrier molecules. Although there are specific transport mechanisms for each of the 20 common amino acids, there is no evidence (yet) that each kind of fatty acid has its own carrier. So the products of lipid digestion are absorbed in the proportions in which they are present in the digested food, but there are complications. The middle fatty acid, which in natural plant lipids is usually unsaturated, is absorbed as a monoacylglycerol, and so forms the basis for the synthesis of new triacylglycerols, while the non-esterified fatty acids from the two outer positions usually follow other pathways: incorporation into structural lipids or oxidation.

Such non-discriminatory mechanisms of lipid absorption mean that lipid-soluble substances can often slip into the body along with the fatty acids and monoacylglycerols. The lipid-soluble vitamins, A, D, K and E, are absorbed in this way, but so are contaminants with similar solubility, notably DDT (dichlorodiphenyltrichloroethane, widely used as an insecticide from the 1940s to the 1970s) and PCBs (polychlorinated biphenyls, formerly used as components of large batteries and other electrical apparatus). As mentioned in Chapter 1, animals take up these substances if they are in or on their food, because the gut 'can't help' absorbing them. Such molecules cannot enter any of the normal pathways that break down lipids, and are much too large and too insoluble in water to be excreted, so they accumulate in lipid-rich tissues, including adipose tissue.

As long as bile and lipases are produced in sufficient quantities, digestion and absorption work so efficiently that nearly all of the common dietary lipids are eventually taken into the lymph or blood. Only a small fraction are not digested efficiently enough to be absorbed if the liver and pancreas are functioning normally, though incomplete digestion and hence poor absorption of lipids is a common symptom of liver diseases. Cholesterol, both that taken in as a component of food and that in spent bile salts, is less efficiently absorbed than the common fatty acids. A proportion, some estimates suggest up to 90%, is absorbed and transported back to the liver, where it is reformed into bile salts or otherwise re-used, but the rest remains in the gut contents and is eventually eliminated.

As well as bile, faeces consist of cellulose and other indigestible carbohydrates, cells shed from the lining of the guts, and the remains of the bacteria that live naturally in the lower gut. The bowels reabsorb much of the water in the faeces, which can become hard if they remain there too long. Ordinary lipids are digested and their fragments absorbed higher up in the gut, but castor oil escapes this fate because the —OH side chains on the ricinoleic acid distort the triacylglycerol molecule so much that the lipase enzymes cannot get close enough to the ester bonds to hydrolyse them. Whole triacylglycerols are too big to be absorbed, so much of a dose of castor oil taken by mouth remains in the gut contents as they pass into the bowels, where it acts as a lubricant to the faeces and possibly also restricts absorption of too much water by forming a fatty barrier between the gut lining and its contents. Other natural lipids that act as laxatives, such as those extracted from the 'castor oil' fish, *Ruvettus*, depend upon similar principles: the molecules are of an unusual configuration, so the lipases cannot reach their ester bonds. The lipid persists intact in the gut contents until it reaches the bowels, and, after doing its work, is eliminated with the faeces.

Another kind of ester bond that human (and rat) gut lipases cannot deal with efficiently is those of waxes. Eating beeswax in honeycombs does people little harm but provides no nourishment: the material simply passes through the gut intact. Very few terrestrial vertebrates try to eat beeswax, but the honey-guides, small Asian and African birds related to woodpeckers feed on large insects including bees, and bees' nests. They have unique micro-organisms in their guts that digest the wax into fatty acids and long-chain alcohols. Many fish eat tiny marine invertebrates, some of which contain a fair proportion of waxes (see Chapters 1 and 3), so their lipases must be able to hydrolyse these lipids effectively.

In developing the new 'fat substitute' called olestra, biochemists exploited the fact that the size and shape of the rest of the molecule can effectively 'protect' its ester bonds from being hydrolysed by pancreatic lipase. Olestra is an entirely artificial substance composed of fatty acids linked by ester bonds (see Figure 2, page 12) to a sucrose molecule. Because it consists mainly of esterified fatty acids, olestra tastes quite like natural triacylglycerols, and it has similar physical properties, so it can be used for frying, baking or making mayonnaise. Its smell and taste when eaten alone and as an ingredient of cooked foods are so convincingly similar to those of triacylglycerols that olestra produces a feeling of satiety nearly as efficiently as natural fats do. But olestra is non-nourishing because, as in the case of castor oil and waxes, human lipases cannot easily reach the ester bonds, so the molecule is hydrolysed only very slowly. Without adequate digestion, there can be no absorption, so slimmers have the flavour and sensation of eating fat without absorbing much of the energy it contains.

Olestra is considered safe because the small amount that is hydrolysed by lipases just breaks down into sucrose and fatty acids, which are then further digested and absorbed in the normal way, exactly as though the person had eaten butter and jam. The main snag is that, like castor oil and other unabsorbed lipids, its persistence in the gut contents can cause diarrhoea, which if intense or prolonged (due to people eating lots of olestra), would lead to dangerous losses of water and minerals. It is mainly because of this hazard that the new slimming aid is not licensed for general use in Europe, though since the mid-1990s, it has been used in the USA as an ingredient of 'fat-free' snack foods such as potato chips.

While olestra might not be perfect, we should admire the ingenuity of biochemists in constructing from natural ingredients a substance that fools the palate and the digestion, if not the bowels. Other pharmacological approaches to limiting lipid uptake are also being developed, including inhibitors of the carrier mechanisms that facilitate the uptake of fatty acids into the gut lining. The physiological mechanisms of lipid digestion and absorption are so efficient that they are not easily disrupted artificially. But these processes are quite simple compared with the problems of dealing with lipids once they are inside the body tissues.

Ruminants

Most animals do not synthesise enzymes that can digest cellulose, the major structural carbohydrate of higher plants. Many micro-organisms and fungi can produce such enzymes (thus many fungi can 'rot' wood) and most herbivorous mammals harbour such organisms in some part of the gut to assist in the digestion of tough plant material. Collaboration between micro-organisms and mammals in the digestion of plants is most sophisticated in ruminants, which have four stomach-like chambers in the gut. Rumination enables mammals to digest foliage from which other herbivores can derive little nourishment and is the main reason for the success of deer, antelope, buffalo, cattle, sheep, goats and their relatives. As well as breaking down complex carbohydrates, the micro-organisms also make major changes to the plant lipids, with important consequences for the lipid metabolism of their hosts.

The first and by far the largest stomach is the rumen, which contains millions of micro-organisms. They break down cellulose into glucose and immediately convert much of it into short-chain soluble fatty acids such as 4-carbon butyric acid, and it is in this form that the products of cellulose digestion reach the intestine and are absorbed.[2] Such fatty acids pass through the cells lining the intestine and thence into the blood, in which they are carried quite easily because they are much more soluble in water than the long-chain fatty acids.

Short-chain fatty acids are not incorporated into storage triacylglycerols but are quickly oxidised to release energy, in much the same way as glucose is in simple-stomached (i.e. non-ruminant) animals. They may also reach the mammary gland, where they are esterified and secreted in the milk, including that of cows, and thence to butter, from which butyric acid was first isolated (the name 'butyric' means 'butter'). Fresh butter, in which all the fatty acids are esterified, is odourless but after boiling or prolonged storage in air, it develops a strong, pungent smell, caused by the release of short-chain fatty acids, among them butyric acid.

Phospholipids in the membranes of cells in grass and leaves contain mostly unsaturated fatty acids, as do the triacylglycerols of many seeds. A herbivorous diet thus includes plenty of polyunsaturates, but micro-organisms in the rumen desaturate and otherwise modify many of them. So most of the long-chain fatty acids actually absorbed by the gut are saturated. Ruminant adipose tissue thus contains a much higher proportion of triacylglycerols with saturated fatty acids. Most of the lipids in the blood, from which foetal calves, lambs or fawns obtain the supplies they need for growth, are also of this composition.

In simple-stomached animals such as pigs, rats, horses and ourselves, digestion does not normally destroy the internal structure of the fatty acids themselves: they are absorbed intact, in all the variety described in the previous chapter. Consequently, the composition of the fatty acids in adipose tissue and milk corresponds approximately to that of the diet. But in ruminants, neither the abundance nor the composition of fatty acids in membranes, adipose tissue or milk lipids corresponds to those of the diet, and cannot readily be altered by changing the lipid content of the food.

Many non-ruminant animals, including rabbits, guinea-pigs and ourselves, harbour micro-organisms in the large intestine which digest complex carbohydrates to short-chain fatty acids in the same way as those of the rumen do. Guinea-pigs can synthesise long-chain fatty acids and hence triacylglycerols from such precursors. But in humans most short-chain fatty acids are quickly oxidised as fuel in the liver, or, particularly in the case of butyric acid, by the cells of the large intestine itself, and they make no contribution to storage lipids. The production of small quantities of short-chain fatty acids in the bowel means that non-ruminants such as ourselves have the biochemical equipment to deal with them. These capacities were simply expanded when humans took to eating ruminant milk and milk products which, as pointed out in Chapter 3, is a recently acquired and highly unusual habit. Being small molecules, short-chain fatty acids evaporate readily, and contribute to the smell of the gases that emerge from the gut.

The milk of all monkeys and apes, including women, contains a lower proportion of lipids in total (and substantially more lactose), than that of almost all other mammals. Although lipids are more abundant in the milk of cows, ewes and other ruminants, a smaller fraction are polyunsaturated fatty acids and far more are short-chain or medium-chain, than is the case in primates' milk. As explained in Chapter 3, some of the polyunsaturates contain *trans* double bonds that may not be suitable for primates. Lambs, calves, kids and fawns have efficient means of obtaining enough polyunsaturated fatty acids from their mothers, both during gestation and while suckling, and of getting them to the growing nervous system and eye where they are most needed. But human infants are not so adapted, and have proportionately larger brains, so problems can arise when they are fed on cows', goats' or ewes' milk.

Regular doses of cod liver and other fish oils help to restore the balance of saturated, monounsaturated and polyunsaturated fatty acids, but the mix still may not supply the growing nervous and immune systems (both major users of specific kinds of polyunsaturated fatty acids) as well as that of human milk. Biochemists have recently developed a means of 'protecting' lipids from the micro-organisms in the rumen: globules of seed oils, or any other lipid, are coated with strands of protein, which are bound together with formaldehyde. Microbial enzymes cannot penetrate this wall of cross-linked protein, so the particles and their precious contents pass unscathed into the cow's fourth stomach. Here, strong acids and enzymes secreted by the stomach lining digest away the protein, releasing the lipids, which are then emulsified, hydrolysed and absorbed in the usual way. Such technology has achieved modest success in increasing the proportion of *cis* polyunsaturated fatty acids in the lipids of cows' milk, thereby improving its nutritional value to humans.

Of the animal milks so far analysed, that of carnivores is closest in fatty acid composition to women's milk. Dog and ferret milk has more fat (7.9% and 6.7% respectively) than human milk (3.6%) but the fatty acid compositions are almost identical, with a high proportion of polyunsaturates, and almost no fatty acids shorter than C14:0. Perhaps Romulus and Remus, the mythical bastard twins who were suckled by a she-wolf and grew up to found the city of Rome, have something to teach us about the ideal baby food.

Lipid transport

Moving lipids around the body is not simple. Non-esterified long-chain fatty acids can't be trusted on their own because they tend to bind to proteins, so once inside the cells lining the intestine, they and the monoacylglycerols are quickly re-esterified into triacylglycerols. Medium-chain and short-chain fatty acids do not esterify readily but instead move across the intestinal cells and into the blood, where they bind to specific carrier proteins called albumin in which form they are safely carried around the body. Triacylglycerols are as insoluble in the blood or lymph as they were in the gut contents, so they can only be transported efficiently when assembled into special structures. These molecular escorts, comparable in some ways to haemoglobin that chaperones oxygen around the body, are called lipoproteins.

A standard method of separating large biological molecules is by spinning solutions of them fast for many minutes in a centrifuge. In such apparatus, the densest molecules sink most rapidly towards the bottom of the tube, while the lighter ones float. Mammalian lipoproteins were identified by such means and so are classified according to their density: those with the highest proportion of protein (50%) and the lowest of lipid are the most dense, around 1.06–1.2 g/ml, and are called 'high density lipoprotein,' while those with only around 10% protein float on top of the others and are called 'low density lipoproteins (LDL)' or 'very low density lipoproteins (VLDL)'. The lightest of all, chylomicrons, have the highest proportion (83%) of triacylglycerols (and only 2% protein).

This classification was constructed for the convenience of biochemists, and does not represent hard and fast differences in the particles' structure or physiological roles, but chylomicrons do appear to be a distinct category of lipoprotein. They are composed of triacylglycerols made from the re-esterification of newly absorbed fatty acids, plus smaller quantities of cholesterol, that are packaged into balls with a few structural proteins. They can be assembled very quickly and in large numbers, and may become huge by biochemical standards. The large ones are up to 0.001 mm in diameter (they average around 70 nm (7×10^{-8} m)), too big to squeeze into the space between adjacent cells of most tissues, but still much smaller than a red blood cell,[3] so they almost touch the walls of the fine capillary vessels as they pass through them.

Chylomicrons appear first in the lymph and thence pass into the blood. After a fatty meal, they may be big enough and numerous enough to produce a noticeable change in the colour and fluidity of the blood. But chylomicrons only last a few minutes: adipose tissue (and some other tissues, such as muscle) synthesises and releases into the tiny blood vessels that permeates it the enzyme lipoprotein lipase that severs the fatty acids from the triacylglycerols, fragmenting the chylomicrons.

Adipocytes can produce lipoprotein lipase in larger quantities than any other protein they export. When still inside the adipocyte, lipoprotein lipase is in an inactive form but it becomes functional when it is expelled through the cell membrane and into adjacent blood vessels. Normally a soluble protein like lipoprotein lipase would be washed away in the blood flow, and some probably does meet this fate, but most is attached by sticky carbohydrate molecules to the walls of the blood vessels, and waits there to serve the adipocytes that made it. Since the largest lipoproteins are so big they almost fill a small blood capillary, each may be attacked on all sides by many lipase enzymes, like maggots on a carcase. The fatty acids so released are taken up by the adipocytes, possibly with the help of carrier molecules in the membrane, and the remaining fragments of the chylomicrons are carried back to the liver which absorbs them entirely.

Once inside the adipocytes, the fatty acids are again re-esterified, and become storage lipids. Like so many biochemical mechanisms, this process is not always 100% efficient. Recent research has revealed that, at least in adipose tissue in the paunch depot on the outer wall of the human belly, not all the fatty acids liberated by lipoprotein lipase are taken into nearby adipocytes. Some remain in the blood, bound to albumin or other kinds of proteins and in this state, they circulate around the body. Thus adipose tissue acts to modify the lipid composition of the blood without actually exchanging materials between itself and the body fluids.[4]

Lipoprotein lipase is present in small quantities on the outer surfaces of many other kinds of cells, including some kinds of immune cells and sperm-forming cells, especially when they are dividing rapidly. It may enable them to extract from passing lipoproteins the fatty acids they need for energy metabolism and for making phospholipids for the membranes of the new cells. The proliferating cells seem to 'take their pick' of the circulating fatty acids and the rest are either broken down to produce energy, or end up in adipose tissue triacylglycerols. Tissues like cardiac and skeletal muscle can produce as much lipoprotein lipase as adipocytes can, but they normally do so only during prolonged fasting and sustained exercise, when supplies of glucose and stores of glycogen are low. Under such conditions, adipose tissue produces much less lipoprotein lipase,[5] thereby avoiding competition for available triacylglycerols with tissues that need to use them as fuels.

Very low density lipoproteins are also eventually broken up by lipoprotein lipase. They are smaller than chylomicrons and 'live' much longer, often for many hours, and they mostly carry triacylglycerols synthesised in the liver (from fatty acids released from adipose tissue, or, less commonly in humans, made anew from glucose) during periods of fasting. On their return to the liver, the lipoprotein remnants are mostly in the 'high density lipoprotein' category, and contain cholesterol esterified to a fatty acid, usually linoleic acid, forming cholesteryl ester.

Once in the liver, the cholesterol may be 'redeployed' as a component of bile salts. Since some bile remains in the faeces, rather than being reabsorbed, the conversion of cholesterol to bile offers, at least in theory, a means of expelling some of it, thereby reducing the total amount in the body. Medical interest in doing so was much enhanced in the 1970s and 1980s, following the discovery that on the whole (there are many individual exceptions), people in populations in which the average blood cholesterol is high die more frequently from ailments caused by blockage of arteries. How far cholesterol is really the culprit in this association, and the contribution of other kinds of lipids, are discussed in Chapter 8.

Lipoproteins seem to be essential to lipid transport in all animals. Although their basic body plans are very different from those of vertebrates, insects and probably many other invertebrates have a comparable range of lipoproteins that convey fatty acids between the gut, storage organs, muscles and elsewhere. Although similar in basic structure, lipoproteins from other kinds of animals differ in detail. Those of most insects contain diacylglycerols, and fatty acids are transported in this form, instead of as triacylglycerols or non-esterified fatty acids. Diacylglycerols are slightly more miscible with water and proteins than triacylglycerols, so off-loading of the lipids onto cells is a little easier. Insect lipoproteins are not re-formed after each passage around the system, so their protein components can act as re-usable shuttles, ferrying diacylglycerols around without being broken down themselves. Lipid transport in insects is certainly efficient: species such as the migratory locust can fly for hours using flight muscles fuelled almost entirely by lipids stored in an abdominal fat body.

These biochemical differences between major groups of animals probably became established long ago, early in their evolutionary history. But insects are nothing if not adaptable and opportunistic. The lipoproteins of those that feed on vertebrate blood as adults, notably *Aedes aegypti*, the mosquito that transmits yellow fever, contain triacylglycerols.[6] Enzymes that assemble and hydrolyse diacylglycerols do not work as well on lipoproteins containing triacylglycerols and vice versa. By modifying their own metabolism, blood-suckers can utilise their hosts' blood lipids with the minimum of biochemical fuss.

Uses of lipoproteins

All but the very smallest eggs contain energy storage molecules, usually triacylglycerols, that fuel the development of the embryo. The eggs of birds and reptiles[7] are among the largest of all and their energy stores are concentrated in the yolk. Modern hens have been selectively bred to lay their eggs even if they are not fertilised (through lack of access to a cockerel). In this state, eggs 'keep' for many weeks, but if fertilised, the embryo would start to develop within a few days of the egg being laid, as long as it is warm. Blood vessels grow out from the embryo into the yolk, and take up the nutrients therein. The yolk contains both high density and low density lipoproteins, presumably to help the embryo to absorb and utilise the lipids in it.

The capacity of yolk lipoproteins to act as an interface between lipids and non-lipids is the main reason for the unique and indispensable role of eggs in foods such as cakes, custards and mayonnaise. Recipes for mayonnaise call for raw egg yolk, which is combined with an impressive quantity (up to twice the volume of the yolks) of salad oil, forming a surprisingly firm emulsion. The mixture must be stirred briskly as the oil is added a little at a time, but not so vigorously as to break the delicate structure of the lipoproteins, upon which the whole confection depends: they expand greatly as they absorb more triacylglycerols into their protein framework, much as they do when transporting lipids after a rich meal.

Like other physiologically active proteins, those of lipoproteins degrade with time, especially if heated or cooled. Stale or improperly stored eggs do not make good mayonnaise, and cooked eggs are useless. Real mayonnaise tends to separate on storage and, because the egg yolks contain proteins and minerals as well as lipid, is susceptible to spoilage by bacteria and fungi such as those that form a bluish mould. As with ice-cream, the industrially produced equivalent contains artificial emulsifying and stabilising agents, and various anti-oxidants and preservatives.

Egg 'white' is almost pure protein and water. Beating it stretches the long protein molecules, enabling them to trap tiny bubbles of air, which expand when heated. The heat also unravels the protein molecules, so they 'set' in the expanded configuration to form a firm framework, as in meringues. Even a small quantity of yolk accidentally spilt into the white on separating eggs can ruin a soufflé: instead of rising to become light and fluffy, it goes flat on cooking, because the fats and lipoproteins from the yolk cause the mesh of protein molecules to collapse, releasing the air trapped in them. For the same reason, beaten whole eggs are never as stiff as beaten egg white alone, and their volume does not expand nearly so much. Nonetheless, whole eggs bind starches (e.g. in flour) and cellulose (e.g. in dried or fresh fruit) to the fatty ingredients, thus determining the texture and friability of cakes and custards. The contrast between the texture of cake and that of water biscuits is due mainly to the presence of egg. No other material, natural or synthetic, quite imitates these properties of eggs, so their role in cookery is secure, at least for the foreseeable future.

Lipoproteins are also the basis for the use of egg yolk in the manufacture of paints. A painting technique known as tempera, in which ground pigments are mixed with some sort of binding substance and applied to well-seasoned wood or other slightly porous material, was developed in ancient Egypt, and widely practised until the invention of modern oil paints in the fifteenth century. Glue made by boiling the skins of cattle, sheep or horses, and even milk were sometimes used as binding agents, but throughout mediaeval Europe, the best quality paintings used hens' eggs. Whole egg, pure egg yolk and yolk mixed with drying oils were used, sometimes different binding agents for different pigments for the same picture.

The key ingredient for making a workable tempera paint is the lipoproteins. They bind strongly to the pigments and form an oil-in-water emulsion that can be diluted to the required consistency with either oil or water (in practice, water was almost always used). As in biological membranes, the lipids spread to form a thin but continuous film that adheres to a slightly porous surface. The 'drying' process owes more to the composition of the yolk triacylglycerols. The evaporation of the water takes only a few minutes, but complete 'drying', the oxidation and polymerisation of the triacylglycerols, takes up to a year. Egg tempera paints eventually form a hard, smooth surface that can be gently polished to a pleasing sheen, and do not flake, darken or otherwise degrade with time.

The egg yolk and the pigment were mixed in roughly equal proportions, so substantial numbers of eggs were required for even quite a small painting. Mediaeval hens laid much smaller eggs than our modern, highly bred birds, and did so abundantly only in the spring. For best results, the eggs must be very fresh – disintegrating lipoproteins in stale eggs do not make smooth, easily applied paints. Once mixed, egg-based paints cannot be stored for longer than a day or two unless a preservative is added, nor can the ingredients be easily extracted and reused. Many of the paints contained extremely expensive pigments such as lapis lazuli, often used for the brilliant blue of saints' robes, and could not be wasted on stale reagents or by prolonged storage.

Artists using this technique had to be supported by several well-organised suppliers and assistants, from poultry keepers to paint mixers, and they had to work as and when the materials could be assembled. Duccio, Piero della Francesca, Fra Angelico, Botticelli, Mantegna and other mediaeval artists achieved masterpieces in what is now regarded as a technically difficult and inconvenient medium. The costs must have seemed enormous compared with ordinary people's earnings. The artists' patrons valued the paintings very highly and many have survived for us to admire. Oil paints, which are water-in-oil emulsions thinned with turpentine or similar solvent, are much simpler to use. Their invention enabled artists to work with fewer assistants and in remote locations instead of being confined to a studio. In spite of these advantages, some effects and colours could at first only be achieved with the old techniques, and many painters of the late fifteenth and early sixteenth centuries used both kinds of paint in the same picture.

Tracer lipids

Lipid digestion and transport mechanisms deliver fatty acids intact from the food to adipose tissue where they remain (although subject to continuous turnover) until they are converted into more complex lipids or are oxidised for energy production. Except in ruminants, the carbon atoms in the fatty acid chain stay together in the same arrangement as they were in the original food. Therefore, in non-ruminants, the relative abundance of different kinds of triacylglycerol fatty acids in storage tissues is similar to that of their diet.

This property explains how the flavour of meat, especially that of birds, fish and single- stomached mammals such as pigs and rodents, is affected by diet: the authentic 'gamy' taste of free-living salmon, pheasants, partridge and wild boar is due largely to the mix of lipids and lipid-soluble compounds derived from the wild plants that they (or in the case of salmon, their prey) eat. Gourmets can identify a 'fishy' taste to the eggs and meat of poultry that have fed on fish-meal. The expression, 'you are what you eat' applies particularly to lipids, a fact that has been exploited to determine the dietary history of certain wild animals.

As mentioned in Chapter 3, certain long-chain polyunsaturated fatty acids are synthesised almost exclusively by marine organisms. Birds and mammals that feed partly or entirely in the marine ecosystem accumulate more of these fatty acids than those that eat from the land. It is often important to know where migratory birds, or coastal and estuarine species are obtaining most of their food in order to understand their habitat requirements and to predict the effects of draining marshes or developing agricultural uses for land.

Analysis of the fatty acid composition of triacylglycerols in a small sample of superficial adipose tissue taken by biopsy can help determine where the birds' food is coming from. In a recent investigation, 10% of the triacylglycerol fatty acids of beach-feeding sandpipers (*Calidris pusilla*) were found to be eicosapentaenoic acid (C20:5n–3) and docosapentaenoic acid (C22:5n–3), fatty acids that are synthesised almost exclusively in the marine ecosystem, while only 5% of those of estuarine plovers (*Charadrius semipalmatus*) and 1% of those of mud-feeding dowitchers (*Limnodromus griseus*) collected from adjacent areas were of these types.[8] Polyunsaturated fatty acids are particularly useful for such studies because the animals' livers can perform only a limited range of transformations and such fatty acids are often selectively conserved.

This kind of detective work can be further refined by studying the isotopes of the carbon atoms in the lipids. Isotopes are alternative forms of the same element that differ in the number of particles in their nuclei. At least two isotopes of most elements occur naturally (though often one isotope is very much more abundant than the others), and in many cases, additional kinds of isotopes, or more of the naturally occurring forms, can be synthesised artificially in a nuclear reactor. Because isotopes of the same element differ only in the composition of their nuclei, their chemical properties, which depend upon the number and arrangement of the electrons, are similar enough for molecules that contain different isotopes to follow the same pathways in biological systems. They are useful to biologists because, with suitable apparatus, isotopes of atoms such as carbon, hydrogen, oxygen or phosphorus can be detected and counted, thus 'labelling' the molecules that contain them.

Unstable, i.e. radioactive, isotopes are more familiar to most people than stable isotopes: fears about the harmful effects of the radiation they emitted kept them in the spotlight. Since the 1950s, radioactive isotopes have been widely used in medical diagnosis and biological research, especially that concerned with investigating where molecules go to in the body and what they become, because the atoms could be located and quantified by the radiation they produced. Stable isotopes do not, by definition, emit radiation, so they cannot harm the organisms being studied or the observers. They also last forever, while radioactive isotopes decay with a half-life ranging from seconds to millennia.

Much more complex and expensive apparatus, called a mass spectrometer, is necessary to distinguish stable isotopes by the small differences in mass that arise from the extra or missing components of the nucleus, and to quantify each one even when only a very little of it is present. Coupled to equipment that separates mixtures of chemicals, mass spectrometers can measure the isotopic composition of the atoms in each compound separately. Stable isotopes are thus becoming increasingly important in physiological and ecological research.

Carbon is by far the most abundant element by weight in most biological molecules, and it occurs naturally as two stable isotopes. In inorganic materials, including atmospheric carbon dioxide, the great majority, 99.11%, of carbon atoms are of the type known as ^{12}C (because they have 12 major particles in the nucleus), and the rest are ^{13}C which are 8% heavier because their nuclei contain one more particle. Slightly different proportions of ^{13}C and ^{12}C are taken up by the tiny single-celled plants that float on the surface of the sea and obtain their carbon in solution as bicarbonate, and terrestrial plants that take in carbon dioxide straight from the atmosphere. The outcome of such fractionation of the mixture of isotopes is that organic materials in marine organisms contain about 5–7 parts per thousand more of the heavier isotope (^{13}C) than similar molecules from animals that have eaten terrestrial foods (though both have fewer ^{13}C atoms than are present in the carbon dioxide in the atmosphere).

Because triacylglycerol fatty acids pass unchanged (apart from several rounds of hydrolysis and re-esterification, which do not affect their carbon atoms) from food to body tissues, the proportions of ^{13}C and ^{12}C isotopes in their carbon skeletons reveal something about the animal's dietary history over previous weeks or months. The new technique of gas chromatography-isotope ratio-mass spectroscopy can measure accurately the ratio of ^{13}C to ^{12}C in each fatty acid separately, thereby indicating whether it originated from the marine or terrestrial ecosystem.

Such methods have been used to investigate food chains and other aspects of ecosystem structure and the dietary histories of wild animals. Arctic foxes (*Alopex lagopus*) are widespread in the Arctic and in high mountains at lower latitudes, in fact, almost anywhere in the northern hemisphere where the climate is too cold for the larger red fox to survive. They are active all winter, and can travel long distances both over land and across frozen seas, so they have colonised all but the smallest islands of the Arctic Ocean and Hudson Bay. On the mainland of Russia, Canada and Alaska, arctic foxes eat mainly lemmings and other small rodents, plus the occasional hare, birds such as ducks and ptarmigan, and fish and carrion if they find it. In many areas, they are so heavily dependent upon lemmings as prey that fox numbers fluctuate from year to year according to the abundance of lemmings.

On Svalbard, there are no lemmings, or any other native rodents, nor hares, no freshwater fish (the species failed to colonise this isolated archipelago after the end of the last glaciation), and ptarmigan are sparse, but there are arctic foxes nonetheless. Being only about the size of a domestic cat, arctic foxes are too small to kill healthy adult reindeer, but they eat the carrion from ones that have died from other causes. They also take the eggs and nestlings of puffins, auks, geese, eider ducks and the many other birds that migrate to Svalbard for a few months in summer to breed. Apart from the geese, these birds feed from the marine food chain, taking fish and invertebrates from the rich coastal waters around the islands. When the fjords are frozen over and the migratory birds have gone, coastal foxes living in areas where reindeer are scarce may follow polar bears onto the sea-ice, and scavenge the remains of the seals that they kill.

Reindeer and ptarmigan eat terrestrial vegetation, so both the chemical composition of triacylglycerols, and the relative abundance of the isotopes of carbon in them should be distinguishable from those of lipids that originated from the sea. Analysis of triacylglycerols in adipose tissue samples from wild-caught foxes[9] showed that palmitic (C16:0) and stearic (C18:0) acids originated from the land: the abundance of these fatty acids was inversely proportional to that of ^{13}C carbon atoms in them. The carbon atoms in the fatty acids had been 'fixed' i.e. turned into an organic compound, by a terrestrial plant. The foxes obtained their supplies of palmitoleic acid (C16:1) almost entirely from marine organisms, while oleic acid (C18:1) came from both terrestrial and marine foods in about equal proportions.

Similar individuals thrived on very different diets: the composition of the fatty acids of some of the foxes caught inland showed that they lived mainly on dead reindeer, while others proved to have been eating large amounts of animals that fed in the sea. Adaptability has long been the key to the success of wolves, foxes and their relatives: food is scarce in the high Arctic, so Svalbard arctic foxes can't afford to be fussy. Nonetheless, the adipose tissue of all individuals studied contained a few fatty acids derived from the marine food chain; for the reasons explained in Chapter 3, such 'fish oils' may be essential nutrients for carnivores, as they are for humans.

The use of stable isotopes for reconstructing contemporary and past diets and ecological relationships between organisms is a fashionable and rapidly expanding area of science. Fatty acids and cholesterol are particularly useful as markers in such studies because they are much more stable than amino acids (the main components of proteins) or carbohydrates. Fatty acids lose their acidic end groups quite quickly, but decomposer organisms such as bacteria or fungi cannot easily break up the remaining hydrocarbon chains any further, so they survive for millennia in the marrow of buried human and some animal bones, and in coral fragments or even just amorphous sediment. After genes, lipids are the most stable of the uniquely biological molecules. Should you ever become a fossil, your fatty acids would reveal more than you might wish about your habits and lifestyle.

Lipids as fuel

Any lipid can be burnt, i.e. oxidised, its component atoms broken from each other and combined with oxygen to form carbon dioxide and water. The latter, of course, is produced as steam, so is no more easily seen than the carbon dioxide at the temperatures of burning. From the development of wick lamps about 40 000 years ago, until the invention of gas and electric lighting just over a century ago, burning to produce light and heat were major uses for waxes and oils. Although less spectacular than axes or arrowheads, the manufacture of portable stone lamps made possible some equally important advances in human culture, including the exploration and subsequent decoration of deep caves, from which dozens of lamps have been recovered, some richly decorated with images of animals. Various other animals use tools for catching or manipulating food or for building nests, but humans are unique in making equipment for illumination.

Some kinds of fuels work better for lighting than others. To be easily absorbed into the wick, the fats should melt at a low temperature, but those that evaporate before they burn are wasteful and generate unpleasant smells. Fuels containing water, protein or other impurities splutter and drip, and the glycerol component of triacylglycerols does not burn completely, producing a smoky, acrid-smelling flame. The highest quality candles and lamp oil were made of purified beeswax and/or certain grades of whale 'oil' (that are mainly waxes); they burn brightly and produce a pleasant smell.[10]

Analysis of the carbon isotope composition of the minute residues left on the Palaeolithic lamps (using variants of the techniques just described) reveals that the fuels were triacylglycerols from wild mammals, probably large terrestrial species such as deer, mammoth, bison and horses. At least in France and Spain where most of the lamps were collected, ancient people did not use vegetable oils or beeswax for lighting. They presumably just put up with the smell. By the time of the Pharaohs (about 5000 years ago), sophisticated Egyptians were burning castor oil, extracted from the seeds of specially cultivated plants, but elsewhere animal fats were widely used for lighting until well into the twentieth century.

The breakdown of lipids in biological systems is basically similar to burning, except that it proceeds more slowly and at a lower temperature, and the energy is released in a form that can be used to drive other chemical reactions in cells. As in burning, oxygen is essential for the production of chemical energy from lipids in biological systems, in contrast to the utilisation of glucose, which can proceed, albeit less efficiently, in the absence of oxygen. The carbon atoms end up in carbon dioxide, which is expelled from the body through the lungs or kidneys, and the hydrogen atoms as water, which may be retained or eliminated as required.

Releasing the stores

One important difference between burning and biological oxidation of fats is that the latter cannot begin until the triacylglycerols are hydrolysed into fatty acids and glycerol. So yet another round of lipolysis takes place, this time inside cells, catalysed by a lipase named, somewhat inappropriately, hormone sensitive lipase. Its production and activity are affected by hormones, notably insulin, and by noradrenalin,[11] but then so are those of the other lipases. The products of lipolysis inside muscle fibres or liver cells are used where they are formed, but those of adipocytes are released into the blood. A small amount of triacylglycerol lipolysis goes on all the time in adipocytes, but most of the resulting fatty acids are re-esterified into the same or adjacent cells before they have a chance to enter the general circulation. The glycerol cannot be reclaimed and a 'new' glycerol molecule is made from glucose for each round of fatty acid/triacylglycerol cycling.

Lipolysis is stimulated by low levels of insulin in the blood, such as those which occur after several hours (at least six in humans) of fasting, and by any kind of arousal, stress, excitement, exposure to cold and exercise that activate the sympathetic nerves to release noradrenalin at their terminals. As described in Chapter 2, fine nerves of the sympathetic nervous system permeate adipose tissue, innervating almost all the adipocytes. Noradrenalin binds to receptors (of which there are several distinct kinds) on the adipocyte surface and thereby quickly (i.e. within seconds) stimulates lipolysis. Activation of the sympathetic nervous system also increases blood flow through adipose tissue, so much of the fatty acids and glycerol so released are carried away and may be taken up by tissues elsewhere in the body.

Noradrenalin is readily available as a laboratory chemical, and adipocytes can quite easily be separated from their framework of connective tissue, fine nerves and blood vessels, so the action of this agent on 'isolated' adipocytes has been extensively studied. It is upon such artificial systems, more than upon the study of intact adipose tissue *in vivo* that most of our knowledge about lipolysis is based. Such methods work well only on mature adipose tissue – very small adipocytes are fragile and break up – and it is difficult to keep isolated cells in good condition for longer than a few hours. But that is long enough to study the basic mechanisms of lipolysis, and much is now known about the molecular steps between the arrival of a few noradrenalin molecules (and it really could be as few as dozens) at the surface of an adipocyte and the increased release of fatty acids and glycerol from the cell.

For a long time, it was assumed that lipolysis produces fatty acids in the proportions in which they occur in the adipocytes' triacylglycerols. But more accurate measurements[12] have recently revealed that adipose tissue releases proportionately more polyunsaturated fatty acids than saturated. Those of shorter chain length (C14 or C16) are more likely to be hydrolysed than longer-chain fatty acids with a similar degree of unsaturation. The biochemical mechanisms behind selective lipolysis are still not clear, nor is it known whether all adipocytes behave similarly in this respect.

In horses and dogs, lipolysis from adipose tissue is sufficient to raise the concentration of both fatty acids and glycerol in venous blood by nearly fourfold after several hours of strenuous exercise. In humans, these concentrations double after about an hour of running, and reach a maximum of three times the resting levels. Such increases might sound quite a lot but, at least in well-fed humans, they represent lipolysis of only a tiny fraction of the total lipids stored in the adipose tissue. When stimulated with maximal doses of noradrenalin, adipocytes in culture hydrolyse the fatty acids from at most 1% of their triacylglycerols per hour of incubation. Lipolysis from intact adipose tissue is probably even slower.

The technical problems associated with monitoring what different adipose depots are doing in the living animal mean that it is still not clear whether all adipocytes everywhere increase their rate of lipolysis a little bit when moderate exercise begins, or a few adipocytes in each depot, or perhaps all of those in just one or two small depots, increase lipolysis a lot, and the rest just 'watch'. There is some indirect evidence that the sympathetic nervous system may selectively stimulate certain adipocytes depending upon the physiological conditions: the pattern of stimulation may be different in fever, chronic nervous stress, brief strenuous exercise, etc.

It is important to answer such questions because too much non-esterified fatty acid in the blood does little to improve the energy supply to muscles and can be positively dangerous. Excess fatty acids bind non-specifically to cell membranes, causing certain kinds of blood cells to stick together in clumps. Such binding also interferes with the transmission of electrical signals within and between cells, and disrupts enzymes. There is more about the implications of excessive lipolysis from adipose tissue in Chapter 8.

Depots differ

Measurements of lipolytic products in the blood reveal nothing about which adipose depots the lipids came from. Figure 13 summarises an attempt to find out, by comparing the action of identical doses of noradrenalin on adipocytes from different sites of the same guinea-pigs. In these experiments,[13] the concentration of glycerol in the solution bathing the adipocytes was measured as an indicator of lipolysis. Adipocytes from all depots of guinea-pigs that had just been running released more glycerol, especially with the higher doses of noradrenalin, than those from comparable depots of resting animals, but there were some important contrasts between depots.

Even large concentrations of noradrenalin do little to stimulate lipolysis from the samples from the perirenal depot on the dorsal wall of the abdomen: glycerol release is high even when the stimulant is completely absent, but compared with the other depots, it responds sluggishly to increased concentrations, and the maximum rate of lipolysis is low. This large depot, accounting for more than a quarter of all the adipose tissue, also has a lower activity of lipoprotein lipase in fasting animals than any of the other depots studied (though its lipoprotein lipase activity does increase substantially within minutes of feeding). Evidently, rapid uptake and release of lipids are not this depot's strengths; it does, however, produce more leptin than any other depot that has been studied. This property, and the fact that it continues to release lipolytic products at a more or less constant rate, regardless of how much noradrenalin is present, suggests that it may be a major source of circulating fatty acids, and the brain's main indicator of total fatness.

The large superficial depots in front of the hindlimb and behind the forelimb respond to noradrenalin in the same way as the perirenal, and together

Figure 13. (opposite) Summary of an experiment to demonstrate site-specific differences in the way that adipose tissue responds to stimulation from the sympathetic nervous system. Adipocytes were prepared from two intra-abdominal depots (dark shading), two intermuscular depots (light shading) and four superficial depots (no shading) of ten sedentary guinea-pigs (solid points on graphs) and eleven exercised animals (open points). The figures on each depot refer to the proportion of the total adipose tissue located in that depot. All the graphs are drawn to the same scale. The vertical axes show the quantity of glycerol released by 1 g of adipocytes in 1 hour, with the scale running from 30 to 70 μmol per ml. The horizontal axes show the concentration of noradrenalin in the incubation medium, from zero (left), 10^{-9},10^{-8}, 10^{-7},10^{-6} and 10^{-5} molar.

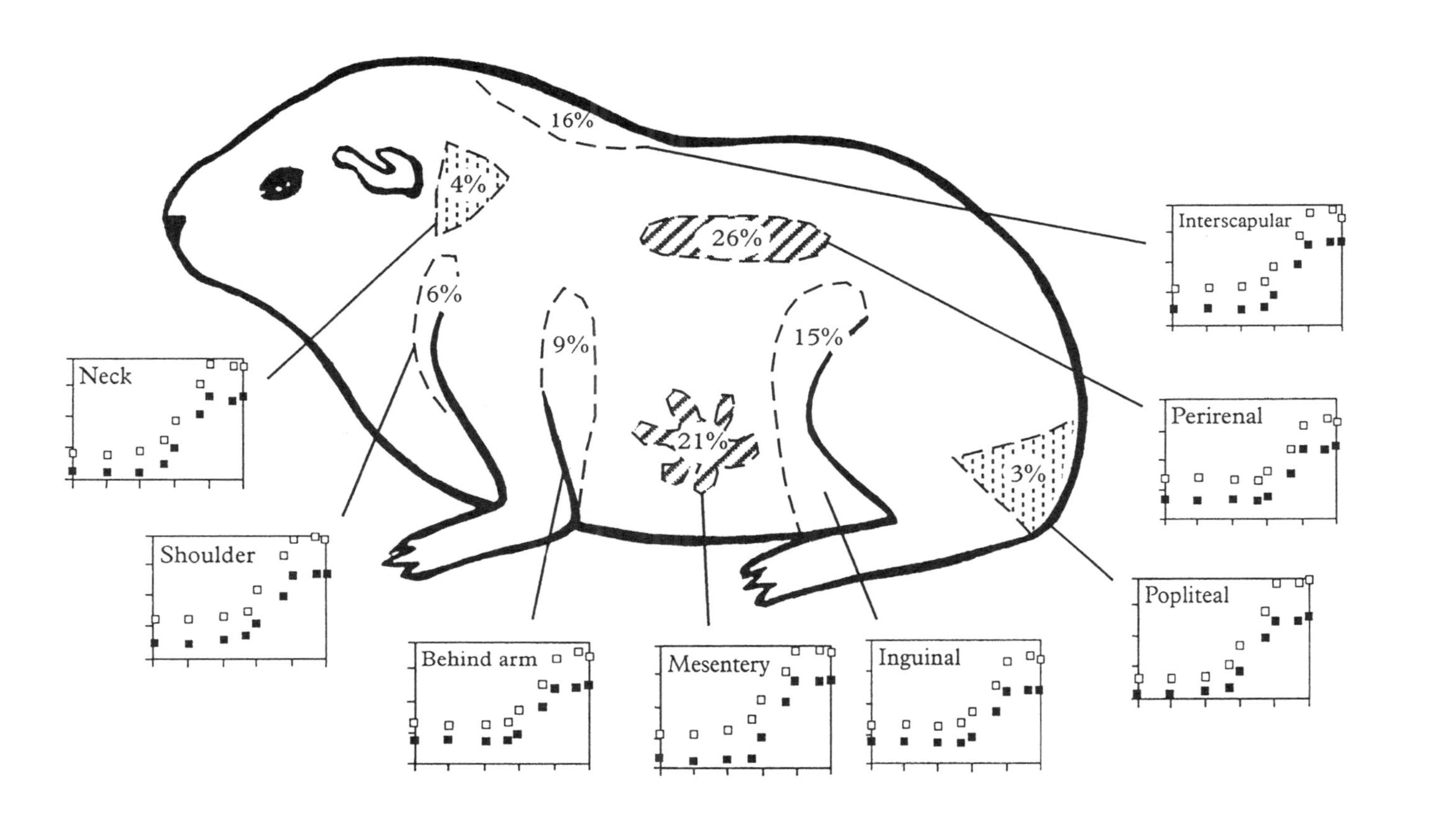
16%
4%
26%
6%
9%
15%
21%
3%
Interscapular
Perirenal
Popliteal
Neck
Shoulder
Behind arm
Mesentery
Inguinal

these depots represent half the total amount of adipose tissue in guinea-pigs. Each of the other depots has its own unique pattern of response to noradrenalin. The two intermuscular depots are the most sensitive to changes in concentration; their adipocytes, especially those from the exercised guinea-pigs, switch abruptly from a very low to a very high rate of lipolysis as the noradrenalin concentration is increased from 50 to 1000 units. The two small superficial depots and the mesentery release glycerol at the same rate when there is little or no noradrenalin, but have a higher maximum.

The site-specific differences shown in Figure 13 may seem quite minor, but the experiments were run for only an hour. Even very slight shifts in sensitivity to nervous stimulation could, over a lifetime, lead to large changes in the depots' mass. Further studies revealed within-depot, as well as between-depot differences. Adipocytes from beside lymph nodes, whether embedded in the mesentery, omentum or intermuscular depots, are slightly more sensitive to noradrenalin than those a few millimetres away from it. There is more about these and other special properties of the adipose tissue around lymph nodes in Chapter 8.

Intermuscular adipose tissue accounts for at best 20% of the total, usually nearer 5% in most mammals, split into several minor depots, many of them too small to provide enough tissue for physiological analysis. We have studied a couple of the larger intermuscular depots in some detail,[14] simply to find out why it is there at all: such a small quantity of lipid could easily be tucked into a larger depot, so why do mammals have adipose tissue in between muscles?

Measurements of the activity of lipoprotein lipase in guinea-pig adipose tissue show that the intermuscular depots have a higher capacity for taking up fatty acids than the larger depots, especially during rest. After a few minutes of exercise, as increased lipolysis in the all adipose depots raises the concentration of fatty acids in the blood, lipoprotein lipase activity in the intermuscular depots falls abruptly: their capacity to remove triacylglycerols from the circulation declines, leaving more available for the muscles. Such properties suggest that depots are biochemically equipped, and anatomically located, to contribute to the nourishment of adjacent muscles, with which they share a blood supply. Such a role may be the physiological reason for the intimate anatomical association between these small adipose depots and muscle.

In spite of being metabolically so active, the intermuscular depots do not undergo large changes in mass in obesity or starvation: on the contrary, they remain remarkably constant in size. The adipose tissue on our lower leg is anatomically an intermuscular depot (the popliteal, see Figures 6 and 7, pages 41 and 43) even though, because people stand on straight knees and the thigh and shin bones are long, much of it appears to be superficial. You may have noticed how little it differs between massively obese and very lean people: the thickness of superficial adipose tissue on the thigh changes enormously, while that around the calf muscle is almost constant.

The adipose tissue associated with the heart also has some highly idiosyncratic properties. The epicardial and pericardial depots can represent a significant fraction, sometimes 10% or more of the total mass of the heart of certain species, although obviously they are small compared to the intra-abdominal or large superficial depots.[15] Although clearly visible in most large mammals (and some large birds, notably swans), as well as around human hearts, the depots are almost absent in laboratory rats and mice. There is a strong tendency among biochemists to believe that if a structure or tissue is absent (without obvious reason) from rats, but is found in humans, its presence in the latter is somehow pathological. Since most humans are fatter than rats, this interpretation is particularly attractive. If tissues are not there, they cannot be studied, so their absence from rats means that almost nothing is known about their metabolic properties.

Butchers and cooks have known for years that cattle, sheep and pigs have cardiac adipose tissue, but there is little incentive to study the composition of a cheap cut of offal in animals that are always slaughtered well before the age at which heart disease presents a significant problem. The epicardial adipose tissue is so firmly stuck onto the heart muscle that dissecting the tissues apart is tedious, and in small animals, difficult to do accurately. So there are very few data on the mass or properties of these depots in domestic livestock or even zoo animals.

Recent measurements from polar bears, arctic foxes and reindeer,[16] show that the cardiac depots can be extensive in some wild specimens. But their masses differ irregularly between apparently similar individuals of the same species, and do not correlate with the masses of any of the larger depots over a very wide range of levels of fatness. Because they are not depleted in starvation, the cardiac depots can appear to be relatively massive in very lean specimens. In this respect, these depots resemble metabolically inert, structural adipose tissue (see Chapter 6) but that is where the resemblance ends.

Biochemical studies on cardiac adipose tissue freshly excised from adult guinea-pigs reveal that the depots are capable of taking up fatty acids, and of making triacylglycerols at a remarkably high rate, higher than either the perirenal or popliteal depots of the same animal. Furthermore, their activities are modulated by feeding the subject on a lipid-enriched diet, but, unusually for metabolically active adipose tissue, not by insulin.[17] The cardiac depots also release fatty acids. Lipolysis from them is more sensitive to noradrenalin than the much larger superficial or perirenal adipose depots, and almost as sensitive as the mesenteric depot.

The heart pumps continuously throughout life, using fuels, especially fatty acids, that could be deployed elsewhere. Any material on the heart muscle simply increases the work of pumping, and is therefore a very silly place to put any adipose tissue, unless its presence there is essential to its physiological role. These points, and its peculiar combination of metabolic and structural properties, make it most unlikely that the cardiac depots are simply part of the general energy storing adipose tissue that happens to have wandered onto the heart. The fact that wild animals have as much, sometimes more cardiac adipose tissue than ourselves and domesticated animals strongly suggests that it is useful, not merely a pathological consequence of obesity or unnatural lifestyle. But it is not at all clear what that useful role is. Possibilities include serving as a local energy source to supply the heart muscle, or perhaps a local sink, that mops up excess lipids released from other depots, during imperfectly regulated lipolysis.

Adipocytes in the structural adipose tissue in joints, feet and the eye-socket (discussed in detail in Chapter 6) seem to lack both the capacity to bind noradrenalin and the biochemical machinery for lipolysis: they are incapable of relinquishing their triacylglycerols, and may be almost unaltered in animals that have died of starvation. Others, notably the adipocytes in bone marrow, are metabolically active and readily take up fatty acids and esterify them. But, at least in rabbits, rats, dogs and probably humans, they do not respond to noradrenalin and do not release their storage lipids into the bloodstream, even in prolonged, severe starvation.

Bone marrow adipocytes are not firmly bound together with collagen as in typical adipose tissue but occur, sometimes in large numbers, as loose cells in close association with the cells that divide to produce blood cells. Until recently, very little was known about their properties or function. One reason for this state of affairs is that rat bones are too small to contain enough marrow for most kinds of assays and experiments. Rabbits or larger animals must be used, and they are expensive to keep. As will be explained in Chapter 8, the role of lipids in protection against disease is now more widely appreciated, and scientists are coming round to the opinion that these adipocytes serve a specific function in supporting the blood cell formation, a role that is probably too important for their lipid to be spared for general consumption by other tissues.

Investigations like the ones described in this section have established beyond doubt that mammalian adipose tissue is far from homogeneous in a wide variety of physiological properties. But with the few exceptions mentioned, we do not know why depots with certain combinations of properties are situated where they are in the body. Relating the anatomy to the physiology is a major question in adipose tissue biology, for which the basic research has hardly begun.

How cells use lipid fuels

Lipids do not decompose by burning, or indeed by most other means, very easily: as just mentioned, some triacylglycerols have survived 40 000 years embedded in crevices of a stone lamp. Oxidation of fatty acids to release energy in a biologically usable form (i.e. as ATP) requires enzymes and coenzymes arranged in highly organised layers of membranes. Animal cells break down fatty acid in mitochondria (see Figures 3 and 10, pages 21 and 54), but in green plants, the process also occurs in smaller bodies called peroxisomes. The abundance of such structures in cells provides a rough indication of their usual sources of energy. Muscles that can make extensive use of lipids as fuel always contain plenty of mitochondria; in some cases, notably the flight muscles of certain insects, there are so many that they occupy as much or more space than the contractile material that does the actual work of producing movement.

Activation of fatty acids begins with yet another round of esterification, this time to a helper molecule called coenzyme A. In a cycle of reactions involving several steps, carbon atoms with their associated hydrogen atoms are removed two at a time. Stearic acid (C18:0), for example, goes round the cycle eight times before it is completely broken into 2-carbon fragments. These small molecules remain in the mitochondria, where they are oxidised to produce ATP, carbon dioxide and water via about a dozen different biochemical steps. Once fatty acids enter a mitochondrion, their oxidation normally goes to completion.

Mitochondria need a large and continuous supply of oxygen to utilise fatty acids efficiently. In the flight muscle of insects, such provision is quite simple because respiration takes place through tiny tubes that open directly to the outside world and insects are so small, the oxygen does not have far to go once inside the body. In many large animals including people, the capacity of the lungs, heart and blood vessels to maintain an adequate supply of oxygen is often limiting: the cardiovascular system has to be very 'fit' to support high rates of oxidation in the muscles for longer than a few minutes. Bouts of running or jumping are usually terminated because we get 'out of breath', not because the energy supply to the muscles is exhausted.

Until recently, it was believed that fatty acids crossed cell membranes by diffusion alone, but various fatty acid carrier proteins have now been discovered, which may detach them from the albumin to which they are bound while in the blood, and facilitate their passage into cells. Whatever the mechanism, fatty acids find their way into the muscles, liver and other lipid-using tissues within seconds of being released from adipose tissue. It is generally assumed that they reach concentrations there similar to those measured in the blood. Glycerol is oxidised in much the same way as glucose, but, although this component accounts for about 10% of the mass of triacylglycerols with long-chain fatty acids, it contains only about 5% of the usable energy because it, like glucose, contains several oxygen atoms (i.e. is already partially oxidised).

It is possible to estimate the overall proportions of lipids and carbohydrates that an animal is using as fuels by measuring the ratio of carbon dioxide released to oxygen taken up. This ratio is about 1 when carbohydrates alone are being oxidised completely, but nearer 0.7 during the breakdown of pure lipids, because they contain more carbon and hydrogen but less oxygen than glucose. Careful monitoring of such respiratory gases over many hours during which subjects eat meals of exactly controlled composition shows that most people oxidise very little lipid while carbohydrate is plentiful. But as soon as supplies fall below a certain level, they switch to using lipids. Eating large, rich meals need not lead to obesity as long as one does not do so too often, and the carbohydrate stores are allowed to run low enough for long enough between such meals to promote lipolysis from adipose tissue and oxidation of the lipids.

Sedentary adults can, if necessary, obtain all the energy they need from carbohydrates, but children, people engaged in strenuous work and those suffering from fever use energy at a high rate and so need lipids as a dense form of metabolic fuel. Supplying glucose continuously as an intravenous drip is not sufficient to meet the energy of young children and they may become fatty acid deficient in as little as 2 weeks unless lipids are added.

Careful monitoring of people being fed for long periods on intravenous drips reveals an essential role of stored as well as exogenously supplied nutrients. To prevent the tubes of the drip from clogging, it is usually necessary to keep the flow going continuously. Glucose is always included in drips, so the apparatus keeps the blood sugar continuously high enough to suppress lipolysis. Adults kept on drips for long periods develop abnormalities of lipid utilisation that can be corrected by reducing the quantity of glucose infused. Some lipolysis seems to be essential, even for very thin, severely ill people, suggesting that some tissues may take up fatty acids from adjacent adipose tissue rather than from the main bloodstream.

Most animals break their daily food intake into just a few discrete 'meals', even when caged with continuous access to food. Animals as diverse as cats and parrots may ignore food for hours then eat steadily for as long as 20 minutes once they settle down to a meal, suggesting to less knowledgeable observers that they must have been very hungry beforehand. If so, why did they not have a snack hours ago (as some modern people would do)? There may be sound metabolic reasons for alternating periods of using mainly glucose as fuel with interludes of fasting, when lipid stores are mobilised.

High performance or maximum economy

The main snag about lipid mobilisation is that, relative to the very high rate of energy utilisation in mammals, it is slow: although even short periods of exercise increase lipolysis from adipose tissue, it can take over 4 hours of continuous running for the fatty acid concentration in human blood to reach a maximum. That is too slow for oxidation of lipids to be of much use in bursts of exercise, such as chasing prey or escaping from predators, which are fuelled by glucose alone. Provided sufficient glucose is available at the same time, fatty acids can contribute up to 60% of the fuel in moderate, prolonged exercise.[18] Stores of free glucose and glycogen are used up in just a few minutes, but animals can make more glucose from other precursors, including glycerol and amino acids. These sources provide more than enough glucose for overnight fasts and similar long intervals between meals, but in humans, the maximum rate of production is only sufficient to sustain continuous running at about half the fastest possible pace. Any faster and demand exceeds supply, depleting blood glucose to a dangerously low level.

Thus the rate of glucose production, rather than the availability of fatty acids, sets the appropriate pace for running marathons and other long distance races: athletes who run too fast, and thus use glucose more rapidly than it can be replenished, eventually feel 'light-headed' (because glucose supplies to the brain become insufficient) and collapse with exhaustion, even though fatty acid levels in the blood are high, and there are plenty more still in the adipose tissue. Even at an ideal pace, rats in exercise wheels and human athletes cannot run for long enough to deplete their adipose tissue stores to anywhere near zero. They have not been adapted by evolution for continuous, strenuous exercise. Chimps, gorillas, orang-utans, many monkeys and most relatives of rats and mice intersperse a few minutes of strenuous climbing, jumping or running with hours of gentle walking, or just pottering about feeding.

Human athletes can extend their capacity for sustained exercise by forcing themselves to eat more starch during the few days before competing, to achieve so-called carbohydrate loading: glycogen stores in the liver and muscles accumulate to higher than normal levels, thereby maintaining supplies of glucose for longer. Similar strategies work for horses but not for dogs, which are thoroughly adapted to a low-carbohydrate, high-fat diet. Experiments on huskies trained to pull sleds show that their performance can be enhanced by feeding extra fat, but not by extra starch. Dog muscles contain proportionately more mitochondria, which might enable them to oxidise fats more efficiently than human muscles can. This species difference means that exercise is likely to be much more effective for slimming dogs than for slimming people.

Many migratory birds and fish such as salmon regulate glucose and lipid supplies so efficiently that they can exercise until their lipid reserves are completely exhausted. Homing pigeons are sometimes found dead with their adipose tissue almost empty of triacylglycerols, having apparently dropped out of the sky with exhaustion. Using lipids rather than glucose has other advantages for long-distance travellers. When lipids are fully oxidised, the carbon atoms become part of carbon dioxide and the hydrogen atoms, water. Usually these end-products leave the body in the expired breath or the urine, the carbon dioxide mainly in the former and the water mainly in the latter, but when an animal is losing more water through evaporation than it can find to drink, the extra produced by the oxidation of lipids can help to balance the loss.

Nearly two thirds of all the atoms in fatty acids are hydrogen (most carbon atoms are attached to two hydrogen atoms – see Figures 1 and 2, pages 10 and 12) compared with half in glucose (formula $C_6H_{12}O_6$). About 1 gram of water is formed for each gram of triacylglycerol oxidised (the exact value depends upon what kinds of fatty acids they contain), compared with only about 0.6 g from the oxidation of one gram of glucose. Migratory birds that use lipids as fuel can thus fly for longer without having to stop for a drink, which can be an important advantage when flying over deserts or the sea.

Even birds can only use large quantities of lipids at moderate rates of energy expenditure: maximal exercise can only be fuelled by carbohydrate. Hovering is one of the most energetically expensive activities known, and only insects and very small birds can do it for longer than a second or two. Hummingbirds hover in front of flowers to feed on nectar, which is a concentrated solution of simple sugars including glucose and, in many plant species, a few amino acids. The birds convert this concentrated food into muscular work at record speed: small hummingbirds weighing about 3 g were found to absorb glucose through the gut seven times faster than non-hovering waxwings of ten times their body mass. The very small amount of waste material was voided less than 50 minutes after a meal. Like butterflies, hummingbirds act as pollinators for many of the species from which they feed, carrying pollen that sticks to their feathers from flower to flower. Some pollen is often knocked into the nectar, and provides the pollinators with a dense source of proteins and polyunsaturated lipids, most of which are used in roles other than as fuel.

In most animals including ourselves, the main use of adipose tissue triacylglycerols as fuels is in fasting. Starvation is metabolically similar to slow exercise, except that the main stimulus to lipolysis is a continuously low level of insulin and a rise in glucagon, another circulating hormone, rather than the release of noradrenalin. In people, attendant feelings of cold, hunger or other kinds of stress or discomfort may activate the sympathetic nervous system so both effects may contribute to lipolysis. Animals that routinely fast for long periods seem to avoid these side-effects of starvation: birds incubating eggs or bears in breeding or overwintering dens show no obvious signs of distress.

In humans, proteins, especially those of muscle, begin to be broken down and their carbon skeletons used to make glucose long before the lipid reserves approach exhaustion, so slimmers find that muscle wasting, and hence physical weakness, are side-effects of trying to reduce fat. In animals adapted to fasting, such processes occur to only a minimal extent until the advanced stages of starvation, thus postponing significant muscle wasting. The importance of such 'fuel management' for natural fasts is discussed in Chapter 5.

Plants

The short-term energy stores of plants are always carbohydrates, usually starches in the form of tiny granules. These reserves maintain the plant during the night and on very cloudy days when photosynthesis is slow or absent. In many plants, the long-term energy stores are also starch, but in some, the need to compress a lot of energy into a small space makes spending the extra energy required to synthesise triacylglycerols worthwhile for seeds. Seedlings utilise their lipid stores to produce both the energy and the structural carbohydrates for growing the roots, shoots and leaves that enable them to synthesise sugars in sunlight.

Green plants have a special biochemical mechanism, not known to be present in animals other than a few kinds of parasitic worms, that permits the incorporation of fatty acid carbon atoms into precursors for the synthesis of glucose, which is the starting point for the formation of the structural carbohydrates that build roots, stems, leaves and, eventually, wood. By converting the fatty acid components of triacylglycerols into carbohydrate, plants can use seed lipids in ways that animals cannot use their adipose tissue stores. The process works equally efficiently from any fatty acid with an even number of carbon atoms, because the carbon atoms are removed from the fatty acid two at a time, and built into 6-carbon glucose. Since it is not tightly limited by the biochemical mechanisms by which the lipids are reclaimed, the exact composition of seed triacylglycerols can be adapted to the storage conditions, temperature or the possibility of being eaten, as described in Chapter 3. Plant seeds have a much wider variety of fatty acids in their triacylglycerols than are found in animal storage lipids.

Mobilisation of lipids begins at germination with the uptake of water into the seeds that swell as the proteins and starches absorb water. Preformed enzymes are activated and new ones synthesised, but, as in the case of animal systems, water-based enzymes cannot easily attack large droplets of lipid. In seeds, the action of triacylglycerol lipases is facilitated by a special protein that mediates the binding of the enzymes to the lipids, in much the same way as lipoproteins promote the action of lipoprotein lipase, and co-lipase assists pancreatic lipase. The triacylglycerols in seeds are always dispersed into tiny compartments called oil bodies, each only a few micrometres (thousandths of a millimetre) across, that are enclosed in membranes, at least some of which contain lipases. Such structures contrast with the (relatively) huge droplet of triacylglycerols in vertebrate adipocytes that may be over 100 μm (0.1 mm) in diameter, or up to a million times the volume of oil bodies in seeds.

The compartmental arrangement of plant storage lipids allows them to be utilised quickly. Similar adaptations are not necessary in white adipocytes because such cells do not metabolise the triacylglycerols themselves to any great extent. They mostly export the lipids to other kinds of cells elsewhere in the body that oxidise them or use them as substrates for the synthesis of other lipid molecules, as the case may be. But as explained in Chapter 2, brown adipose tissue oxidises its own lipids to generate heat, although when such supplies run low, it can, like most other animal cells, take up lipids and glucose from the blood circulation. The similarity in size and general appearance between plant oil bodies and brown adipocytes may arise from the fact that in both cases, storage lipids are utilised in the cells in which they are sequestered.

Plants do not eat meals or run away from predators, and their growth proceeds at a fairly steady rate, so they do not have to cope with abrupt changes in the supply of or demand for glucose and other energy-providing molecules, in the way that vertebrates, insects and other complex animals do. Plants can match the rate of mobilisation of seed lipids to their requirements without having to resort to agents like insulin, which acts to prevent excesses or deficits in the availability of circulating lipids and glucose by adjusting the activities of liver, muscle and adipose tissue on a time scale of a few minutes or hours.

Differences in the quantity of storage tissue is the main cause of the enormous range of sizes of seeds, from coconuts that produce just a few seeds each weighing several kilograms, to herbaceous weeds and orchids that produce millions of tiny seeds. Seedlings from the latter must establish themselves quickly or perish, but it may take weeks or months for young coconuts or avocado pears to exhaust all the energy reserves in their large, nutritious seed.

The capacity to sequester and reclaim lipids in this way seems to be confined to seeds and seedlings, for which compactness and low density are important qualities in storage materials. The energy stores of the vegetative stages of plants, which are not so constrained by size and weight, are almost always carbohydrates. Biennial and perennial plants sequester energy between growing seasons as starches or sugars often in the underground parts of the plant. Potatoes, parsnips, carrots, beetroot and dandelion roots are familiar examples of such storage organs. Such materials are mobilised early in the spring of the plant's second or later year, giving its growth of roots and shoots a 'headstart' over that of annual plants. We harvest the storage tissues as food in the autumn or winter of the first year: if you delay until the plants sprout in spring, they become flabby and tasteless.

In contrast to leaves, from which many valuable nutrients, especially those of the pigments essential for photosynthesis, are withdrawn before they fall, plants cannot reclaim the nutrients in seeds and fruits, if (as often happens) an insect establishes itself inside a seed or fruit, and eats the embryo. The most it can do is to abort the doomed structures, thereby preventing them from accumulating even more nutrients. Eliminating diseased or damaged tissues means that more nutrients can be directed into the stores of viable seeds. The plant 'gives them up for lost' and a special layer of cells in their stem weakens, so they are blown off in the next gale. A large proportion of the apples, peaches and other fruits that fall prematurely are infested with caterpillars (which produce a distasteful brown dross) or other seed predators. The uninfected fruits and seeds remain firmly attached, and their storage tissues expand, taking the share of nutrients that might otherwise have gone to those that were aborted. These seeds thus become bigger and would (if humans did not eat them first) produce healthy seedlings when they germinate.

Animals eating plants

The contrast between the microanatomy of oil bodies in seeds and lipid droplets in white adipocytes also helps to explain why peanuts, cashew nuts, sesame seeds, sunflower seeds and walnuts, do not taste 'greasy', even though they may be over 50% lipid (as much as some 'light' margarines), while chunks of vertebrate adipose tissue, and the purified oils extracted from the nuts, are unappetising to many people because of their greasy taste and texture.

The partitioning of seed lipids into tiny oil bodies also makes them more digestible for animals because less mechanical emulsification is required. Among higher vertebrates, digestion is further speeded up by preliminary processing in the mouth. Parrots, finches and other seed-eating birds use their powerful, finely controlled beaks to shred nuts and seeds into tiny fragments before swallowing them, carefully spitting out the shell and any other indigestible components. Seed-eating mammals such as rodents (rats, mice, squirrels and hamsters) gnaw or crush seeds with their hard, continuously growing teeth, often holding the food in their front paws for a really neat job. Gnawing and shredding nuts may take some time, but the digestibility and high nutritional quality of the food so obtained make it worthwhile, especially for flying birds, which cannot afford to carry around heavy gut contents for long periods of slow digestion.

People also extract most plant oils by crushing or grinding the seeds, which breaks up the tiny oil-containing compartments, and squeezes the lipids out. Powerful, sustained forces and hard grinding surfaces are necessary to obtain a high yield: oil presses had to be so strongly built that many, including some very ancient ones, have survived in the archaeological record. In only a few large seeds, such as coconuts, is the oil concentrated enough to be extracted efficiently just by baking or boiling. Such strenuous methods are not necessary to obtain animal triacylglycerols from adipose tissue. As Palaeolithic people first discovered, they can easily be obtained in a highly pure form by 'rendering': pieces of adipose tissue are heated and the melted lipids simply flow out as the collagen shrinks and the enclosed adipocytes burst.

Building lipids

The membranes in and around plant cells and oily fruits and seeds are the original sources of most of the world's biological lipids. Higher plants synthesise fatty acids inside chloroplasts[19], or, in non-green tissues such as seeds, in similar membranous structures, starting from small molecules produced directly from photosynthesis. In animals, fatty acids are synthesised in the semi-fluid cytoplasm of cells, not in mitochondria or other membranous structures, but the necessary enzymes are usually grouped tightly together.

Although the starting points differ, the basic biochemical steps for synthesising fatty acids are similar in all plants, animals and microorganisms. Carbon atoms, with their attendant hydrogen atoms, are added two at a time in a cyclical process that involves the coordinated activity of several enzymes. This mechanism of fatty acid synthesis usually stops at 16 carbon atoms, i.e. with the formation of palmitic acid, after which additional enzymes are needed to make further modifications. The process requires significant amounts of energy in the form of ATP: in humans, converting dietary glucose into fatty acids for storage as triacylglycerols uses about 25% of the energy in the glucose, compared with only 7% for the synthesis of glycogen.[20]

Whether built *de novo* or absorbed from the diet, saturated fatty acids may be desaturated by desaturase enzymes, and all may be further elongated by elongases (see Chapter 3). Their activities, and hence the nature and extent of fatty acid modifications depend upon environmental factors, notably temperature. There are no genes for lipids *per se*, just genes for enzymes (i.e. proteins) that can take up, build or modify lipids for which the diet provides the precursors. So which lipids occur where in plants or animals, and in what quantities depend both upon the enzymes present (and hence upon which genes are active in producing them) and upon the organism's circumstances, including in the case of animals, what it can find to eat.

Most animal and plant cells can synthesise small quantities of fatty acids for their own use, for example, as components of membrane lipids. As mentioned in Chapter 1, cells of the outer layers of the vertebrate skin synthesise a variety of complex lipids, most of which contain fatty acids. During lactation, the mammary gland (breast) makes fatty acids for secretion as triacylglycerols into the milk, sometimes at an impressively high rate. In a few mammals, including rats, mice and guinea-pigs, the adipose tissue makes substantial quantities of lipids but in general, only the liver (or the liver-like organs of invertebrates) is capable of large-scale synthesis of lipids for export to other kinds of cells elsewhere in the body.

Although they may have the capacity for synthesising lipids, many animals rarely use their biochemical apparatus for lipid synthesis if their diet includes sufficient fats. Our modern western diet contains plenty of lipids, and as long as we can digest them efficiently, supplies are more than sufficient to keep the adipose tissue replete with triacylglycerols and provide enough precursors for other purposes, such as making phospholipids for membranes. Human adipose tissue and liver can synthesise fatty acids from glucose, but they do so only very slowly. Lipogenic enzymes in the human liver are less than a quarter, sometimes nearer 1%, as active as those of rats or birds. The formation of lipids from glucose makes a very small contribution to adipocytes' contents unless we eat carbohydrates in enormous quantities for long periods.

The main effect of frequent meals and sugary snacks is to suppress the utilisation of lipids, so any fats in the food end up in the adipose tissue. Almost all meals contain some of fat, which is normally burnt by light exercise, but if carbohydrate is regularly eaten to excess, dietary lipids gradually accumulate in adipose tissue, leading over months and years to obesity. Older readers may be surprised by this statement. Until the 1970s, sugar, bread, potatoes and other starches were regarded as 'fattening' because human tissues were thought to resemble those of rats, mice and many other herbivorous animals in readily converting the carbohydrates into fatty acids. Recent research has overturned this advice. It is now believed that fats, and, within reason, fats alone fatten people. Starch-reduced slimming products such as Energen bread have disappeared, to be replaced by skimmed milk, cottage cheese and low-fat cakes.

Carnivores such as cats and dogs, and probably snakes, eagles and vultures, also make few lipids for themselves. Although they often fast for long periods (months in the case of large snakes) between irregular meals, reclaiming the adipose tissue triacylglycerols, their diet always provides plenty of lipids and not much carbohydrate. There would be no point in making triacylglycerols from proteins or carbohydrates unless adipose tissue reserves had run exceptionally low, or the meal was exceptionally lean. Many parasitic animals also make little or no lipid, but for a slightly different reason: while living in or on their host, parasites have continuous access to food, so storage is unnecessary.

In animals, endogenously produced fatty acids are mixed up with those from the diet, so nearly all complex lipids are the products of both the animals' synthetic capacities and its dietary experiences. This statement is as true of invertebrates as it is of mammals and birds: animals can make their own lipids, but do so only to a modest extent if sufficient can be obtained from the diet, and fatty acids from the two sources may end up in the same triacylglycerol or phospholipid molecule. Experiments in which the fatty acid composition of the diet was changed in young, growing guinea-pigs revealed that triacylglycerols with fatty acids corresponding to those of the new diet appeared first in the inguinal and perirenal depots.[21] The simplest explanation is that these depots obtain most of their fatty acids from the blood which in these well-fed animals means straight from the diet. Other depots, notably the popliteal, if they expand at all, do so mainly by making their own fatty acids, so the composition of their triacylglycerols only slowly follows that of the diet.

In insects, as in vertebrates, the contributions of endogenous synthesis and diet differ greatly between species. The majority of insects, including aphids, most caterpillars and many kinds of beetles are herbivores. Except for those that eat lipid-rich seeds (as many do – most kinds of nuts and seeds are eaten by at least one kind of insect), the diet is low in lipids, so the capacity for synthesis is well developed, especially in females while they are building up nutrients in preparation for laying their eggs. But certain seed-borers, insects that live as parasites on other insects or on vertebrates, or which, like mosquitoes, suck vertebrate blood, rely almost entirely upon dietary lipids, and retain little of the biochemical apparatus for synthesising their own. In a word, such insects are like most modern humans: their diet includes plenty of lipids so they don't make more unless they have to.

The same principles hold for the synthesis of sterols. Cells of insects and plants, and most kinds of mammalian cells, can synthesise sterols in quantities more than sufficient to meet all their requirements. Plants have to make their own, of course, but vertebrates eating vertebrates (e.g. cats, ourselves), and insects eating insects (e.g. tiger beetles) take in plenty of the appropriate sterol with their food. In such cases, endogenous synthesis of sterols is down-regulated: sterols are large, complex molecules whose synthesis requires many different enzymes, and there is every reason to avoid making them if there is enough in the food.

One of the most concentrated sources of cholesterol in the human diet is the yolk of hens' eggs, and persons in whose blood it is present in excess are advised to avoid such food. However, reducing dietary intake may do little more than lift the suppression of the endogenous synthesis of cholesterol, so the total amount in the blood and other tissues may be unaltered. Most omnivorous animals, including most kinds of monkeys, eat any egg they find if they can crack it. One family of snakes (Dasypeltinae) has special adaptations for cracking hard-shelled birds' eggs, and eats almost nothing else. Its cholesterol metabolism has never been studied.

Regulating synthesis

Lipids require special enzymes to oxidise them and all the paraphernalia of lipoproteins to carry them around, and they cost energy to make. So perhaps it is not surprising that the storage materials of many plants are mainly or entirely carbohydrates, with lipids synthesised only for indispensable roles such as in membranes. If starches work so well as long-term energy stores for the humble potato, why cannot animals do likewise? We find great many starchy plants, and a few sugary plants (e.g. many fruits, sugar beet) but no sugary animals.

Almost all tissues can use glucose as fuel and many do so in preference to using fatty acids. Some animal and plant (but not bacterial) proteins, notably those of connective tissues such as cartilage and the lens and cornea of the eye, have components derived from glucose that are added in a process called glycosylation, after the standard steps of protein synthesis are completed. Although indispensable in these roles, glucose is harmful at high concentrations – it binds to proteins in ways that make them useless as enzymes or structural materials. Lipids, especially non-esterified fatty acids, are also harmful in excess, but as already explained, if there is more than enough glucose around to meet current needs for energy, adipose tissue quickly takes up and sequesters the superfluous lipids and stores them safely as triacylglycerols. Prompt, efficient mopping up of excess fatty acids may be an even more important function than supplying stored lipids for some adipose depots, especially the small ones.

Tidying away excess glucose after a large meal is just as essential, but involves different tissues. Insulin is a major player in the regulation of glucose metabolism. Within minutes of eating food containing sugars or starch, the pancreas secretes more insulin and its concentration in the blood rises, in normal humans by around fivefold. It acts on receptors and transporters in cell membranes, prompting tissues to take up glucose from the blood and use it for energy production. Glucose that cannot be oxidised at once is taken into cells, especially, but not exclusively, liver and muscle, and converted into an insoluble glycogen. Only a little glucose can be stored as glycogen; the maximum in humans is 200–500 g (less than 1% of the body mass), or up to 8% of the mass of the liver, and 2% of the skeletal muscles.

These mechanisms keep going for as long as glucose is being absorbed from the gut, but once glucose is 'cleared' from the circulation, insulin concentration in the blood declines. The liver switches from removing glucose to releasing it back into the blood, from which it is extracted and used by other tissues as they exhaust their own reserves. After a large meal, stores of glycogen in the liver and muscles may be sufficient to last adult humans for about 24 hours of sedentary living, but people (and rats) normally keep such short-term reserves at well below the maximum capacity.

Any excess glucose in the blood after glycogen deposits are filled up is converted into lipid, as a means of getting rid of it from the circulation. To put it simply, animals are adjusted to run on low (but not too low) levels of circulating glucose and do not have the means of storing large quantities as starch in the way that plants can. They have to turn excess glucose into lipid, but such emergency measures are rarely necessary. As explained in Chapter 8, according to some current theories, appetite (i.e. signals from the brain that promote or suppress feeding) is controlled mainly by blood glucose and tissue glycogen levels, so people, rats and presumably other animals do not normally eat more glucose-generating foods than they can safely handle. Just before hibernation, migration, breeding fasts or other special circumstances, high rates of lipid synthesis may be appropriate: the necessary enzymes are activated, and circulating glucose as well as fatty acids are turned into storage lipids before the glycogen stores are filled up, thereby preventing suppression of appetite and allowing feeding to continue.

For reasons to be explained in the following chapter, very large flying birds such as swans and geese are strictly limited in how much fat they can carry in flight. So they do not naturally overeat, no matter how delicious the food they are offered. To produce the French gourmet dish, *foie gras* ('fat liver', often turned into *pâté de foie gras*), captive geese have to be force-fed for many days. Their large wing feathers are usually removed, so they cannot burn off the excess energy by flying, and they are kept in small pens that prevent much being consumed by walking around. Once the adipose tissue is 'full up', the lipids accumulate in the liver, which may weigh up to 1 kg in a specially fattened Alsace goose, compared with 100 g in normal specimen.

Built-in lipids

In much of Chapter 3, polyunsaturated fatty acids are presented as multipotent, valuable, and in many animals, scarce lipids, while the saturates are seen as common, easily synthesised and not good for much except energy production. The majority of fatty acids of most membrane phospholipids are unsaturated, but those in the cells that line the lungs have two saturated fatty acids, usually palmitic acid (C16:0). These special phospholipids act as surfactants, enabling the lungs to inflate properly for only modest increases in air pressure, thereby minimising the muscular work required to breathe efficiently.[22] Phospholipids with straight chain saturated fatty acids pack neatly into a small space when the lungs are deflated, but re-spread readily on inflation. They serve as a sort of anti-glue, preventing the walls of all the tiny pockets of the lungs from sticking together, enabling them to fill easily with air. The proper composition and arrangement of these surfactants are particularly important in neonates. The lungs do not mature until just before birth, and inappropriate or insufficient phospholipids in their lining are a major cause of breathing difficulties in premature babies.

For many years, membranes were believed to be the main forum for physiologically important structural relationships between fatty acids and proteins, but recently another kind of association, called acylation has come to light. Many fatty acids stick to proteins in a non-specific way – that is one of the reasons why they are always esterified or bound to albumin when out in the blood – but myristic (C14:0) and palmitic (C16:0) acid form specific, functionally important attachments to a wide range of different proteins in both micro-organisms and multicellular animals. In spite of their chemical similarity, they link up with proteins in contrasting ways.[23] Myristic acid is always attached to one particular kind of amino acid, and the association is forged soon after the protein is synthesised. The fatty acid and proteinaceous components stay together for as long as the molecule remains functional. Palmitic acid is attached to a different type of amino acid, and seems to be more labile, separating from its original partner and attaching to other proteins quite frequently.

It is certain that the partnership between lipid and protein is essential to their function – the proteins are defective in mutant bacteria in which the attachment fails – but in many cases, it is not very clear exactly how the fatty acids help. Some acylated proteins, such as receptor molecules for hormones, are associated with membranes. Having lipid appendages may enable them to position themselves correctly on or in the membrane, but explanations along these lines do not obviously apply to many others. Recent reports[24] suggest that cholesterol links up with certain mammalian proteins in much the same way as fatty acids do, and that the union is essential to normal function.

We know so little about acylated proteins that the importance of the link with fatty acids is probably underestimated. From the biological point of view, the big difference between these structural roles of saturated fatty acids, and those of polyunsaturates, is that animals can make as much of the saturated lipids as they need, wherever they need them. No protein need go short of a lipid partner. In contrast, building phospholipids correctly may depend upon dietary supplies of essential polyunsaturated fatty acids, of which shortages are a real possibility.

Lipids as messengers

Certain lipids also serve as messengers within and between cells. The steroid sex hormones, testosterone and oestrogen, were the first lipid hormones to be recognised as such, initially in mammals, then in a wide range of other vertebrates. Other steroid hormones have since been identified, including ecdysone, the hormone that controls the moulting of insect cuticle. Steroid hormones are synthesised from cholesterol or other sterols and, like cholesterol, readily mix with triacylglycerols and phospholipids. They permeate cell membranes so easily that their receptors are sometimes located inside the nucleus as well as, or instead of, on the cell surface. They circulate through the body, coordinating the many different tissues that contribute to fertility and breeding. Intensive research since the beginning of the twentieth century, culminating in the development of the contraceptive pill for women (and now modified for use on many other domesticated and zoo mammals), has made the vertebrate steroids, testosterone and oestrogen, among the best known of all hormones.

Until 1970, steroid hormones were believed to be the prerogative of vertebrates and insects. Then a steroid that acted as an essential and unique growth promoter was isolated from cabbage pollen. Several such substances have now been identified in a wide range of plants, and named brassinosteroids after the cabbage family (Brassicaceae) in which they were first discovered. Plants that lack the genes for producing them are stunted, and their failure to grow cannot be corrected by any other growth hormone. The fact that steroids act as regulators in green plants as well as in widely different lineages of animals suggests that this role is a very ancient feature of multicellular organisms.

Prostaglandins, leukotrienes and thromboxanes are families of messenger molecules derived from fatty acids. Prostaglandins are so named because they were first isolated from the prostate gland that surrounds the male urinary duct. (It tends to enlarge in elderly men, constricting the flow of urine, and so it may be surgically removed or its growth suppressed.) Prostaglandins have since been found in almost all tissues of mammals, and in other kinds of animals, and have a variety of roles. The most thoroughly studied in fish, amphibians, reptiles and birds are the maturation of eggs and sperm and other aspects of the reproductive system. In mammals, the best-known function of lipid-derived messenger molecules is the coordination of the cellular processes that contribute to tissue inflammation. This complicated sequence of events occurs over several days in response to almost any kind of injury, including cuts, bruises, burns, 'bites' from insects and venomous snakes, and the introduction of foreign proteins (e.g. for therapeutic or preventative immunisation) and bacteria.

These messenger molecules are formed from long-chain fatty acids with several *cis* double bonds (see Figure 12, page 98), such as arachidonic acid (C20:4*n*–6). Its four *cis* double bonds combine to bend the chain of carbon atoms into a tight curve, like a hairpin, so a ring structure readily forms in the middle of the molecule. Specific enzymes add or remove further groups of atoms, converting arachidonic acid into a prostaglandin or a leukotriene. The tendency to form such curved configurations may be one reason why the *cis* arrangement is so much more common than *trans* in animal fatty acids. Plants and micro-organisms do not have such messenger molecules, so they can afford to be less particular about whether their desaturase enzymes form *cis* or *trans* bonds in fatty acids.

Arachidonic acid is widespread and often abundant in the membrane phospholipids of many kinds of mammalian cells, and is released from them by a specific phospholipase. So in a way, the membrane phospholipids act as a store for the small quantities of these fatty acids that are used to make messenger, rather than energy-generating, molecules. As well as the messenger molecules derived from fatty acids of the *n*-6 family, there are other classes of prostaglandins, leukotrienes and thromboxanes with different suffixes and subscripts, derived from the *n*–3 family, notably eicosapentaenoic acid (C20:5*n*–3).

Dozens of such lipid-based messenger molecules have been isolated and more are still being described. Their syntheses and properties are complicated and their interactions with each other and different kinds of cells are the subject of much current research. All messenger molecules bind to specific receptors and hence have a unique and invariant chemical structure. So, in contrast to storage and most structural lipids, for which there are a range of substitutable fatty acids, most lipid-derived messenger molecules can be synthesised only from a particular kind of fatty acid, or members of the same family of fatty acids. Lipid messengers from different families of fatty acids often have opposite effects, and their relative abundance at particular sites controls a range of processes from the contraction of visceral muscles to dilating or constricting blood vessels, and the aggregation of certain blood cells.

Most lipid-based messenger molecules work at very low concentration, sometimes less than one part in a billion (10^9), which in the blood translates to fewer than a thousand molecules for each cell that lines the small blood vessels. They are produced in tiny amounts, of the order of 1 mg (10^{-3} g) per day, usually close to their site of action, from locally available precursors and are broken down, often within seconds of their synthesis, to inactive residues (which eventually appear in the urine). Their short 'life' and low concentration make them difficult to study even in cultured cells, and even harder to monitor in whole animals.

Their discovery in the 1960s and 1970s coincided with the peak of research that led to our current understanding of gene action and protein synthesis: during this period, lipids and carbohydrates were regarded as merely fuels or secondary products. The revelation that certain fatty acids may be specific building blocks in the synthesis (rather than simply providing the fuel) of particular messenger molecules that regulate important cellular processes has altered physiologists' perception of fatty acid metabolism and the organisation of cells and cell components that contain them.

The synthesis and roles of such messenger molecules are one of the fastest moving areas of study.

As well as being the raw material for the synthesis of prostaglandins and leukotrienes, certain fatty acids themselves are now viewed as possible messenger molecules, perhaps conveying signals locally between adjacent cells, either alone or in association with members of a family of proteinaceous messenger molecules called cytokines. Certain polyunsaturated fatty acids may act directly on genes, promoting or inhibiting their action. Little is known about how they work, or in what concentrations, but the important point is that carrying signals between cells is no longer seen as the prerogative of proteins or large, complex sterols. The prospect of finding new and better ways to control gene action always excites biochemists, especially those involved in the drug industry, so the interactions between genes and fatty acids and fatty acid-derived messenger molecules are being actively investigated.

The importance of certain polyunsaturated fatty acids as precursors for synthesis of lipid-derived messenger molecules, or as messenger molecules in their own right, has raised the possibility that certain depots of adipose tissue may actively sequester rare, 'valuable' fatty acids for adjacent tissues that need them, rather than releasing them into the blood circulation for general consumption. Such notions, which suggest that adipose tissue actively selects and organises its lipid stores, rather than simply serving as a general repository, are highly controversial. We return to this topic in Chapter 8, in the light of information about the internal organisation of adipose tissue that surrounds lymph nodes.

And in the end . . .

'We fat all creatures else to fat us, and we fat ourselves for maggots.' remarked Hamlet[25] about the stout, avuncular Polonius whom he had accidentally killed. Under natural conditions, the carcasses of freshly dead animals are picked clean, often within hours, by the combined efforts of scavengers, from vultures and hyenas to blowfly maggots and burying beetles. Fresh adipose tissue is nutritious, especially to birds and mammals that can digest concentrated lipids efficiently, and so its disposal in natural ecosystems presents no problem. We do not normally see large chunks of adipose tissue lying around, as we find bones, hair, horns and wood, all of which decay only slowly by the action of fungi and bacteria.

Lipid breakdown becomes inefficient when, for whatever reason, the bodies of large vertebrates end up where the scavengers cannot reach them. The proteinaceous components of adipose tissue are quickly broken down by bacteria (assisted by the body's own enzymes). The triacylglycerols leak out of the disintegrating adipocytes, and in a damp, well-aerated environment such as a shallow grave, are hydrolysed by bacterial action, or, slowly, just split up spontaneously. The water-soluble glycerol is leached away, leaving a greasy, foul-smelling residue known as adipocere, that consists mainly of non-esterified fatty acids and hydrocarbon chains. Such material is insoluble in water and is not easily broken down further by bacteria, fungi or any other kinds of scavenging organisms, so it persists for months or years. Being less dense than water, the fatty acids ooze towards the surface, forming a whitish scum.

This distressing sequence of events is more likely to happen to human corpses because they generally contain far more fat, and the skin is weaker, than those of horses or other deliberately buried domestic animals. Tracker dogs can smell adipocere at very low concentrations, and its formation has betrayed the presence of many an illegally buried body that has been hastily or inexpertly disposed of.

The formation of adipocere requires water, and so can be prevented by keeping the body very dry. The option was exploited by ancient Egyptian embalmers who supplemented the advantages of the very dry climate in the desert just a few kilometres from the Nile flood plain by applying salts that draw out and absorb water from the tissues. The eviscerated body was packed in crystals of salt (sodium chloride) or natron (sodium carbonate), probably for several weeks. When thoroughly desiccated, the mummy was coated with 'drying oils', which further inhibited the normal decay mechanisms (though after thousands of years, certain species of fungi have managed to inflict significant damage on some mummies).

The brain is too inaccessible to be preserved easily by such means, but, being mainly lipid, decays in much the same way as adipose tissue. The Egyptian embalmers' first act was to remove the entire organ by drawing it out through holes drilled in the back of the nose. Several thousand years elapsed before anyone seriously believed that the brain was involved in sensations or behaviour. To the Pharaohs' technicians, it was a greasy nuisance that spoilt a good mummy. We do not know whether the standard procedure had to be modified for severely obese cadavers; perhaps it was not attempted on such material. Egyptian art almost always portrays people as slim. Whether we should believe that they invariably were so is discussed in Chapter 7.

Final words

We read a great deal these days about genes, their natural replication, inheritance and mutation, and artificial deletion, insertion, substitution or transfer. They have such a high profile in contemporary biology that some people have forgotten that there are interesting and important processes in plants and animals in which genes are involved only indirectly. Fat, in all senses of the term, is among them.

The basic biochemical mechanisms for the synthesis, digestion, absorption, transport and utilisation of lipids were established early in evolution, because, although rarely cast in starring roles, lipids are essential to life. The biochemical apparatus for dealing with lipids is present in one form or another in all organisms, although the capacity to synthesise or use lipids owes much to habits and opportunities as well as to ancestry, and may almost disappear if it is not required. So while certain features of proteins may reveal something about the genes of the organism that synthesised them, the structure, arrangement and abundance of its lipids can tell us something about the life history and experience of individuals. Rapid, efficient uptake and release of storage triacylglycerols are not easily achieved, but higher vertebrates have evolved adipose tissue, a tissue that is devoted mainly to the regulation of lipid storage. Its natural ecological and physiological roles are the subject of the next chapter.

5

The functions of fattening

Functionality has been a guiding theme in biology for much of the second half of the twentieth century. If a structure is there, the evolutionary biologists claim, it must have, or at least have had, a function. The relationship between structure and function has been thoroughly explored for organ systems such as sense organs, limbs and feeding structures. Detailed study of their structure and operation in a few laboratory species is combined with observations from a range of wild animals to produce a comprehensive picture of the basic mechanisms and how they are adapted to the special habits of each species. Storage tissues have not been subjected to similar scrutiny. Until very recently, biologists studying wild animals measured only the quantity of storage lipids, overlooking the possibility that their anatomical location or chemical composition may also be important for certain roles. The availability of better equipment for separating and identifying lipids, and of wildlife biologists able and willing to use it, has revealed some unforeseen subtleties in the sequestration and mobilisation of energy stores.

Fat for travelling

The accumulation of fuel reserves in preparation for long-distance migration was among the first forms of natural obesity to be studied scientifically, and is still among the most familiar. Animals accumulate lipids in adipose tissue (or, in the case of fish, in liver or muscle) before migration and use the fuel on the journey, during which they eat very little, sometimes nothing at all. Some limitations on the chemical composition of such fuels were mentioned in Chapter 3, and some aspects of the way lipid is used in strenuous or prolonged exercise were discussed in Chapter 4. Here we examine how animals adjust their lipid stores to routes, weather conditions, and their own abilities, so as to maximise the chances of completing the journey successfully.

Size and habits

For swimming and walking or running animals, the energy required to transport each kilogram of body mass through a kilometre decreases with increasing body mass.[1] Because locomotion by these means uses less energy per kilogram for larger animals, they can carry proportionately more fuel, i.e. they can be fatter at the start of long journeys or long periods of fasting, than smaller ones, and their reserves last longer. For terrestrial locomotion, the gait used and the amount of fuel carried are also limited by the strength of the skeleton. Smaller animals generally have stronger skeletons in proportion to their body mass, and so they can perform athletic feats such as jumping and climbing that impose large mechanical forces on the skeleton with a much lower risk of injuring themselves than large animals can. The skeleton is strong relative to their size, so small animals can carry large 'payloads' and still safely run fast: their offspring can be large and/or numerous at birth, and they may become fat.

Very large animals such as elephants and adult rhinos cannot gallop or jump because the skeleton cannot support the weight of the body on just a single leg. They do not live in mountainous country and have to be careful on steep, slippery river banks, preferring to bathe in pools with gently sloping sides that they can reach via well-worn, if lengthy, paths. Zoo-keepers exploit this limitation: elephants can be confined in pens surrounded by a shallow, steep-sided moat, across which a smaller animal would have no difficulty jumping or scrambling. So, although the energetic cost of moving while obese or heavily pregnant decreases with increasing size, very heavy animals have to move cautiously to avoid injury. The longest journeys on foot are usually undertaken by medium-sized animals travelling over fairly even terrain, such as reindeer, wildebeest and, of course, people.

In contrast to walking, running and swimming, the muscle power required to stay airborne by active flapping flight increases disproportionately with increasing body mass.[2] So the ability to fly upwards decreases with size, and birds heavier than about 13.5 kg cannot fly at all (though some larger birds can glide for short distances, launching themselves off cliffs). Very large birds such as swans cannot perform the steep climbs or controlled flight at the low speeds needed for near-vertical take-offs and landings. Swans have to take off and land on water, using their feet to accelerate to flying speed, and as brakes when landing. Many smaller birds can generate substantially more power than is needed to keep airborne, so they can fly steeply upwards and can carry proportionately more fuel, up to 50% of the body mass if necessary.

The relationship between body size, flight speed, distance travelled and fuel loads has been studied in some detail in migratory birds,[3] which, as pointed out in Chapter 4, may run their adipose tissue triacylglycerols almost to zero on long flights. Small birds of body mass around 100 g that set out with adipose tissue triacylglycerols amounting to 100% of lean tissue mass (i.e. half the body mass is fuel) can fly for 3–4 days non-stop or 3000–4000 km (depending greatly upon wind direction and other weather conditions). Flying time can be prolonged by slowing down, or shortened by speeding up or by flying into a headwind. In general, larger birds can fly faster, and so get there sooner, but because they can carry less fuel, they often cannot go as far as medium-sized birds. Birds with a lean body mass of around 100 g lose about 0.7% of the body mass per hour while migrating, but larger birds of around 500 g flying similarly loaded lose 1–1.5% per hour. The longest non-stop journeys are undertaken by medium-sized birds, such as sandpipers, knots, turnstones, curlews and godwits of body mass around 100–800 g.

The rate of energy expenditure decreases as the birds become lighter through oxidisation of fat. As they set out fully loaded, each gram of triacylglycerols takes them less than half as far as at the end of their journey, when their fuel is almost exhausted, because flying with half the body mass as fat uses three times as much muscle power as flying 'on empty'. Calculations suggest that a bird that sets out with half its total body mass as fuel and flies to exhaustion uses 40% more energy for the whole journey than one that sets out with a fuel load of 10% of the body mass, flies as far as it can, stops to feed, and continues on the next lap, repeating the cycle until it reaches its destination. Such a journey is slower, of course, because even with unlimited food, birds cannot fatten faster than about 10% of the body mass per day, usually nearer 3–6% per day. So at least a week of steady feasting is necessary to increase the body mass by 25%, 3–4 weeks to accumulate maximal fuel reserves.

Journey plans

While on migration, birds may take different food from that which they normally eat as adults, often reverting to the diet that enabled them to grow very fast as nestlings. For example, queleas, sparrow-like birds that swoop around in amazingly dense flocks over grasslands in Africa, normally eat small seeds, but they feed their chicks on termites, grasshoppers and other large insects, and when on migration, the adults try to find such food for themselves.

Migrating birds rely upon food being available in plenty as soon as they arrive at their stopovers. Mud-flats and meadows are important foraging grounds for many northern hemisphere migrants, so draining a marsh or ploughing a meadow for agriculture can invalidate a migration route for long-distance travellers. Migratory birds have probably always had to cope with making adjustments to routes and fuel loads, often within a single lifetime: as geological features go, coastlines, estuaries and lakes change very rapidly.

Different populations of the same species use different routes and stopover points, and may adjust their fattening to the contrasting 'journey plans'. Some of the most demanding routes for European birds are those across the Sahara and the Mediterranean Sea. Garden warblers (*Sylvia borin*) which breed in Britain and northern Europe during the summer, but winter in Nigeria, Botswana and Zimbabwe, have different fat loads at departure when travelling in opposite directions, presumably because winds are more favourable in one direction than the other. What is interesting from a physiological point of view is how the birds achieve a remarkably exact match between distance to be travelled and the amount of food they eat before setting out.

The amount of fuel a large bird can carry is strictly limited,[2] and most journeys begin with a full load and end with little to spare. Whooper swans (*Cygnus cygnus*) are among the largest of all migratory birds. One major population migrates between north-west Scotland and its breeding grounds in central Iceland. The swans set out with adipose tissue triacylglycerols amounting to about 20–25% of their lean body mass; calculations suggest that this quantity is barely sufficient, but the swans would be unable to take off with a larger 'payload'. Rather than waste fuel flying into the wind, if they encounter bad weather, they land on the sea and wait, sometimes for 30 hours or more, until travelling conditions improve and they take off again.

Birds have both superficial and intra-abdominal adipose tissue, but they lack true intermuscular adipose tissue. The main superficial depots are at the base of the neck around the furcula ('wishbone'), on the anterior surface of the thigh, over the breast muscle and on the back near the tail. Biochemical differences between adipose depots in birds have not been as thoroughly studied as those of mammals, but as in mammals, the superficial depots usually enlarge more than the internal ones. They expand laterally as well as thicken, forming an almost continuous layer that is thickest over the back, breast and the anterior surface of the thigh. The rates at which the lipid is withdrawn from different depots during a long journey have not been studied. Aircraft fuel tanks are depleted simultaneously, not one at a time, to avoid creating asymmetries that would cause tilting or rolling in mid-air. Birds probably do the same.

Bird migration is the most spectacular and best-studied form of animal travel, but other kinds of animals, notably fish such as eels and salmon, and various mammals, including some species of whales, bats and antelopes also travel long distances. Most mammalian journeys, like those of birds, are seasonal and follow established routes, but a few species are nomadic, roaming unpredictably over large areas in search of food.

Almost all large solitary carnivores, including most species of bear, are territorial: a mature male and one or more females hunt and breed in territory that they defend against adult intruders. The male is the father of cubs born in his territory, from which he expels his sons as soon as they grow large enough to pose a threat to his dominance. Such arrangements are impractical for polar bears, living as large carnivores in a climate as erratic as that of the Arctic: the food supply is too unpredictable in space and in time for animals to confine themselves to a permanent territory. Polar bears of all ages travel huge distances between places where seals are abundant and accessible. Satellite tracking and recapture of marked specimens show that they can cover at least 30 km a day for many days.

Polar bears do not feed when on the march but live off their fat, as migrating birds do. But unlike birds, the frequency of journeys and the distances travelled vary unpredictably, being determined by when and where the weather and currents make for good seal hunting. To be sure of having sufficient reserves to last until the next meal, polar bears have to be fatter than most other itinerant animals. In fact, most adults are moderately fat almost continuously.

Somehow the bears must know where they are and where they are going. Tracking data show that, for all their wanderings, each animal remains within its native area, which may be the size of France and Germany combined. Females build their maternity dens in the same place, even though successive pregnancies may be as much as 4 years apart. For reasons that are still not clear, large numbers of male bears of all ages assemble in certain places, among them the north-western shore of Hudson Bay around Cape Churchill, where scores can be seen together in late October and November. This former trading post where, since the early eighteenth century, Eskimos[4] sold animal skins, narwhal ivory and reindeer meat for guns, knives and other manufactured goods has become the main centre for polar bear tourism.

Fat for mating

Energy reserves are essential to some aspect of breeding in nearly all organisms. In animals, nutrient storage plays a part in attracting mates, formation of eggs or sperm, gestation of foetuses, nourishment of nestlings and survival of the young after parental care ends. Fatness indicates success in finding food, and in deterring others from stealing their prey until they have had their fill. Fatter animals may be stronger, more successful as foragers, or more socially dominant, all qualities that, if inherited, make them desirable breeding partners who would endow his or her offspring with favourable genes. Fatter individuals are also better provisioned for spending more time incubating eggs or patrolling territories without having to take time off to feed. They thus make more diligent parents, able to raise better nurtured young. But there are other less obvious advantages to preferring fitter, fatter mates.

Healthy mates

The life cycle of a great many parasites depends upon one of its hosts eating another. Dogs, cats, snakes and many other carnivores including humans can acquire tapeworms, liver flukes and certain threadworms from eating other animals. The gut lining and associated tissues are constantly 'on patrol' to attack such invaders, and mounting a swift, thorough defence can be metabolically expensive. In mammals, defences against invasion through the gut involve special immune tissue known as 'gut-associated lymphoid tissue' as well as numerous large lymph nodes in the mesentery and omentum. Parasites that evade these defences and establish themselves successfully can be a significant drain on their host's energy reserves, and often produce substances that are actually toxic to the host.

People who are infected with parasites often complain of lethargy, skin irritation or chronic pain, and animals thus debilitated are far more likely to fall prey to predators. The best explanation for why parasites make their host feel weak and depressed, but not ill enough to die, is that lethargic or inattentive animals are much more likely to fall prey to predators. If the predator happens to be one in which the parasite can continue its life cycle, so much the better for it. But if the host dies quietly in a corner, the potential predator or scavenger may not find it soon enough to rescue the parasite before it is killed by deterioration of its host.

In 1982, Hamilton and Zuk[5] suggested a link between the brightly coloured, well-kept plumage that so often characterised birds during the mating season, and resistance to parasites. They proposed that, in the long term, a bird's chance of perpetuating its genes was best served by choosing a partner with proven capacity to resist disease and parasitism. The problem is, how to determine which individuals have done well in the war against disease. Discriminating against lethargy and weakness in favour of vigour and strength is clearly one way, but that may not always be easy to do where courtship rituals are highly stereotyped.

Bright, immaculate plumage or pelt indicates both that the bearer was sufficiently well nourished to grow it, and that its skin is not irritated by parasites or diseases that would make it scratch. Poor health, the theory goes, would show up in dull, worn or untidy plumage. By the same argument, larger or fatter individuals should also be favoured in the competition for mates. Observations on wild ducks including mallards, pintails and shovellers show that fatter individuals win fatter mates, and once the pair has formed, they maintain the similarity in fatness by feeding together and assuming similar social status. There is some evidence that rival stags assess each other's size before testing their strength in direct antler-to-antler combat, but even the most observant courtiers cannot easily estimate the fatness of a potential rival or mate under a thick coat of fur or feathers. Hair reduction, as in humans, removes this obstacle. The implications of this evolutionary change for sex differences in the distribution and abundance of human adipose tissue are discussed in Chapter 7.

Freedom from disease not only indicates that an animal has sound, efficient genes to pass on to its descendants. It also bodes well for the successful development of healthy offspring from fertilised eggs, but complete success can never be assured, and other measures to protect the developing offspring may be necessary. Eggs, sperm, seedlings and embryos seem to be more susceptible to poisoning than adult tissues. During the early stages of pregnancy, it may be easier and safer for the mother to reclaim her energy reserves than tackle the demanding and potentially dangerous business of feeding. When living on adipose tissue and other reserves, the job is done. The health risks and metabolic expense of feeding are removed, at least temporarily.

Concepts along these lines have been proposed as the functional basis for such familiar effects as nausea, anorexia and aversions to certain food that women (and possibly other female mammals, but we don't know) often experience during the early stages of pregnancy. Highly complex tissue such as the brain and sense organs may be particularly at risk, and any damage to them is irreparable because they cannot regenerate. These tissues are among the first to develop and most of their formation and growth take place mainly during the first 4 months. Mothers usually feel better during the second half of pregnancy, and their appetite recovers sufficiently for them to fatten themselves, as well as to sustain the growth of the foetus.

Time to spare

In many animals, mating and breeding are limited to a short season, often determined, as explained in the next section, by the availability of food or safe nesting areas. Animals whose courtship and mating are restricted to a brief period cannot afford to waste time on eating even if there is food available. As well as growing magnificent antlers and a glossy coat, mature stags fatten before the rut, and live off their reserves during several weeks of roaring, showing off to, and sometimes actually fighting with, rivals, and maintaining and servicing a harem of females. Fatter males can devote more time and energy to these evolutionarily important activities than those who have to take time off to graze. Many male seals, notably the huge elephant seals, also fast throughout the mating season, devoting themselves entirely to mating with the much smaller females, and to repelling rivals.

These habits often evolve in species in which a few dominant males father a large proportion of the young, and lower ranking males do not get a chance to breed at all. Harem formation and polygynous mating are widespread among mammals, because the females can raise the young without the male's direct or frequent assistance. Such social situations are rare among birds, but when they occur, the results are spectacular.

In most species of birds, both parents contribute to the onerous task of obtaining and delivering food to the nestlings. They may also share other jobs associated with raising their offspring, such as nest-building and incubating the eggs. In habitats that provide little food, the older offspring from previous clutches may assist their parents in raising later broods, rather than find a mate and try to breed themselves. One of the best-studied examples is the Florida scrub jay (*Aphelocoma coerulescens coerulescens*), a subspecies of a species that breeds as monogamous pairs over most of its range from Oregon to Colorado. In the small, isolated population that breeds in oak scrub vegetation in Florida, a habitat that offers very few insects and other invertebrate foods, as many as six adults may collaborate to raise a brood of chicks.

Where food is plentiful, the opposite happens, and the female can find enough to raise her chicks, unaided even by their father. The males' contribution to the next generation is reduced to providing sperm: they do not help with nest building, incubating eggs or feeding nestlings because the females can manage on their own. In many cases, they do not even have to spend much time foraging for themselves. Such conditions allow evolution under sexual selection, a form of natural selection determined by rivalry for and choice of mates, rather than by climate, predators, finding food and all the other problems that animals have to contend with.

Bower birds and birds of paradise have lived in the forests of New Guinea for long enough for rampant sexual selection to have promoted the evolution of characters that at first sight seem to be irrelevant to the serious business of survival. Few places in the world offer such easy pickings for birds – usually other species of birds, mammals or insects move in to take advantage of the plenty – but New Guinea's peculiar geographical position and ecological conditions have apparently maintained this situation for a long time. The females are busier with chores such as building a nest and bringing food for the chicks, but they do have time to inspect the appearance and performances of a range of possible mates, and to make their choice according to some special feature. Sexual selection has promoted the evolution among birds of paradise of an immense range of brilliantly coloured and bizarrely shaped plumage, displayed by elaborate, sometimes very energetic, movements. Such delicate, trailing feathers could not be kept in perfect display condition if the males undertook hard work like collecting nest-building materials, or crawling in and out of holes or burrows.

There are dozens of species of birds of paradise and bower birds in New Guinea and adjacent islands, each with different plumage, display routines or other habits. Male bower birds are not so brilliantly coloured as male birds of paradise, in some species, hardly different from their females, but their emancipation from parental duties leaves them free to spend many hours bringing 'treasures' – brightly coloured stones, flowers, inedible fruits – to build and decorate a bower. High-quality food is readily available, and energy storage mechanisms are efficient enough for the males to be able to obtain all the food they need in only a few minutes of foraging away from the bower. The builder patrols his bower continually, ready to lure any passing female into it, and to protect it from theft and vandalism by neighbouring males.

Elaborate dances and aerobatics that display spectacular feathers or flying skills are the prerogative of medium-sized birds, which, as explained in connection with migration, can perform energetically expensive flight manoeuvres and carry a substantial 'payload', as either adipose tissue or ornaments. The most spectacular feats of load-carrying, including lifting freshly killed rabbits or fish, and airborne courtship, are performed by birds such as eagles, ospreys and hawks that weigh around 1–5 kg. Very large birds (weighing over 10 kg), such as swans, geese, albatrosses and condors, do not perform elaborate aerobatics and do not carry prey or heavy nest-building material. In such species, the sexes usually have similar plumage and courtship displays are performed on the ground or on water.

On the basis of diet and skeletal characters, ornithologists believe that bower birds and birds of paradise share recent ancestors with the families Sturnidae (starlings and oxpeckers) and Corvidae (crows, magpies, jays and rooks) and most species are of a similar range of sizes.

Fat for breeding

Lipids are the main agents of transfer of energy from mother to offspring in most vertebrates and a great many other kinds of animals, including insects. Mammals are somewhat unusual in that their eggs are very small and contain almost no yolk. Mammalian eggs do not need energy stores because soon after fertilisation they attach themselves to the lining of the uterus, forming a placenta, through which the embryo obtains all the nutrients it needs. Lipids cross the placenta and enter the foetus's blood, but most seem to be deployed in structural roles, particularly the formation of phospholipids in the membranes of the brain and nervous system, rather than being broken down for energy production.

The situation changes abruptly at birth, when the neonate starts to suckle. The milk of most mammals (humans are exceptional in this respect) is rich in triacylglycerols but contains only small quantities of carbohydrate as lactose. During suckling, carbohydrates and lipids almost swap the roles they had in the foetus: suckling mammals convert lactose into glucose, most of which is incorporated into structural materials, often as 'finishing touches' to proteins destined to move between cells or to attach to cell surfaces, and lipids become the main source of fuel.

Milk

Almost all mammals grow much faster while suckling than during gestation, and so need much more food. The nutrients still come from the same source: the mother's body. Many species give birth during periods in which food is abundant and easily digested, and are thus able to meet the demands of lactating by eating more. Some small mammals that bear large litters eat as much as twice the average intake in the non-breeding state. Sheep and deer mate only in the autumn, are pregnant during the winter months when food can be very scarce, and give birth just before spring brings new grass and leaves. They exploit the food bonanza as efficiently as possible, often spending the whole day and much of the night eating or ruminating.

Fatty acids in milk are the main agents of transfer of energy-providing material from mother to offspring. The mammary gland takes up nutrients from the blood, synthesises milk and secretes it to the young at an impressively high rate. Some milk lipids are made from triacylglycerols that come directly from the gut in lipoproteins, others are synthesised in the mammary gland from glucose or other precursors, and, especially during fasting, some incorporate fatty acids released from the adipose tissue.

Although adipocyte triacylglycerols contain only long-chain fatty acids, the enzymes of the mammary gland are less particular and readily esterify medium-chain and even short-chain fatty acids into triacylglycerols. So the triacylglycerols of milk from animals that are eating enough for most or all of the precursors for lipid synthesis to come straight from the gut without going via the adipose tissue may include fatty acids that are never incorporated into storage triacylglycerols. But during fasting, fatty acids released from adipose tissue become the major source of raw materials for lipid synthesis, so the milk triacylglycerols from starving animals always contain many more long-chain fatty acids.

The composition of fatty acids in milk triacylglycerols is thus very variable in many mammals, depending upon what the mother has just eaten, and how much is derived from her adipose tissue, or obtained directly from the diet. Although a nuisance for quality controllers in the dairy industry, such variability is probably inconsequential for most wild animals. But it can matter if the mothers take in lipid-soluble pollutants: those eaten during lactation may go straight into the milk, and in fasting mothers such as bears, those that have accumulated in adipose tissue during the previous months get released with the fatty acids and so transferred into the milk, sometimes at alarmingly high concentrations.

One of the advantages of the mammalian nervous system completing so much of its growth during gestation is that the fatty acids incorporated into its phospholipids are accumulated selectively over a long period. Milk lipids are used mainly as fuels, a role for which fatty acid composition is almost irrelevant, and contribute little to brain development. So sucklings of one species, humans for example, grow quite well when fed on the milk of another, such as cows or goats,[6] as long as the gross proportions of lipid, water and lactose are adjusted to avoid osmotic problems. Recent research suggests that human infants, especially premature ones, may be unable to elongate and desaturate linoleic and linolenic acids efficiently enough to meet their needs. Human milk contains significant quantities of the necessary polyunsaturated fatty acids, and their presence may explain why breast-fed babies enjoy better health, and may have slightly more precocious intellectual development than bottle-fed infants of the same age.

For some species, food supplies are too erratic, or obtaining them requires too many long absences from the nest, for lactation to be fuelled just by eating more. Milk production has to draw upon the body's reserves of lipid from adipose tissue, calcium and other minerals from the skeleton and protein from it is not clear where, probably mainly muscle and liver. So many female mammals fatten during pregnancy, and deplete their adipose tissue lipids during lactation. The requirement to be simultaneously obese and pregnant is a major departure from the role of storage tissues to reproduction in other kinds of vertebrates. Normally, mothers fatten before mating and growth of the eggs or embryos begin, and since the processes are fuelled by storage lipids, the adipose tissue shrinks as the reproductive tissues expand. Such animals can have both fat bodies and reproductive organs inside the abdomen without the need for its massive expansion. Similar arrangements are clearly less feasible for mammals, which have to be obese and gravid simultaneously, in many cases for several months at a time.

Polar bears[7] mate in the spring, but development of the fertilised eggs does not begin until autumn, and the cubs are born in midwinter. In the intervening time, the mated females feed voraciously, first on seal pups that make easy prey as they remain in shelters on the ice while their mothers are feeding in the water below. After the pups are weaned, they moult their white fluffy coats and go to sea. So the bears turn to catching seals as they come to breathe at holes in the ice. The hunting technique works well only when the sea is almost completely frozen over, and the seals are limited to surfacing at small holes or cracks in the ice. Summer and autumn are lean periods for bears: the seals may be around, but they swim too fast for the bears to catch them. But by this time, the females have fattened to up to twice their non-pregnant body mass.

As gestation begins, the pregnant females walk inland and excavate a den in a bank of snow formed by a river bank or hillside. The cubs, usually twins, are born in the den, weighing only about 1 kg each. The female stays with them continuously for several months, suckling them on rich, creamy milk synthesised entirely from her own body reserves. She may take the odd mouthful of snow, but she certainly does not eat – there is no suitable food nearby. In March or April, she walks back to the coast with her offspring, where she has her first meal for many months, and sets about the arduous task of teaching the cubs how to catch and dismember seals. She may continue to suckle her cubs for a further 2, or at high latitudes, up to 3 years.

Lactation during fasting depends upon the mammary glands being supplied from the large quantities of adipose tissue, and calcium and other minerals being mobilised from the skeleton. Carrying so much storage tissue around is only feasible because bears have no natural predators (other than humans) and hunt by stealth and skill rather than by speed and agility. Storage tissues enable female polar bears to separate feeding and mothering to a greater extent than is allowed to almost all other warm-blooded animals, thereby emancipating their families from the vagaries of food supplies in the Arctic.

Marine mammals can afford to be fatter, at least for short periods, than terrestrial species because they do not attempt to move far or fast on land, and so are less constrained by the mechanical limitations to fatness. There are no large terrestrial predators in the Antarctic, so seals breeding on beaches or on sea-ice can afford to suckle their young for several weeks. But those living in and around the Arctic are obliged to share their habitat with polar bears, and they have evolved ways of transferring lipid from mother to young at a spectacular rate.

Harp seals (*Phoca groenlandica*) that breed on pack-ice throughout the Canadian and Russian Arctic are only 3% by weight lipid at birth but reach 47% lipid at weaning 13 days later.[8] The single pup drinks an average of 3.7 kg of milk each day, gaining weight at an average rate of 2.3 kg per day. Transfer of lipid from mother to offspring gathers pace as the young matures: the lipid content of the milk increases from 36% at the start of lactation, to 57% (similar to rich cream) just before weaning. The activity of lipoprotein lipase reaches record levels in both the mother's mammary glands, where it hydrolyses the triacylglycerols in circulating lipoproteins so the mammary cells can take up the liberated fatty acids, and in the pup's adipose tissue, into which 70% of the milk lipids are deposited.

Lactation in the hooded seal (*Cystophora cristata*) which breeds on temporary ice-floes in the North Atlantic, is even more compressed. The single pup weighs about 22 kg at birth, large relative to adult size, but is only slightly fatter than harp seal pups. It almost doubles its body mass to around 43 kg in only 3–5 days of suckling. Over 70% of the increase in body mass is blubber, which holds enough lipid to sustain the pup during a post-weaning fast that can last several weeks. During the 4 days of lactation, the mother's body mass declines from 179 kg to 150 kg, over 80% of the loss being from the blubber. She thus synthesises milk at 2.5–6 times the rate of other seals that have been studied, probably a record for any mammal. Such haste may be necessary to minimise the risk that mother and pup become separated by storms and currents that move ice-floes around, as well as that of predators.

In such seals, the transfer of nutrients via milk seems to be the primary purpose of the maintenance of contact between mother and pup. Unlike bears (and most other carnivores), seal mothers do not teach their offspring to hunt for food, though some species coax the pup into the sea and swim with them at around the time of weaning.

Babies

Many features of neonatal and adult mammals can be attributed to the radical alterations in the organisation of growth that are made possible by lactation.

Suckling, by definition, is powered by the facial and respiration musculature, leaving the teeth and jaws to grow unimpeded by the need for efficient function. Because they can rely upon the mother to provide them with liquid food, young mammals do not need to chew 'real food' until they are as much as two-thirds adult size. Parental feeding thus facilitated the evolution of exactly opposable teeth for efficient chewing. Teeth are not needed at all until weaning, then the 'milk' teeth enable the young to feed itself while the skull completes its growth. With a few exceptions such as elephants and whales, the jaws are almost full-size by the time the adult set of teeth erupt. Opposing teeth match each other exactly, permitting sophisticated jaw actions such as gnawing (as in rats and beavers) and chewing the cud (as in cattle, antelope and deer).

Even so, most mammals are weaned with their metabolic reserves as high as possible, especially carnivores and other species for which skill and experience are essential for finding food. Much of the triacylglycerols transferred to suckling seals is stored in their own blubber, and utilised during the long period between weaning and the pups becoming proficient at finding food for themselves. Baby seals spend their time lying on the ice wrapped in thick fur, but many other suckling mammals not only move around much more than they did *in utero*, they also keep warm by non-shivering thermogenesis.

Brown adipose tissue has no known role in the foetus, which is as warm as it could possibly want to be inside its mother's uterus. In most species, the tissue expands and matures in the final stages of gestation, and can be activated from the moment of birth. Brown adipose tissue can oxidise glucose but rarely does so after the glycogen stores with which the neonate is born have been used up, because fatty acids, derived from the mother's milk if it has just been fed, or released from its own white adipose tissue during fasting, are usually plentiful in the blood.

Although their brown adipose tissue is extensive and mature, almost all mammals have very little white adipose tissue at birth, including bears, reindeer and muskoxen that are born into very cold climates, and bare-skinned species such as pigs. White adipose tissue grows very rapidly during the suckling period. Existing adipocytes fill with lipid, and new ones form by division of pre-adipocytes (see Chapter 2) at faster rates than at any other time of life. In rats, almost the entire complement of adipocytes, around 10^7 cells, develops during suckling, though overfeeding later in life can promote the formation of a few more. Experiments with pre-adipocytes extracted from neonatal rats and maintained in culture indicate that the main factor promoting proliferation and maturation of adipocytes is the high-fat diet of milk.

For mammals other than carnivores (whose diet is mostly proteins and lipids) and ruminants (in which rumen micro-organisms convert dietary carbohydrates into short-chain fatty acids), weaning entails a transition from lipid to carbohydrate as the main source of energy. A single glucose-rich meal fed to suckling rat pups aged about 3 weeks (i.e. close to weaning age) can, within a few hours, turn on or off certain genes in the cells of the liver and pancreas, enabling them to make the enzymes needed to handle glucose. The body's energy metabolism is thereby abruptly converted from using mainly fat to using mainly carbohydrate.[9] Unfortunately, similar investigations have not been carried out in carnivores, in which weaning is usually more gradual, and the chemical composition of the adult diet more similar to that of milk.

The diet of most animals that are large as adults changes as they grow: many lizards and snakes eat insects, slugs and other small prey when young, and, for obvious reasons, defer tackling larger prey until they are large and experienced. But the composition of the diet of weanling mammals changes more radically and abruptly. The adjustment is just one of the many physiological adaptations of both mother and offspring to this uniquely mammalian means of nurturing the young. The lactation habit has the enormous advantage that it emancipates mammals from the limitation of breeding only where and when suitable food is available for the young, and permits rapid growth of the neonate. Triacylglycerols and adipose tissue are central to this breeding strategy. Without them, the mothers could not store enough materials to synthesise milk at a high rate, and the young could not take full advantage of the nutrients their mother provides during the often brief period that they are together.

Seals give their young the best possible start in life, at least as far as size and energy reserves are concerned, by combining the advantages of producing a single relatively large but lean offspring at each birth, and sustaining a very high rate of postnatal growth on huge quantities of rich milk. Almost all of the rapid increase in body mass is due to faster growth of the adipose tissue: a recent investigation found that none of the organs studied – intestine, liver, pancreas, spleen and heart – grew faster in seals than they did in ordinary domestic pigs, and some, especially those associated with the gut, grew significantly more slowly. The pups eat hardly anything for a month after their mothers abandon them: they remain on the ice until spring brings more food for them. So just as adipose tissue emancipates the mother from the need to breed where and when suitable food is available, it also permits temporal separation between birth, weaning and nutritional self-sufficiency for the young.

The energetics of human reproduction contrast with those of seals: babies are born quite fat, fatter than nearly all other mammalian neonates except guinea-pigs. Women's milk is exceptionally rich in the sugar, lactose, but low in triacylglycerols (only 4%), so adipose tissue lipids can make at most only a small contribution to milk synthesis. Human infants grow more slowly than those of almost all other kinds of mammals, wild or domestic. At around 1 kg, newborn piglets are less than half the size of human babies at birth, but modern breeds of pig are heavier than a man (and suitable for use as pork) by the age of 7 months. Sows regularly raise 15–20 piglets in each litter, while women manage only one child at a time, occasionally twins. Humans thus avail themselves of the capacity to fuel rapid infant growth from adipose reserves less than most other mammals do. On the other hand, human infants remain associated with, and partially dependent upon, the mother for longer than almost any other mammal. Some implications of these peculiarities of our species are discussed in Chapter 7.

There are obvious advantages to being relatively large and mature at birth: heat loss is less of a problem, the young can last longer between suckling periods, and can be weaned sooner, etc. Although their cubs are born during the winter, all members of the family Ursidae are smaller at birth, relative to adult size, than any other mammal.[10] The cubs of species that breed in very cold climates are, if anything, even smaller than those native to lower latitudes. In view of the advantages of larger size in cold weather, why are bears so small at birth?

Several suggestions have been made, but the most plausible hinge on the fact that mother bears are usually fasting for much of pregnancy. Female polar bears make snow dens, and brown bears, black bears and other omnivorous ursids of both sexes enter holes and caves when fresh plant food becomes scarce in autumn. They are too big to hibernate in the same way as squirrels, hedgehogs and other small mammals can. If they let their body temperature fall as low as these animals do, rewarming would take too long, and use too much energy. The bears' body temperature falls by only a few degrees and they sleep, sometimes for several months.

The mother's tissues can manage quite well using lipids as the main energy source, but those of the foetus cannot. They need carbohydrate, of which supplies are very short. If glucose made from glycerol is not sufficient to meet the demand, more is made by breaking down proteins. The mother can ill-afford to use proteins for energy production: the limited reserves are needed to make milk proteins that the cub uses to build its growing tissues. After birth, easier access to oxygen and other metabolic changes remove this problem, and the suckling cubs can get almost all the energy they need from lipids. The greater the proportion of their growth that takes place after birth, the more bear cubs can be supported by nutrients from the mother's adipose tissue. The best explanation for bears' exceptionally small size at birth is that it enables the cubs to use their mothers' lipid reserves efficiently.

Eggs

The formation of large, yolky eggs that nourish the embryo until it is at an advanced stage of development is a very ancient means of reproduction among vertebrates. Most sharks, skates and rays, whose ancestors were abundant in Devonian seas 400 million years ago, produce such eggs, which hatch into miniatures of the adults, capable of feeding for themselves.

One third of the fresh mass of the yolk of birds' eggs is lipids, a proportion that may represent the maximum that can form a stable emulsion, holding the lipids in a state that the embryo or its mother can handle. The relative size of the yolk and the 'white' part of eggs differs greatly between species, being lowest in species that feed their young on the nest for some time after hatching, such as pelicans (17% of the mass of the egg), gannets (18%) and crows (19%). Yolk content is highest in species whose relatively large eggs hatch to produce large, mature chicks that can walk well and feed themselves. The yolkiest eggs are those of kiwis (65%), megapodes, also known as brush turkeys or moundbuilders (over 51%), ducks (40%) and, of course, species of the order Galliformes, which includes turkeys, peafowl, pheasants, guineafowl, quail and domestic poultry, whose eggs we eat.

The final phase of egg maturation, when lipids enter the yolk and the shell and membranes are added, is fuelled entirely by adipose tissue, which, unlike the gut, can be tucked away in parts of the body where it causes little impediment to other organs. Although reptile eggs are usually smaller relative to the size of the adults than those of birds, they are often laid in big clutches, several score at a time in the cases of large marine turtles, crocodiles and some snakes. Scope for expanding the abdomen is limited in reptiles: turtles and tortoises have to fit themselves into their inextensible shell, and a swollen abdomen limits locomotion and, when massive, respiration. Many large reptiles stop eating before laying, sometimes for several weeks, and the gut regresses, leaving space for the eggs and associated reproductive organs to expand.

Some reptilian mothers regain an appetite within a few days of laying their eggs, but others, notably crocodiles and alligators, remain near their nests, repelling egg-eating predators, and eat little or nothing. Some species dig out the hatching young and guard them for a few weeks before they resume feeding. Large reptiles can easily sustain several months of fasting, even at high temperatures and while remaining active. Adult crocodiles and alligators can afford to be fat: their diet of other vertebrates is highly nutritious, and except when making or guarding the nest, they spend most of their time floating in water, occasionally crawling just a few metres up the beach to bask in the sun. Some birds, especially large species, also eat less while breeding, but are too small, and use energy at too high a rate, to fast for long periods.

The composition as well as the abundance of egg lipids is important to their viability. Egg formation has been studied in detail only in domestic fowl. All the lipid, 5–6 g in the eggs of modern hens, is deposited in the single large cell while it is still in the ovary. Special features of the egg membranes and the tiny blood vessels around them enable triacylglycerols to move straight from the blood, where they are bound in very low density lipoproteins, into the yolk, without lipolysis and re-esterification. The fatty acid composition of the yolk thus exactly matches that of the lipids in the hen's blood, and any lipid-soluble toxins, such as pesticides, that have found their way into her tissues, pass unhindered into the eggs.

Large reptiles and birds have long been hunted for sport and for their valuable skins, meat or feathers, but the decline of such species in the wild has stimulated interest in breeding them in captivity. Crocodiles, alligators, marine turtles, and 'exotic' birds such as ostriches are now 'farmed' in various parts of the world for their meat and leather. Such enterprises are frustrated by the finding that even when kept clean and at ideal temperatures, the hatching rate of eggs laid by captive females is often substantially lower than that observed for the same species in the wild. It is becoming increasingly clear that the fatty acid composition of yolk lipids is at the root of the problem.[11] The principal fatty acid in the eggs of domestic fowl and some other galliform birds, including turkeys, peafowl, guineafowl, pheasants and quail, is linoleic (C18:2*n*–6) and in ducks, arachidonic acid (C20:4*n*–6). Although ostriches are, like these birds, mainly herbivorous, they belong to a completely different, and possibly more primitive order (Struthiformes). In wild-laid eggs, *n*–3 fatty acids, especially linolenic (C18:3*n*–3), are much more abundant, around 22% of the total in the yolk triacylglycerols, but those laid by captive ostriches are more similar in composition to hen's eggs.

The diet fed to the animals in captivity may provide enough essential fatty acids to sustain the adults, at least for long enough for them to be harvested, but it is apparently insufficient for the formation of viable eggs. Breeding success can be improved by correcting the lipid composition of the diet, but it is no good feeding a diet enriched with *n*–3 fatty acids only while the birds are laying. They have to eat lipids approximating to those of their natural diet almost continuously, presumably because much of the yolk lipids are derived from adipose tissue. Problems with spontaneous oxidation (see Chapter 3) may set an upper limit on the proportion of polyunsaturated fatty acids that birds or their eggs can handle.

Bird embryos oxidise mostly carbohydrate in the early stages of development, later switching to lipids. In domestic chickens, the cells lining the embryonic gut start 'eating' droplets of yolk around the twelfth day of incubation, and pass its lipids into the blood as lipoproteins. At the same time, the adipose tissue matures and starts to produce lipoprotein lipase in large quantities, enabling it to take up the yolk lipids. Both the number and the mean size of adipocytes increase until at least the nineteenth day of incubation, and continue after hatching on the twenty-first day. The adipose tissue serves as more than just a temporary store: as a cellular tissue rather than just a sack of yolk, its adipocytes and the lipoprotein lipase they produce discriminate between triacylglycerols of different fatty acid composition, and thereby 'ration' the embryo's irreplaceable lipid provisions to ensure that it develops properly.

As mentioned in Chapter 4, birds' eggs are rich in cholesterol, and are a major source of this lipid in the human diet. Recently, genetic engineering techniques have been used to manipulate domestic hens into laying low-cholesterol eggs. When fertilised, such eggs are non-viable: development gets started, but the embryo soon dies. So the genetically modified hens are unable to pass on the trait to future generations. A substantial amount of pre-formed cholesterol, as well as an appropriate mix of fatty acids in triacylglycerols, must be essential for normal development of chicks, and probably that of all other vertebrates.

Developing brains

Highly unsaturated fatty acids, especially arachidonic acid (C20:4*n*–6), docosahexaenoic acid (C22:6*n*–3) and eicosapentaenoic acid (C20:5*n*–3), are essential for the formation of the phospholipids in nerve membranes. Quite minor imbalances in their supply can distort the normal development of the brain and eyes, and the chick dies before hatching. In spite of their importance, these polyunsaturates are not abundant: for example, only about 0.5% of the fatty acids in yolk are docosahexaenoic acid, mostly in phospholipids. The embryo's adipose tissue incorporates this and other polyunsaturates into its triacylglycerols so effectively that the proportion of docosahexaenoic acid reaches an unprecedented 20% during the third week of development (less than 1% is usual in adult adipose tissue).

Formation of phospholipids for the brain takes priority, and selective lipolysis and uptake of these long-chain polyunsaturates direct almost all of them into the growing nervous system. At the same time, over 80% of the much more plentiful saturated and monounsaturated C16 and C18 fatty acids are oxidised for energy production. Selective deposition of rare polyunsaturates into the nervous system is so efficient that the composition of brain fatty acids is almost identical in all birds, regardless of their diet: the wide variation in the composition of their yolk fatty acids is reflected in that of the heart and liver membranes, and of course, the adipose tissue triacylglycerols.

Polyunsaturated fatty acids are known to be important structural materials for these tissues in all vertebrates. In general, embryos of fish, reptiles and birds seem to be more severely affected by insufficiency of *n*–3 fatty acids than mammals, in which *n*–6 can apparently be substituted, at least for some roles. In foetal mammals, white adipose tissue does not appear at all until development of the brain and sense organs is far advanced, and it is among the last of the major tissues to mature. The main reason for the much earlier maturation of bird adipose tissue may be its role in managing fatty acid supplies to ensure unhindered growth of the nervous system and sense organs.

The need for *n*–3 polyunsaturated fatty acids may underlie the fact that, on reaching a mammalian carcass, scavenging birds such as crows and storks often eat the eyes first, before tackling apparently more nutritious tissues. The cornea (transparent outer surface), lens and most of the eyeball consist mainly of tough collagen or watery jelly, but the retina and optic nerve at the back of the eye are a concentrated source of long-chain *n*–3 polyunsaturated fatty acids. They are much more accessible to small, sharp beaks than the brain and spinal cord, that are encased in hard, tough bone. This macabre habit was well known to our ancestors, who would sometimes deliberately leave the bodies of executed criminals and battle casualties exposed for long enough for the wild birds to inflict this final humiliation on them. For terrestrial birds without access to fish or aquatic worms, mammalian retinas may be a valuable source of nutrients that are essential to the viability of their eggs.

There is little reason to doubt that dinosaurs depended as much upon dietary sources of specific fatty acids as modern birds and reptiles do. Their decline towards the end of the Cretaceous has been linked to the rapid increase in flowering plants and other changes in the flora. The greater variety of toxins in such foliage is usually blamed, but reduced availability of lipids and other essential nutrients during egg formation could also have contributed.

The dependence of female birds upon adequate supplies of these fatty acids to produce viable eggs may help to explain why it is so much easier to breed most kinds of mammals in captivity on a less than completely natural diet than it is to breed tortoises, iguanas and other plant-eating reptiles, and herbivorous birds such as parrots, on an artificial diet. Mammalian mothers nourish their offspring during gestation and lactation, and so have longer in which to supply them with these and other essential nutrients. On a highly abnormal diet, reduced availability of polyunsaturated fatty acids can also impair breeding in mammals.

Until the 1950s, zoos fed lions, tigers and other big cats on butchers' meat, mostly from cattle, sheep and other ruminants and, although the adults remained healthy enough to survive, attempts to breed them were very rarely successful. Once they were fed whole carcases, including the gut, its contents and the long bones containing fatty marrow, lions and tigers started breeding like domestic cats. There are now so many captive-bred big cats that vasectomy and other contraceptive procedures are necessary to curtail breeding. The key ingredients of the more natural diet were vitamins associated with the vegetable matter in the gut contents, minerals in the bones, and the high concentration of polyunsaturated fatty acids in the marrow fat. Ruminant adipose tissue is particularly low in such lipids, and over a long period, a diet consisting almost entirely of beef and mutton did not meet the big cats' needs.

Raising chicks

Neonatal mammals and most hatchling birds are fed by their parents, but whereas all mammals do so with milk, which can be synthesised from nutrient reserves in the body, the food the parent birds bring to the chicks[12] is obtained in the vicinity of the nest and within hours, often within minutes, of being collected. If supplies are insufficient due to exceptional weather or other hazard, many parents abandon their nests and the chicks starve. Most birds are thus limited to breeding where and when sufficient food is available for them to bring to their young. Some species migrate long distances to take advantage of a temporary glut of food, often newly emerged insects or fish fry, on which they can feed their young.

The main exceptions are pigeons, which eat a very diverse diet, and penguins, some species of which breed far away from their feeding grounds. These birds feed their chicks on a 'milk' formed from deciduous tissues at the back of the throat. Several days, and long distances, can separate the collection of the food and its nutrients being passed on to the young. The largest living species of penguin, the emperor penguin (*Aptenodytes forsteri*) feeds on fish, squid and large crustaceans that they catch by diving and chasing the prey underwater. Like all penguins, they do not fly so, in spite of their size, they can carry large quantities of storage lipids in adipose tissue, which proves to be essential to breeding in the hostile climate of Antarctica.[13] Emperor penguins mate and raise their chicks on traditional breeding grounds that may be as far as a hundred kilometres from open water.

The adult males are fat when they leave the feeding areas early in April (autumn in Antarctica) to walk to the breeding grounds inland They gradually lose weight as they fast during 6 weeks of vigorous courtship and for a further 2 months while brooding their eggs. As in all birds, the internal adipose depots fill first, followed by superficial depots which enlarge to form a subcutaneous layer over much of the back, thighs, neck and breast. The latter depots are visible under the large patch of white feathers on the front, and can be effectively displayed by flipper flapping and energetic calling with the head thrust upwards. The females choose as mates males with deep, loud voices and plenty of rippling adipose tissue on the breast and neck. They present their partner with a single egg, then return to the open sea to feed.

The father carries the egg on his feet and broods it in a special flap of feathered skin that extends from his abdomen. Brooding penguins keep close together in large groups and walk an average of only 30 m a day, thereby minimising energy expenditure. If his mate has not returned by the time the chick hatches, the male feeds his offspring on 'curds' formed from deciduous tissue in the oesophagus. Large males can produce several hundred grams of such material, enough to sustain a chick for up to 2 weeks.

As soon as they are relieved by the return of their mates, the fathers walk back to the open water in what is by then midwinter, continuously dark and very cold. By this time, the males' average body mass has fallen from around 40 kg at the start of courtship, to around 23 kg when they leave the colony.

Only about 2 kg of the latter is triacylglycerols in adipose tissue, just sufficient to sustain the penguins as they walk, using energy at 2.8–4.5 times the resting rate, back to the open sea, where there is food to be caught. Under normal circumstances, the birds complete their long journey and begin feeding just before their adipose tissue lipids are exhausted. If the weather is unusually severe, or the sea-ice is exceptionally extensive, or stocks of fish at the feeding grounds are low, penguins that were even slightly underweight at the start of the breeding season may not survive.

The females also fast during courtship, but they return to the sea after presenting their mates with an egg that is large relative to their own size. The females fatten quickly while at sea, eating 6–8 kg per day and increasing the mass of their adipose tissue by up to 200 g per day. The females' body mass rises by about one third before they return inland to take their turn to feed their chicks on curds and partially digested food regurgitated from the stomach.

From hatching onwards, penguin chicks have to cope with large but infrequent meals; king penguin chicks are fed on average once every 39 days – a far cry from the chicks of most garden birds, whose parents may feed them dozens of times a day in good weather. Fattening is thus a priority from a very early age, and penguins may be fatter as chicks than at any other time of their life. If their parents' return is delayed, king penguin chicks can fast for up to 5 months. They stop growing and their basal metabolic rate falls to only 42% of the normal value as they wait, sometimes in vain, for their parent to come back and feed them. Almost all nestling birds have ways of reducing energy expenditure when bad weather or other hazards prevent their parents from bringing food, but those of penguins, and of other large seabirds such as gannets that forage far afield, are much more spectacular.

Mothers

A hungry mother mouse that was about to wean a large litter came to the traps with which Frances and George Hoggan collected their research material (see Chapter 2). They noted that her adipocytes were depleted almost to exhaustion, and concluded that, at least in small mammals, adipose tissue lipids make a major contribution to lactation. The adjustments needed to direct lipids, minerals and other nutrients from the storage tissues to the mammary glands have been intensively studied because of their relevance to milk yields in dairy cattle. As soon as the mother gives birth, an array of changes to hormones, their receptors and the production of enzymes and lipoproteins gives the mammary gland preferential access to circulating lipids. The hormones that prompt the maturation of the mammary gland during the final stages of gestation and trigger milk secretion after birth also act on adipose tissue. As lipoprotein lipase activity in the mammary gland increases greatly, that of adipose tissue is sharply depressed, thereby disabling its uptake of triacylglycerols from the blood, and lipolysis is stimulated.

These changes direct fatty acids from the adipose tissue and/or the gut to the mammary gland, where they are incorporated into milk triacylglycerols. While lactation is in progress, the adipose tissue temporarily stands back from competing with the mammary glands for the ingredients needed to build lipids: in cows, fatty acid synthesis in adipose tissue falls to less than 1% of its maximum capacity. The mammary gland also acquires the ability to synthesise fatty acids from glucose, and may, during peak lactation, do so at a higher rate than any other tissue. In small mammals that have large litters, such as the mouse, the mammary glands can synthesise fatty acids nearly ten times as fast as the adipose tissue is ever capable of doing.

In many mammals including women, the mammary glands are embedded in superficial adipose tissue (much of the human breast is adipose tissue, even during heavy lactation), but it does not seem to be equipped to respond preferentially to the hormones or other signals that stimulate milk production. There is no evidence that mammary adipose tissue contributes more to milk synthesis than any other depot. When animals fatten before breeding and deplete their adipose tissue during lactation, the depot on the inner dorsal wall of the abdomen and the large superficial depots undergo the largest changes, while the smaller, metabolically more specialised intermuscular and cardiac depots remain more or less constant. With a few exceptions, notably humans, there is no difference in the relative masses and anatomical arrangement of adipose depots between adult males, which never lactate, and females which do.

Pregnancy and lactation stretch muscles and skin, and women tend to get fatter with age, but breeding *per se* does not change the distribution of adipose tissue. Nor is there any obvious relationship between species differences in the distribution of adipose tissue and the extent of reliance upon stored lipids to fuel lactation. Lactation requires rapid, bulk transfer of lipids into milk, entails all the major adipose depots and is coordinated by circulating hormones. The distribution of adipose tissue seems to be determined by factors other than the capacity to take up and release lipids into the general blood circulation during lactation or fasting.

While feeding their young, the parents have to support both themselves and the juveniles, whose nutritional requirements become greater and greater as they grow larger. So while parental feeding enables the young to grow faster, and to become fully functional adults more quickly, gathering special food and/or milk production can impose enormous metabolic demands on the parents. During lactation, the mother's own requirements and those of her offspring are in competition for her finite resources. In the past 20 years, much attention has been devoted to the implications of such 'parent–offspring conflicts'. The theory is simple: the offspring give themselves the best possible start in independent life by growing and fattening as much as possible at the expense of their parents. But the adults' evolutionary fitness is improved by retaining enough metabolic resources to enable them to breed again as soon as possible, perhaps with a different partner whose genes may prove to be superior. Because most evolutionary theorists are more familiar with animal behaviour than with biochemistry, most of the examples of 'weaning conflicts' have been behavioural. Thus the theory is exemplified by tales of parents ignoring, or actively repelling, offspring that are old enough to feed themselves, but beg to be fed or try to suckle.

In the case of mammals, the physiological aspects of 'parent–offspring conflicts' focus almost entirely on the mother, and are not confined to weaning but extend throughout lactation: by definition, the young is not viable without the mother until weaning, so deterioration in her health prejudices the offspring's chances of attaining adulthood as well as her prospects for reproducing again. Adipose tissue resources must be divided between the nutritional demands of the offspring, which could be almost infinite, and the need to sustain the mother in good health. The quantity and composition of milk produced from a mammary gland are partially determined by the pattern of the offspring's suckling from its teat, especially in kangaroos (and other marsupials) that can suckle two joeys of very different ages at the same time.

The evolution of the lactation habit may have been a major factor promoting faster, more efficient metabolism of adipose tissue and more elaborate control of its activities. Advances in the regulation of adipose tissue include both more signalling and messenger molecules for more sensitive temporal control, and improved spatial organisation of the tissue by partitioning it into depots. It is perhaps no accident that lactation and substantial, intricate site-specific properties of adipose tissue are both unique and universal features of all living species of mammals. And that includes the duck-billed platypus and the spiny echidna, the monotremes of Australia and New Guinea that lay large, yolky eggs similar to those of birds and reptiles, but suckle the hatchlings on milk that is remarkably similar to that of other mammals.

Fat for playing

Being able to sustain oneself on stored nutrients between occasional, highly nutritious meals means that there is plenty of time for other activities. Snakes, it is true, spend much of their 'spare' time resting in places that offer suitable combinations of warmth and protection from disturbance. Birds and mammals in similarly fortunate nutritional situations use the time in ways that, from an evolutionary point of view, prove to be more profitable. As pointed out earlier, prolonged, strenuous courtship is possible only when supported by easy access to nutritious food or, more often, by energy reserves that were accumulated before the mating season began. Similar principles apply to the evolution of other 'spare time' activities, notably social behaviour, which, because it entails sophisticated communication between animals, promotes the evolution of what we call intelligence.

Most primates seem to have plenty of spare time, and now that we have the money to pay scientists to watch wild troops for weeks at a stretch, there is much more information about how such animals spend their days. Many monkeys and apes, especially chimpanzees and gorillas, spend a large proportion of their waking hours grooming each other, establishing dominance ranking and in other ways servicing social relationships. The juveniles play with each other and with the adults, often in energetic, even dangerous, ways. As food becomes harder and harder to find, a smaller proportion of each day remains available for play. Such activities seem to be the most elastic, being curtailed when scarce food supplies require long hours of foraging but expanding (together with ample snoozes) to fill the time when life is easy.

If hard times persist, the troop becomes dysfunctional and may split into several smaller groups. Juveniles raised under conditions in which there was little opportunity for playing prove to be less effective as adults than those that grew up while food was plentiful enough to allow for such activities. When food becomes scarce, baboons and other primates spend more time foraging, but there is no evidence that easier pickings lead to excess consumption and hence to obesity. Wild primates seem to know when they have had enough.

Humans take this trend further, and many of us spend a huge proportion of the day engaged in activities unrelated to obtaining food. Such 'leisure' time, and time for energetically expensive spectacles such as the mating displays of birds of paradise and bower birds are only made possible by the evolution of the capacity to obtain, handle and digest plenty of nutritious food and of efficient storage mechanisms, especially adipose tissue.

Fat for eating

One might imagine that continuous access to small meals would be an ideal situation, and would eliminate the need for storage tissues. So it does for animals such as internal parasites and for mammalian foetuses, but the food of most free-living animals is widely and variably distributed in space and in time, and requires processing: catching, chewing, digesting etc. A branch of the study of animal behaviour known as 'optimal foraging' has revealed that animals regulate, and to some extent plan, the quality and quantity of food eaten. They set out at certain times of day to find particular foods, and if the required nourishment is insufficient in quality, or cannot be harvested efficiently, foragers may switch to alternative sources of food.

Even when exploiting a satisfactory resource, the exact timing and quantity of food eaten often cannot be predicted accurately. Observations on large carnivores such as lions and cheetahs hunting in the wild show that even experienced individuals are successful in only about one in five attempts. So most animals are 'hungry' i.e. they could and would eat if food were found, for much longer than they are actually eating. Keeping the gut in a state of readiness for digestion and absorption, as well as the actual processes themselves, are metabolically expensive.

The energetic cost of eating has been most thoroughly investigated in snakes, whose special habits make them particularly suited to such studies. All snakes are predators, and, although the hatchlings may eat large insects, snails etc., the adults of all the large species live almost entirely on other vertebrates. Such food is both very concentrated, and of almost exactly the same chemical composition as the snake itself. A few, very large meals thoroughly digested provide adequate nutrition and generate very little waste. Even under ideal circumstances, large pythons, boas and vipers eat only about every 1–2 months, and can fast for up to 18 months without irreversible consequences. The time courses of the metabolic processes involved in digesting prey have been studied[14] in the sidewinder rattlesnake

(*Crotalus cerastes*) a venomous species whose diet consists largely of rodents, and the Burmese python (*Python molurus*) that feeds mainly on mammals and large birds when adult, and kills its prey by constriction.

Young, partially grown specimens were kept in captivity and fed on laboratory rats or mice. Within 3 days of eating a meal amounting to 25% of its body mass, the snakes' oxygen consumption increased up to 17-fold, and remained continuously high for several days, before gradually declining over the following 2 weeks. This increase in oxygen uptake is greater than that observed when a person goes from lying in bed to fast running, and means that glucose and/or fat is being consumed at a very high rate. These snakes remained inactive while digesting their food, so the energy must have been used for the synthesis of tissues and secretions.

Examination of specimens killed at various times after feeding showed that much of the extra energy was being devoted to fuelling reconstruction and modification of the internal organs. The stomach, intestine, liver and heart enlarged greatly, starting within a few hours of the prey being eaten, and the gut enormously increased its capacity to digest and absorb nutrients.

The highest oxygen consumption and the fastest changes in tissue mass and properties occurred before digestion was sufficiently far advanced for significant quantities of any nutrients to be absorbed, so the adaptations to feeding must have been fuelled by stored energy. In other words, the animals need sufficient reserves to fuel the modifications to their intestine and other viscera that enable them to digest the food effectively. Big snakes can become quite fat: their fat bodies expand to form thick, bolster-like masses that extend the whole length of the abdomen. Unfortunately, the investigators did not study adipose tissue or lipid metabolism, but there seems little doubt that it contributed to fuelling growth of the viscera as well as to sequestering the lipids after they were digested and absorbed.

Big snakes apparently alternate brief periods of high energy expenditure during feeding, digestion and absorption, with long intervals in which the gut and other viscera are reduced in size and metabolic capacity. During fasting, the body is maintained at a much lower rate of energy expenditure by lipids released slowly from the adipose tissue. Of course, some energy is expended in converting the nutrients absorbed from the food into storage triacylglycerols, and in breaking them down again when they are needed, but the quantity is small compared with the cost of maintaining the gut in a state of continuous readiness for frequent, small meals. Such very long fasts are easier for large snakes than for most other animals because, being cold-blooded, their resting energy expenditure can be quite low. They also travel around very little, and rarely have to move fast, as few predators are likely to tackle them.

Rattlesnakes and their relatives are even better equipped to keep energy expenditure low by avoiding unnecessary exercise. When disturbed, they vibrate their rattle (formed from loose caudal vertebrae enclosed in segments of dead skin at the end of the tail), which warns potential predators and unwary passers-by of their presence, so they, and not the snake itself, do the running away. Moving fast enough to produce an audible sound is usually energetically expensive, but the tail-rattling muscle has special biochemical and structural features[15] that enable it to generate a loud, fairly high-pitched and above all, distinctive noise remarkably cheaply. Although large snakes are often spectacular and generate much fear among potential prey and bystanders (e.g. ourselves), their metabolic economies mean that the amount that they actually eat is tiny compared with that needed to sustain birds or mammals of similar size.

During severe illness or other circumstances in which eating is impractical, many other animals close down the gut and rely upon reserves in much the same way as large snakes do. Elderly animals and those weakened by chronic illness or defects of the mouth or gut (e.g. tooth wear or decay) reach a point beyond which they cannot muster enough reserves to fuel digestion. Without the means of obtaining further sustenance, their decline continues until they die, usually, in the case of small animals, within a few hours.

In large animals with substantial energy reserves, the interval between the loss of capacity to feed and death may last some time, long enough to be noticed by human observers. Wild animals that do not die from predation – the fate of the majority – often appear to die of starvation, a conclusion based on the fact that post-mortem examination usually reveals an empty gut and little or no adipose tissue. In many such cases, starvation may have been due more to loss of the ability to handle, digest and absorb food than because none was available. The animal 'lost its appetite', but lived on its adipose reserves for days or weeks, or in the case of large snakes, terrapins and tortoises, as long as several months.

These days, intravenous drips and other forms of artificial feeding enable physicians and vets to provide adequate nourishment to people and domestic animals whose health has deteriorated far beyond the state in which they would be able to digest and absorb food naturally. If such treatment is prolonged, patients may reach a more advanced state of decline than would ever occur in wild animals.

Fat for sleeping

Hibernation and its hot weather equivalent, aestivation, are periods of inactivity and seclusion, usually accompanied by cooler than normal body temperature, which enable animals to pass through seasons when food is scarce or inaccessible. Hibernation was among the first physiological states in which adipose tissue of wild mammals, reptiles and amphibians was studied thoroughly. They depend upon their fat reserves while their body temperature is too low to allow the collection and digestion of food; but reclaiming the stores is not as straightforward as it first seemed.

The enzymatic processes involved in fasting and starvation are essentially similar to those of slow exercise, but there is a critical difference: during exercise, the body is warm, often slightly warmer than when sedentary, but in hibernation, the body temperature is low, sometimes falling by 35 °C to close to 0 °C. Enzymes do not work on solidified fats, any more than they function in frozen water. Animals must still be able to metabolise the triacylglycerols in their adipose tissue, albeit much more slowly than when they are fully active. The fatty acid composition of triacylglycerols is largely irrelevant to their role as fuel when animals are warm, but it is crucial for their use during hibernation.

Experiments[16] on captive chipmunks (*Eutamias amoenus*) and golden-mantle ground squirrels (*Spermophilus lateralis*) show that they enter hibernation more readily, remain cooler for longer and are better able to survive long winters when plenty of polyunsaturated lipids are included in their diet during the weeks preceding hibernation than when they are fed on saturated fats of similar calorific value. One reason could be the contribution of unsaturated fatty acids to lowering the melting point of the triacylglycerols and the structural phospholipids in membranes. The lipids remain more fluid, and hence retain their proper affinity for carrier molecules and enzymes, at cooler temperatures. The chemical composition of the storage lipids, together with other aspects of adipose tissue, is thus adapted to the physical conditions and its role in whole-body metabolism.

Squirrels obtain many such unsaturated fatty acids from the seeds and other plant parts that they eat. Hibernation is an active, physiologically controlled process, and especially in mammals, metabolic preparations can be identified days or weeks before the animal actually allows its body to cool. As the weather becomes cooler and the days shorten in autumn, squirrels and other hibernatory rodents actively seek nuts and other foods that contain these lipids. The woodchuck or marmot (*Marmota flaviventris*) selectively retains linoleic acid (C18:2) before hibernation: the saturated fatty acids are released and oxidised by the muscles, liver, etc. while the animal is warm and active, but the polyunsaturates remain in the adipose tissue, for use when the body is cold. The obvious conclusion is that these herbivores depend upon the increased availability in autumn of seeds rich in polyunsaturated lipids. Failure of a seed crop could prevent successful hibernation and thus lead to death from cold or starvation, even if plenty of other foods were available.

The rate of spontaneous oxidation of lipids is believed to impose limits on the maximum proportion of polyunsaturates that biological lipids can contain. In the investigation of the effects of diet on hibernation, it was found that feeding unnaturally high proportions of polyunsaturated fatty acids reduced the time for which ground squirrels remained dormant nearly as effectively as feeding too few such lipids. The most likely explanation is that these unstable lipids readily oxidise to form toxins that disrupt hibernation (see Chapter 3).

The most obvious ways of reducing energy expenditure are to stop running about and lie down quietly out of sight of predators, and, weather permitting, to cool down. Less obvious, though equally effective, is to get rid of inessential, energetically expensive tissues. The brain uses a lot of energy but cannot easily be rebuilt when normal activities resume. The digestive system also uses a lot of energy, and it can be reconstituted efficiently. Hibernation is economical for amphibians, reptiles and mammals not only because they are cooler and less active – that has been generally understood for a long time – but also because they are fasting. To economise on the energy required for its maintenance, the gut closes down, much as it does between meals in snakes. Secretion of stomach acid and digestive enzymes and renewal of the gut lining almost stop, thus reducing the energy required to maintain it. In ground squirrels that arouse themselves and feed every few weeks during the hibernating season, the lining of the intestine becomes measurably thinner between feeding periods.

There is another reason for avoiding food and living on energy stores while the body is cool. Like all chemical processes, digestion is slower at lower temperatures, which might not matter if the body's rate of using energy is similarly slowed, but it would give intestinal parasites and harmful bacteria more time in which to breach the gut lining. Mounting effective immune responses would also be slow at low temperatures, so the risk of acquiring diseases from food could be dangerously increased. If they are below their critical body temperature, most fish, frogs, snakes, tortoises and lizards refuse to eat, even if suitable food is put into their mouths. What that temperature is, of course, differs widely from species to species, and can sometimes be adjusted within narrow limits by acclimatisation, and, as mentioned in Chapter 3, may be affected by the lipid composition of the diet.

It is important to realise just how economical sleeping undisturbed in a cool, secluded place can be: small tortoises may lose less than 5%, and rarely lose over 15%, of the body mass during 5 months of hibernation at about 5 °C, and much of that is water lost through evaporation from the lungs. On emergence during the first warm spell of spring, they potter about for a few days. Once the gut is fully regenerated, they are ready to start eating and drinking, and may regain the body mass they have lost in less than a month. Many reptiles and amphibians emerge from hibernation with a surprisingly large proportion of their lipid reserves still remaining. These stores not only come in handy during spells of cold weather when food is scarce, but are often important for fuelling mating and egg production. The less lipid a hibernator uses during the winter, the more it has left to fuel a 'head-start' in the all-important business of breeding the following spring.

Fat redundant

When adipose tissue is not needed for any of these roles, it disappears, not completely of course, because the mature adipocytes remain intact, but they may lose almost all their lipid and become almost indistinguishable from other connective tissue cells. So far as it is known, they also become metabolically inert, and may remain in this state for months. The start of the breeding season, eating a large meal or other metabolic stimulus re-activates the dormant adipocytes: they synthesise and secrete lipoprotein lipase, take up lipid from the blood and are back in business within hours.

We think of autumn as a time for feasting and fattening before the onset of winter, and so it is for hibernating animals. But several species of small rodents, among them the dwarf hamsters *Phodopus*, do the opposite: they lose weight as winter approaches. Male *P. campbelli* that weigh around 40 g in summer shrink to little over half that size in winter, while the smaller females lose one third of their body mass in the same period. The adipose tissue undergoes the largest change in mass, decreasing from a maximum of 35% of the body mass to less than 5%, but some of the lean tissues, especially the reproductive organs, also shrink.

These little hamsters are native to Siberia and Mongolia, where they build tunnels under vegetation, or, for much of the year, under snow. They can eat almost any plant material, but during the long, cold winters, they live mainly on the seeds of grasses and annual herbs. Apart from brief periods of torpor, the hamsters remain awake all winter and forage under the snow. Sexual behaviour is switched off completely – males and females ignore each other – but they come into breeding condition as soon as the days become substantially longer than the nights. Under favourable conditions, a pair can raise several litters of over a dozen young in a single season. Like lemmings, their numbers fluctuate enormously from year to year.

Losing weight seems to be another energy-saving strategy that enables the hamsters to survive the winter on meagre rations: smaller bodies require less energy to move around, so less food is needed to keep going, and supplies last longer. In summer there is plenty of food, but not enough time to eat it and to mate, defend territories and attend to all the other duties of breeding. So the first response to lengthening days is to eat more and grow fatter, fat enough to devote much of their time and energy to reproduction. In winter, these provisions are not necessary: undistracted by the demands of sex and breeding, the hamsters can spend more time, all day if need be, finding enough food to keep going. Heavy adipose tissue adds to the cost of foraging, so, quite sensibly, the hamsters shed the excess lipid, if necessary by eating less for several weeks. Experiments[17] in which adult *Phodopus* were kept in warm or cold conditions, and exposed to artificial long or short days show that reduced daylength, not air temperature, triggers these changes in body size and fatness. Perhaps the weather in Siberia and Mongolia is too unpredictable for temperature to be as reliable a guide to season as daylength.

Although most of the adipose tissue of very obese hamsters is superficial, and the depots may abut to form an almost continuous layer, there is no evidence that its contribution to insulation is important to thermoregulation. As shortening days prompt depletion of adipose tissue, the grey summer fur is replaced with a paler, thicker winter coat that seems to be more than adequate to keep the hamsters warm under the Siberian snow.

Final words

Many of the concepts described in Chapter 4 arose from laboratory experiments on isolated molecules and cells, or whole animals under tightly controlled conditions, but the conclusions of this chapter are based upon a different kind of evidence: comparative studies of wild animals. Such data are difficult to interpret. There are not only many practical difficulties in finding out what free-living animals actually do, but one cannot perform experiments in the sense of inventing new kinds of animals with specified combinations of characters. It is a matter of putting diverse observations together and drawing the most logical conclusions from them.

Storage tissues play a fundamental role in underpinning spectacular activities such as long-distance travel, elaborate courtship and parental care. Many of these aspects of animals' lives are interesting in their own right, and have attracted much scientific and popular attention, but in marvelling at the impressive activities and structures, we should not forget the physiological mechanisms that make them possible. Lipid storage expands the range of ways in which animals can live and so creates a wider variety of species. Many areas that provide adequate food are infested with predators, have a severe climate or in other ways are unsuitable for breeding. Efficient adipose tissue enables animals to migrate seasonally to places that offer different combinations of amenities and hazards, or to commute between them, as penguins do.

6

The functions of fat

Adipose tissue and its contents could perform all the biochemical roles so far discussed wherever it is located in the body. As pointed out in Chapter 2, the adipose tissue of most amphibians and reptiles is organised exactly as would be expected for a tissue whose functions are only whole-body metabolic regulation and energy storage. But in mammals and birds, adipose tissue is always distributed, partitioned into several large and numerous small depots that are scattered around the body, and associated with the skin, muscles, lymphatic system, gut and other internal organs. Why the tissue is arranged in this curious way in higher vertebrates has led to much armchair theorising, but regrettably little practical research.

Especially in cases where the adipose depots are massive, additional functions are proposed, among them that the tissue acts as an insulator, impeding heat transfer from the warm interior to the cool skin, and that it protects vital organs from mechanical damage. The notions that the superficial adipose tissue insulates the body against heat loss and internal adipose depots protect associated tissues from mechanical damage are some of the most firmly established of all dogmas in twentieth century biology. Thermal insulation is widely stated as the reason for the association between adipose tissue and the skin in both learned and elementary biological literature, almost invariably without any supporting evidence.

A coat of adipose tissue

The principal mechanism of thermal insulation in most mammals and birds is the trapping of a layer of stagnant air or water around the body by fur or feathers. Very fine underfur beneath stout guard hairs, or the fine down feathers that grow under the stiff flight feathers, form the best insulators. They trap millions of very tiny pockets of air, which are warmed by the body but cannot escape to mix with the colder outside air. The downy feathers of the eider duck, which breeds beside fjords and lakes in Iceland and Svalbard but overwinters in northern Europe, have long been regarded as the best material for eiderdowns and other high quality bedding and outdoor clothing. Air conducts heat much less efficiently than water, but as divers using wet suits know well, even a thin layer of water held in place by tight clothing provides useful insulation in very cold water.

Small animals cannot have very long fur, and have a proportionately larger surface area, so they risk losing heat at a high rate unless their insulation is very efficient. The furs of small animals have a higher capacity for insulation per unit mass than those of larger species that live in similar climates. The winter furs from ermine (adult body mass 50–250 g), mink (0.5–2 kg), muskrats (0.7–2 kg) or arctic foxes (average 2–4 kg) are lighter, softer and warmer to wear than those of larger animals such as bears. All three qualities can be measured scientifically. Insulation, or its converse, conduction of heat, can be quantified by wrapping an object of standard size and shape in a layer of the material to be tested and measuring its rate of cooling under standard conditions.

The same concepts and techniques can be applied to assessing the contribution, if any, of adipose tissue to insulating mammalian and avian bodies. The rate of heat conduction through adipose tissue taken from human cadavers turns out to be only a little less than half that of muscle or stagnant water, but much more than that of pelts of fur or feathers. Of course, in living animals, blood flowing through the superficial muscle or adipose tissue acts like water in a radiator, bringing warmth from deep inside the body towards the surface. Blood flow through muscles is usually higher than that through adipose tissue, especially during vigorous exercise, so more body heat would be lost from superficial muscles than from superficial adipose tissue. Such measurements suggest that adipose tissue would limit the transfer of heat from the body core to the surface more efficiently than muscle or other highly vascularized tissue, but its capacity to do so is nowhere near as good as that of external coverings such as fur or feathers. The finding that a structure has properties that *enable* it to serve in a certain role, does not justify the conclusion that it is *adapted* to that purpose.

Keeping warm

The hypothesis that adipose tissue is *adapted* to act as an insulator can be tested by determining whether its distribution and anatomical arrangement have evolved to carry out such a role more efficiently in animals that need more thermal insulation, for example, species that remain active all winter in cold climates, or habitually swim in very cold water. The identification of characters related to a particular habit – such as living in a cold climate – is easier if the species to be compared are similar in other ways, either through having common ancestry, or though shared diets and other habits. The mammalian order Carnivora fits these criteria better than any other: species occur in all habitats from hot tropics (e.g. desert foxes, jackals) to the Arctic (e.g. polar bears, wolverines, arctic foxes) and almost all are predators, at least to some extent. A further advantage is that Carnivora occur in an enormous range of sizes, from stoats and weasels that weigh only a few hundred grams as adults, to bears and tigers that can be several thousand times larger. Thus it is possible to compare two alternative explanations for species differences in the relative abundance of superficial adipose tissue: are they determined by habits and habitat, or merely by body size?

The opportunity came in 1988, when the Canadian polar bear expert, Dr Malcolm Ramsay, invited me to help with a study of adipose tissue in polar bears. We dissected out and weighed all the superficial and intra-abdominal adipose tissue of 13 wild polar bears that had been shot by native Eskimo hunters, and compared the observations with similar measurements from various wild and zoo-bred carnivores that I had dissected over the previous 10 years.[1] Although they were collected in early November, when the sea-ice was just beginning to freeze over and seal hunting was again becoming efficient, all the polar bears were between 10 and 21% dissectible fat, which

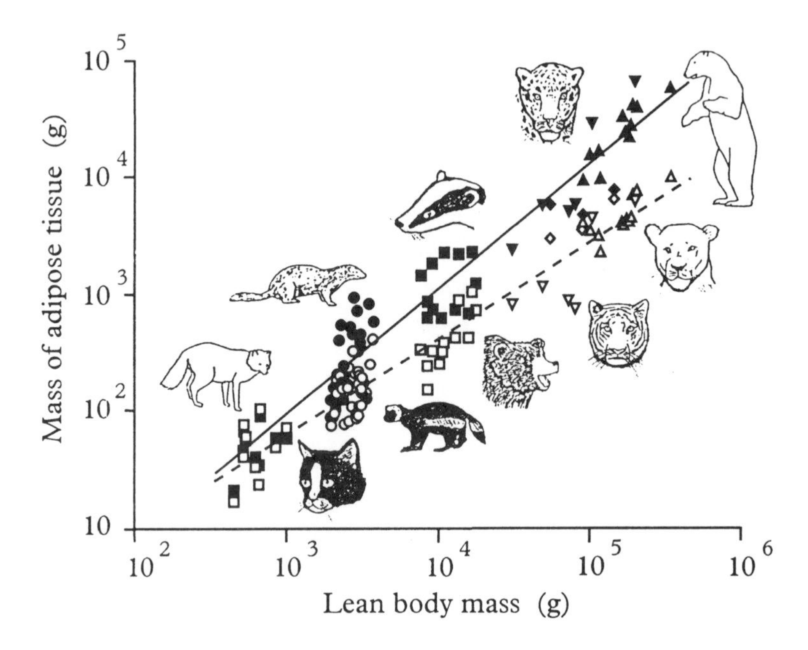

Figure 14. Comparison of the masses of superficial (represented by solid symbols) and intra-abdominal (represented by open symbols) adipose depots in various species of the order Carnivora. Squares show data from species of the weasel family (Mustelidae) including mink, badgers and wolverines; diamonds show the cat family (Felidae), cats, jaguars, tigers and a lion; circles show arctic foxes (the dog family, Canidae), and triangles show bears (Ursidae). The lines are fitted to all the data *except* those from the polar bears (upright triangles), although the measurements from these partially aquatic arctic bears fall very close to the extensions to these lines.

amounted to 73 kg of adipose tissue for the largest bear whose total body mass was over 400 kg. Fatness in this range is high for a wild mammal, and a substantial proportion of the adipose tissue is subcutaneous, so it could in theory contribute to thermal insulation. But is the adipose tissue so distributed *in order to* act as insulation?

Figure 14 shows some measurements of the masses of all the intra-abdominal, and all superficial adipose tissue from a variety of temperate, tropical and arctic carnivores. The Carnivora are almost certainly descended from a common ancestor, and so they share similarities in body shape and proportions as well as in diet. To ensure comparability, the sample included only specimens that were of similar fatness to the bears. Using logarithmic scales enables data from small mink of body mass less than 1 kg, and huge bears weighing up to 500 kg to be shown on a single graph, and facilitates the mathematical analysis.

There is a fair amount of inter-individual variation in the partitioning of adipose tissue between internal and superficial depots even in these genetically uniform populations, but overall the mass of superficial depots increases in simple proportion to body mass. In contrast, the mass of the intra-abdominal depots becomes proportionately smaller in larger mammals, but the trend is exactly the same whether the animals are native to the Arctic or to warmer habitats. The measurements from the wild polar bears fall very near to the lines fitted to the data obtained from tropical and temperate-zone carnivores, so this comparison provides no evidence for any major differences in the distribution of adipose tissue in arctic bears and that of related animals from warmer climates.

The partitioning of adipose tissue between internal and superficial sites arises entirely from the fact that polar bears are very large and become very obese. It is not necessary to invoke a net shift in the partitioning of adipose tissue between internal and superficial depots to explain the thick layers of subcutaneous fat in polar bears. The extra fat usually goes into superficial depots, whether or not the animals are partially aquatic or live in cold climates. In other words, an apparent change in the distribution of adipose tissue arises as a direct consequence of altered body proportions due to larger overall size and higher fatness, and has nothing to do with thermal insulation. Polar bears have other adaptations to cold climate, such as reduced ears and tail, and thicker fur that extends over part of the soles of the feet, but the arrangement of adipose tissue can be explained by the facts that they are very large, and very fat.

In small mammals such as mink, about half the adipose tissue is internal and half superficial, although in all species studied, the proportion in superficial depots increases with increasing fatness. The proportion inside the abdomen decreases with increasing body size (independent of fatness) so in large bears, less than 15% of all adipose tissue is internal. As well as having proportionately more adipose tissue in superficial depots, the surface area of larger animals is also smaller, relative to their volume or mass. The body surface area scales as body mass$^{0.66}$, i.e. it becomes proportionately smaller with increasing body size. With a smaller area to cover, the superficial depots are thicker in larger animals, even if the specimens are not fatter (i.e. if the ratio of the masses of lean and fat tissues is unchanged). The superficial adipose tissue is ten times thicker in a 500-kg bear than in a 500-g rat or weasel from this effect alone, quite apart from the fact that in larger animals, a greater proportion of the adipose tissue is superficial, and bears are often fatter than rat-sized animals. The adipocytes of superficial depots, especially those around the hind legs and tail, are less responsive to noradrenalin and hormones such as insulin (see Chapter 4), so they may be metabolically as well as anatomically the best place for long-term storage of large quantities of potentially dangerous lipid.

This partitioning of adipose tissue is particularly clear among the Carnivora, possibly because the abdomen is relatively small in these mammals, with the liver and the guts each amounting to only about 2% of the body mass in large species such as polar bears. There would not be room for adipose stores amounting to 50% of the body mass in an abdomen that normally accommodates less than 5% of the body mass. If all, or even a large fraction of the adipose tissue were intra-abdominal, the bear's abdomen would protrude and sag.

This analysis highlights some pitfalls of using measurements of the thickness of superficial adipose tissue as a means of estimating fatness, as described in Chapter 2. Fatness is a ratio of masses, the mass of energy stores as a percentage of total body mass, but adipose tissue thickness is a linear measurement. With increasing body mass, the superficial adipose tissue becomes thicker for animals of the same total body composition, because the ratio of their surface area to mass decreases. The superficial adipose tissue becomes thicker in larger animals, and depots that are adjacent but distinct in small species and the juveniles of large species merge in large adults to form a continuous layer, albeit of rather variable thickness. An almost continuous layer of superficial adipose tissue can create the impression that the specimen is obese, even when the layer represents only a small fraction of the total body mass.

The depots are discrete in young bear cubs (which, being small, have a proportionately larger surface area) and very lean specimens, but in large, moderately fat adults, the superficial depots form a continuous layer of 'subcutaneous' adipose tissue covering the whole body except parts of the head and forearms. This arrangement has a casual resemblance to blubber in whales and seals, which, together with their coastal habitat, has given rise to the notion that polar bear adipose tissue has evolved into blubber. But in fact, as the comparison in Figure 14 shows, the distribution of adipose tissue in these animals, which in evolutionary terms are new recruits to aquatic life, is not fundamentally different from that of their terrestrial relations.

Although the relative abundance of adipose tissue in the major depots differs greatly between species, it is difficult to account for most arrangements in terms of adaptation to thermal insulation. There is no evidence for reorganisation of adipose tissue in the smaller species of mammals that have evolved to become aquatic. Otters, for example, have many adaptations of the trunk, tail and sense organs to their life as predators in lakes and rivers, but the arrangement of their adipose tissue is not different from that of their terrestrial relatives, stoats, weasels, wolverines and badgers. Otter fur is thicker and less compressible than that of the terrestrial species, so it retains its insulating properties when in water. The warmth and water-repellent properties of otter fur are the main reasons why the European otter (*Lutra lutra*) and the sea otter (*Enhydra lutris*) of Alaska and coasts around the northern Pacific Ocean were hunted for their pelts both by native peoples and, more intensively, by westerners.

So why do superficial adipose depots become so massive in mammals? There is no agreed single answer. One reason might be the demands of lactation immediately following gestation. Reptiles are usually fattest just before maturation of the eggs begins, and the intra-abdominal fat bodies are much depleted by the time the clutch is ready to be laid. Heavily gravid females are often fasting, so the empty, inactive gut frees up yet more space inside the abdomen. But most mammals feed normally while they are pregnant and accumulate lipid in preparation for lactation. Fattening, feeding and gestation all take up space inside the abdomen. The reproductive organs and the gut cannot be excluded from the abdomen, but adipose tissue can. The subcutaneous depots may simply be the most convenient place to put a lot of adipose tissue.

Like all birds, penguins have both intra-abdominal and superficial depots of adipose tissue. They fast for impressively long periods, while remaining quite active, so they need large quantities of storage lipid. As they enlarge, the superficial depots expand laterally as well as thickening, and eventually form an almost continuous layer of adipose tissue over the thorax, abdomen and much of the neck, though not over the flippers, legs or head. But it would be rash to conclude that its function in such a position is as an insulator. In swimming animals, indefinite expansion of the abdomen is not compatible with streamlining while in water, but a sleek profile can be maintained if the additional fat is spread out over much of the body.

The same arrangement is found in migratory birds. Birds normally have very little superficial adipose tissue except for small depots around the thigh and wishbone (furcula), but just before migration, the subcutaneous depots enlarge, especially those over the back and pelvis, and may become several millimetres thick. They also expand laterally so the depot around the wishbone extends over the anterior part of the flight muscles and along the neck. It is the only way that adipose tissue amounting to 50% of the total body mass can be accommodated (see Chapter 5). Assessing the fatness of small birds by inspection and/or palpation of the superficial depots has the advantage of being able to take account, at least in a qualitative way, of changes in both the thickness and the area of the adipose tissue.

Keeping cool

Thus living in polar regions does not necessarily promote changes in the distribution of superficial adipose tissue in birds and mammals, beyond those associated with being larger and often fatter, but what about maximising heat dissipation in hot climates? The comparison is difficult because where the climate is wet as well as warm, and in tropical regions with little seasonal variation, the food supply is sufficiently reliable for most large animals to be continuously lean. Only in hot deserts is moderate obesity combined with tolerance of very high temperatures. Most desert animals avoid the heat by spending the days in underground burrows and foraging on the surface only at night, but camels are often out in the sun all day long.

Camels' main claims to physiological superiority are the capacity to recycle much of their body protein, thus enabling them to manage on forage of very low quality, and the ability to withstand the loss of a larger proportion of the body's water than most mammals could survive. Their blood cells can cope with drinking huge quantities of water, if necessary as much as 100 litres or 22% of body mass, in just a few minutes. They do not squander precious water on sweating or salivating as a means of disposing of excess heat: when exposed to full sunshine in hot air, they allow their body temperature to rise by as much as 7 °C (enough to kill a man) before evaporative cooling increases. Adults are so big and their thick fur impedes heat uptake so effectively that usually they do not reach peak temperature until just before evening. They can cool off during the night, before heating up again the next day.

Camels are not exceptionally obese: an intact adult is normally around 5–10% by weight dissectible adipose tissue, about average for a large mammal. Even geldings that have been overfed in captivity yield only about 100 kg of lipid, about 13% of the live body mass. Camels are most famous for having humps, which contain not water, as is sometimes supposed, but adipose tissue, of which they hold about a third of the total (see Figure 7, page 43). Whether as two separate mounds, as in bactrian camels, or just one as in dromedaries, the relative size of their adipocytes indicates that the humps are just greatly enlarged interscapular depots, and consist, as these depots always do, of relatively small adipocytes. The other superficial depots are reduced in camels, especially that on the anterior surface of the thigh, which amounts to less than 1% of the total in a bactrian camel, compared with 15–30% in most other mammals (see Figures 6, 7 and 13, pages 41, 43 and 135). Like other terrestrial mammals, camels have large quantities of adipose tissue inside the abdomen, partitioned into the usual four major depots.

So humps do not enable camels to be substantially fatter, and the basic pattern of site-specific differences in adipocyte volume is unaltered. The presence of large humps is associated with a reduction in the relative masses of most of the other superficial adipose depots, while the internal ones have remained much as they are in other animals of similar size and dietary habits. One possible explanation for this rearrangement of camels' adipose tissue is that it allows the body to remain tall and narrow, thereby minimising the amount of solar heat that falls on it from above. Thinner superficial adipose tissue over the rest of the body facilitates heat loss by simple conduction through the skin.

Blubber

Three separate lineages of mammals have become fully committed to aquatic habitats: cetaceans (dolphins, porpoises and whales) and sirenians (dugongs and sea-cows) spend their entire lives in water, while the pinnipeds (seals and walruses), emerge only to mate, give birth and suckle their young on beaches or icebergs. These marine mammals (only a small minority live in freshwater) have specialised adipose tissue called blubber that forms in the inner layers of the skin. Blubber looks and feels tougher than typical adipose tissue because it contains much more collagen and other extracellular material. But the limited information available – it is obviously very difficult to obtain tissue from such animals in fresh enough condition for the physiological studies described in Chapters 2 and 4 – suggests that blubber adipocytes have similar properties to those of adipose tissue in terrestrial mammals.

Smaller cetaceans and pinnipeds have almost no intra-abdominal adipose tissue: in dolphins, porpoises and pilot whales, the perirenal depot around the kidneys and dorsal wall of the abdomen is reduced to a few tiny patches, and apart from several large veins and many smaller blood vessels, the mesentery is a thin sheet of tissue so transparent one can read the newspaper through it. Almost all the storage tissue is displaced to the superficial depots, where it contributes to insulation, at least in adults.

Blubber that is full of lipid is a better insulator (i.e. conducts heat less efficiently) than that depleted of lipid, but the improvement is not very great. In tests[2] using samples from the harbour porpoise caught in the North Atlantic off Nova Scotia and spotted dolphins from the Pacific Ocean off Central America, insulation provided by a similar thickness of material was found to be about three times better for blubber from the fattest specimens that were 90% by weight lipid than for depleted blubber (5% lipid). These two cetaceans were compared because they are about the same size (body mass under 90 kg, about the same size range as people) but the harbour porpoise (*Phocoena phocoena*) lives in much colder water than the dolphins (*Stenella attenuata*). Factors other than insulation per unit mass of the blubber itself turned out to make a greater contribution to differences in heat loss between the species. *Phocoena* has a smaller surface area than *Stenella*, mainly due to its shorter flippers, tail and dorsal fin. At least in midsummer when the samples were collected, the porpoise's blubber is about twice as thick as that of the dolphin.

As well as differences between species and between individuals of the same species, there are also site-specific differences in the properties of the blubber. In minke whales,[3] insulative capacity was found to be highest in the blubber over the ventral wall of the abdomen. There was no general relationship between the thickness of blubber at any particular site on the body, and its insulative capacity per unit thickness, which suggested that one or other parameter, perhaps both, are determined by some physiological or biomechanical requirement unrelated to thermoregulation. The possibilities that spring to mind are the need to streamline the body shape for swimming, or something to do with the role of the blubber as an energy store.

Measurements of the fatty acid composition of triacylglycerols in the blubber of ringed seals (*Phoca hispida*) provide direct evidence that there is normally a thermal gradient across it. Samples taken from near the skin on the trunk, or from anywhere on the flippers, had a greater proportion of monounsaturated fatty acids, especially palmitoleic (C16:1) and oleic (C18:1), and fewer saturates than those from the inner side of the blubber near the muscles.[4] This property would allow the triacylglycerols near exposed surfaces to remain liquid even though they are always cooler than the core and often close to the temperature of the surrounding water. The gradient of fatty acid composition was more pronounced in fatter, older seals than in pups, suggesting that insulation becomes more efficient, i.e. less heat is allowed to reach the surface, as animals grow and by itself provides a good reason for growing large.

In mammals at least, adipose tissue can be almost anywhere in the body, but foetuses have to be inside the abdomen: that is where the uterus is. Although cetaceans and sirenians never have twins (they are among the few kinds of mammals to have eliminated multiple births entirely),[5] even a single foetus and its placenta produce a visible bulge in the mother's sleek, streamlined shape as birth approaches. Females usually fatten during pregnancy and use the stored lipid to fuel lactation. If adipose tissue were inside the abdomen, such fattening would add to the bulge, but if it is distributed around the body, the animal can expand uniformly while retaining its streamlined shape. The removal of adipose tissue from the abdomen to the superficial depots may have evolved to maintain streamlining, as much as to maximise blubber thickness.

Because blubber is the first tissue biologists meet on looking inside a dead specimen, and because it can sometimes be many centimetres thick, there is a general belief that marine mammals are considerably fatter, in the sense that they carry larger lipid stores, than most wild terrestrial animals. But in fact, because the ratio of surface area to body mass is so low in large animals with reduced limbs, the overall proportion of adipose tissue is often not very different from that of terrestrial mammals. It just seems more because the distribution resembles that of very fat terrestrial mammals.

As pointed out in Chapter 2, naturally obese animals are most common at high latitudes, where large seasonal changes in plant productivity and an unpredictable, severe climate lead to irregular food supplies. Warm-blooded animals in such habitats might also be expected to benefit from the insulation produced by superficial adipose tissue. Those in the warm, equable tropics have less need to be fat, and, other things being equal, need less insulation. Blubber as an insulator always presents a paradox: should its lipid be withdrawn for use as fuel when food becomes scarce, thereby weakening insulation and increasing the amount of energy required just to keep warm? Grey seals (*Halichoerus grypus*) breed on North Atlantic coasts, including western Britain. The pups fast for up to 10 weeks after being abandoned by the mother, and measurements of internal body temperature show that superficial muscles do indeed become cooler as the blubber layer shrinks due to the lipid in its adipocytes being withdrawn.

Blubber has to be thick to insulate thoroughly. As just explained, superficial structures can only be thick if they represent a large proportion of the total body mass (i.e. the animal is fat) or they are spread over a proportionately small area (i.e. the animal has a low ratio of surface area to body mass). So in general, blubber is only satisfactory as the sole form of insulation for large animals. Most of the smallest cetaceans, porpoises and dolphins, are restricted to temperate and tropical regions, and many of the great whales, e.g. humpbacks and fin whales, that feed in polar seas migrate to warmer waters to give birth to calves that are only a twentieth of their size. Many such places provide little or nothing for adult whales to eat, so the mothers fast for several weeks while lactating, drawing on their lipid reserves, as female polar bears do. The whales do not return to cooler, food-rich waters until the calves have grown large enough and fat enough for blubber insulation to be efficient.

As well as a lower limit in the size of a marine mammal, retention and dissipation of heat also impose an upper limit on the thickness that blubber can be. Marine mammals dissipate excess heat mainly through the blubber-free skin of the flippers and tail. Blood flow through these appendages greatly increases as soon as the core temperature rises above normal. On hot days, seals on land sometimes hold up a flipper to the breeze to accelerate cooling.

Seals are born on beaches or ice floes and most species do not go to sea until they are several weeks old. Neonates have thick coats of fluffy fur but little adipose tissue anywhere, although they fatten rapidly while suckling their mothers' rich, creamy milk (see Chapter 5). At about the time the seal pups are weaned, the neonatal fur is replaced by shorter, stiffer fur. But fur of some sort continues to make an essential contribution to insulation in all seals except the adults of the largest species, the elephant seals (*Mirounga* spp.). Walruses (*Odobenus rosmarus*) are an exception to these generalisations. The hair is very sparse throughout life, but the pups are as heavy as a woman (45–70 kg) at birth, up to twice the size of neonatal elephant seals, although as adults, the latter are about twice the size of the former. So the single walrus pup is very large at birth compared with its mother.

Both these very large seals occur in cold seas, *Odobenus* in the Arctic Ocean, and one species of *Mirounga* in the northern Pacific including around Alaska, and the other in the Southern Ocean. Walrus blubber can be up to 15 cm thick and blood flow though it and the skin can apparently be adjusted rapidly and efficiently. Biologists prefer to visit the Arctic during summer, and observe walruses most frequently when they are hauled out on beaches basking in sunshine. Under these conditions, their skin is noticeably pink, due to the profuse flow of blood through it and the underlying blubber. But when in the sea or exposed to cold weather, walruses are a ghostly pale grey: the blood is withdrawn from the skin and blubber to the muscles and other warm internal tissues, and so contributes nothing to the superficial colour. Blubber can work as an adjustable insulator in this way because blood flow through it and the skin can be shut down almost completely for long periods without ill-effects. Cutting off supplies of oxygen and nutrients for a long time is only possible for tissues like adipose tissue that use such materials at a very low rate: muscles, kidney or brain would suffer irreversible damage if subjected to similar treatment.

Like most large, long-lived marine animals, walruses are plagued by skin parasites and encrusting organisms such as barnacles, which are kept in check by regular shedding of the outer layer of skin, parasites and all. The blood supply to the skin is much enriched while its outer layer is regrowing. The walruses spend most of the moulting period on beaches and are reluctant to enter the water, perhaps to avoid curtailing superficial blood flow.

If necessary, marine mammals can be very fat. Many of the factors that limit fatness in terrestrial animals do not apply to aquatic mammals. As in the case of penguins, body bulk is only a slight impediment to swimming, and since seals and whales are nearly neutrally buoyant, they weigh very little in water. The maximum amount of blubber that whales or seals can have without risking overheating during vigorous swimming depends upon body shape, the temperature of the water they are living in and how much heat they generate. A thickness of about 20 centimetres seems to be the upper limit. Blubber of that thickness apparently does not hold enough lipid to meet the energy storage needs of the very largest whales, fin whales, blue whales and their relations. They have adipose tissue in the mesentery that supports the gut and around the kidneys, and some intermuscular adipose tissue in certain muscles, just as terrestrial mammals do.

A cushion of adipose tissue

As well as having some physical properties that lend themselves to a role in thermal insulation, adipose tissue also has features that would suit it for mechanical roles. An almost incompressible, moderately viscous fluid (i.e. triacylglycerols at body temperature) held in small, uniform pockets (adipocytes) bound firmly together with a light, tough but flexible material (extracellular collagen) absorbs impacts very well – indeed, 'bubble' wrapping materials and other kinds of padding have a similar structure. Some of the energy of deformation may be stored, enabling the material to recoil to its original shape when released, like a tennis ball. Depots of adipose tissue can be of any shape, so in principle, such a material could mould contours and fill spaces between tissues.

'Structural' adipose tissue consists of collagen and adipocytes in different proportions, and is arranged in a wide variety of different ways according to its anatomical location and the species. Its adipocytes are generally smaller than those of storage depots, and they do not hydrolyse triacylglycerols, take up glucose or fatty acids or respond to hormones to any measurable extent. So they contribute little or nothing to the uptake or storage of lipids for the rest of the body, and do not expand in obesity, nor shrink during fasting. The lack of involvement in lipid metabolism provides biochemists with the excuse they need to ignore structural adipose tissue, but anatomists and engineers study its mechanical properties. Their measurements have shown that, while it may not supply any energy-yielding materials, structural adipose tissue in limbs can reduce total energy expenditure during normal movement, and protect delicate or brittle tissues from the large mechanical forces that inevitably accompany fast movement.

Heels and toes

Structural adipose depots are components of the limbs of many mammals. The most familiar are soft pads under the tips of our fingers and toes that consist of adipose tissue embedded in a dense network of collagen. The heel pad covering the outer faces of the calcaneus, or heel bone, appears to be similar. But while most larger primates have distinct toe and finger pads, that of the heel is far more extensive in humans than in any other species. That fact alone suggests that its properties may be linked to the uniquely human erect posture and bipedal locomotion.

At its peak, which lasts less than a twentieth of a second, the force exerted on the hard ground while a young man runs at 6.5 metres per second is nearly three times his body weight. At each stride, the heel pad is squashed from a resting thickness of about 1.5 cm to about 1 cm, but it does not bulge very much. The fatty components are encased between firm bands of collagen, forming 'compartments' that are tightly bound to both the skin and the heel bone. Mechanical tests of the kind used by engineers to characterise inanimate materials show that this structure is bouncy, absorbing the energy of impact between the heel and the ground, and recoiling within milliseconds to its original shape when the load is removed. These properties protect the bone and its joints from the shock of impact at every stride, and the recoil may help to place the bones in the correct position as the body weight is transferred from heel to ball of the foot.[6]

Only about 70% of the energy that was used to deform the heel pad appears in the recoil. The rest is degraded into heat. That heat can be useful, of course, and its absence is one reason why one's feet get colder faster when standing or sitting in a cold climate than when walking briskly or running under similar conditions. Perhaps surprisingly, in view of the temperature sensitivity of so many other properties of adipose tissue, percentage energy loss was found to increase only slightly as the specimens were cooled from 37 °C to below 10 °C. In other words, our heel pads are slightly less bouncy when we are walking barefoot on ice than in an indoor gymnasium. Corresponding measurements on the paw pads of other mammals reveal similar independence of mechanical properties and temperature.

One's feet feel less supple and springy than normal when one tries to walk on hard rocks after sitting or lying for some time. This effect is not due only to muscle stiffness, though it plays a part, but to the fact that the mechanical properties of heel pads change slightly over minutes and hours. In tissues from middle-aged and elderly people, bounciness was most efficient when about 0.6 s elapsed between each cycle of squeezing and releasing the heel pad, a regime that simulates brisk walking or gentle running. About 3.7% more of the energy absorbed by the heel pad was lost as heat (i.e. did not contribute to recoil) for each tenfold increase in rest time. So the message from this structural adipose tissue, as from the metabolically active depots, is that people are built for walking, and that's what they should do – far and often.

Sensations received from joints, muscles and skin of the feet are immediately integrated with the commands that the nervous system sends to the muscles, adjusting each stride to unforeseen irregularities in terrain. Many different sensations are encoded, including pain and measures of substrate texture and temperature, but the most abundant and important report joint position and tissue stretch, features that are due directly to the mechanical properties of the structural adipose tissue in the foot. Most of this neural feedback is handled reflexively, so we are usually only very dimly aware of it, but injury or disease (e.g. degeneration of the peripheral nerves) in any of the structural tissues, including the fat pads, distort this neural feedback, thereby impairing accurate stride-by-stride control of movement, and greatly increasing the risk of falling, especially during fast walking over steep or rough surfaces. These structural depots do not need, so do not have, more than a meagre blood supply, but poor perfusion means that they are slow to regenerate after a major injury. Perfect repair is so important to comfort and safety, so we should be glad that surgeons and engineers take an interest in them.

The heel and toe pads not only protect the bones and joints from immediate impact, but, by ensuring that the sensations from the feet are prompt, exact and reliable, they unobtrusively prevent embarrassing and potentially dangerous accidents. Wearing shoes distorts their roles in several ways. All footwear acts as an insulator, sometimes so effectively that the heat produced by muscle contraction and the compression and release of foot-pads cannot be dissipated fast enough, and the feet become uncomfortably hot during running. Walking boots and sports shoes with thick, springy soles supplement the natural resilience of the fat pads, which both protects the feet from impact damage, and enables athletes to run a little faster. But all stiff footwear impairs the accuracy of neural sensations received through the feet, which does not matter much when walking on a smooth pavement, but can be a serious handicap in rock-climbing.

The paws correspond to our fingers and toes, but their mechanical properties, like those of the heel pad are intricately associated with their owner's locomotory habits. The paw pads of domestic cats and all the larger members of the cat family (lions, tigers, leopards, etc.) are larger but softer and more compliant than those of dogs, foxes or other fast running carnivores such as cheetahs. Soft paws absorb the energy released by the impact of the foot into the ground, and, especially when combined with retractable claws (a special feature of all members of the cat family) enable the animal to walk silently even across hard surfaces. Cats and their relatives hunt by stealth, for which silent movement is obviously important; they extend their sharp claws only to climb trees or grip prey. Dogs and cheetahs do not climb and rely more on speed to catch prey. Their narrow, firm feet have smaller paw pads that contain more tough collagen and fewer adipocytes than those of soft-pawed cats, and bounce off the ground after each stride with minimum energy loss, more like a cricket ball (or a baseball) than a football.

Bears, especially polar bears, have huge feet, much larger in proportion to their body mass than those of the cat family. Each finger or toe has a wad of structural adipose tissue underneath it, and an even larger one forms the central paw pad. Bear tracks always include clear claw marks because they cannot retract their long, stout claws, which nonetheless remain sharp enough to be useful for subduing prey and, in forest-living species, for climbing trees.

The bare surface of the footpads is covered with firm conical projections, each about a millimetre in diameter, formed from the thick outer layer of the skin. This material has 'non-slip' properties, that together with the underlying pads of soft, flexible adipose tissue, enable the feet to grip the slippery ice over as wide an area as possible. The rough, thick hair protruding from between the toepads and around the edge of the main sole probably also helps to prevent slipping on very smooth surfaces. Bears are primarily animals of the cold forests and northern tundra. Large, non-slippery feet may be an adaptation to moving safely across snow and ice. Those of polar bears work better than any snowboots humans have been able to construct. On meeting a bear on the arctic ice, one is well aware that it can run fast on such a slippery surface, and people can't.

The mechanical properties of different kinds of paw pads are due mainly to the quantity and arrangement of the collagen around the adipocytes, but it is also important that the lipids in them remain liquid. If they freeze, the animal would be walking on blocks. The extremities are nearly always cooler than the body core, so the triacylglycerols in paw pads, and those in the marrow of peripheral bones, include a much greater proportion of unsaturated fatty acids than those of other adipose depots. The composition of adipose tissue and its perfusion with warm blood ensure that the paws' temperature remains within the limits for maintaining the appropriate mechanical properties. The paw pads of cats and dogs are bare of fur and, like our own feet, may serve as surfaces through which excess heat is dissipated during strenuous exercise or in very hot climates. Those of animals that live in very cold climates, such as arctic foxes and hares, are covered with fur.

Eyes and cheeks

Vertebrate eyes are mounted in bony sockets in the skull. Their exact position within the socket is determined by the size and shape of a mass of structural adipose tissue behind and around the eyeball and enclosing the large optic nerve. The depot looks and feels like ordinary storage adipose tissue, but attempts to stimulate lipolysis show that it behaves like a structural depot: the response of isolated adipocytes to noradrenalin is so slow, it cannot be detected during the few hours normally allowed for such measurements in the laboratory.

In mammals, the cellular structure of eyesocket adipose tissue shares some features with the storage depots.[7] Adipocytes in the eyesockets are larger in larger mammals, ranging in size from about 10 picolitres (10^{-11} litre) in shrews, to over a nanolitre (10^{-9} litre) in fin whales (see Chapter 2). The association cannot be explained only by the fact that whale eyes and whale eyesockets are larger. Species such as camels, that have relatively large quantities of fat around the eye, simply have more adipocytes of the appropriate size. Within a single species such as guinea-pigs, individuals whose storage adipose tissue consists of a relatively large number of relatively small adipocytes also have smaller adipocytes in their eyesockets. So whatever determines the basic cellular structure of storage adipose tissue, also determines that of this structural depot. The matter has not been investigated in any other structural depot.

Except during hibernation, the brain is one of the warmest parts of the body, and its temperature is tightly regulated. In Svalbard reindeer, the triacylglycerols in the eyesocket adipose depot located beside the brain contain slightly more polyunsaturates than those in the storage depots, but many fewer than those in the bone marrow of the lower limbs. Nothing is known about how these structural depots obtain mixtures of fatty acids appropriate to their operating temperatures.

Most mammals have long, narrow snouts and extensive lips, making the mouth well-suited to sucking. So, although mammalian neonates usually have relatively shorter mouths than adults (due to the precocious development of the eyes, ears and brain, that enlarge the back of the skull), they are still well equipped to suck from mammary teats. In primates, the brain and eyes are larger still, and the adult face is foreshortened, making the mouth short and wide, a shape that is less well suited to sucking. Sucking in newly born primates is facilitated by a pad of adipose tissue, called the corpus adiposum buccae, situated between the muscles of the cheek.[8] As well as filling the spaces between the jaw muscles, the buccal pad's mechanical resilience helps to re-extend the muscles at the end of each sucking cycle. Its rhythmic movements can be clearly seen in the cheeks of many young primates, especially apes and humans.

The adipocytes that form the corpus adiposum buccae proliferate and start to fill with triacylglycerols from early in the second half of gestation, at a stage when most other adipose depots are hardly developed at all. They contribute greatly to the chubby-cheek appearance of human infants, and are apparent even in those who are otherwise rather lean. The buccal fat pads reach their maximum size at birth, but persist long after weaning, becoming relatively smaller with age as the adjacent muscles and skeleton grow. In dissections of cadavers aged 60 years or more, each buccal fat pad was found to weigh 8–11.5 g, even in people who had died from wasting diseases such as cancer, although they sometimes shrink extensively in very old people, producing the characteristic hollow-cheeked appearance.

Most mammals recognise each other's sex, reproductive status and individual identity mainly by smell or voice, but for monkeys and apes, with their superior visual capacities, appearance is also important. Structural adipose tissue contributes to features that indicate an individual's sexual and social status to its conspecifics. Mature male orang-utans have cheek pouches, that seem to be larger and more conspicuous in high ranking individuals (see Figure 8, page 44). Senior male gorillas also have a small but conspicuous lump of adipose tissue on top of the skull, making the head appear larger, like wearing a helmet. Birds also recognise each other by sight, and although plumage is the main distinguishing feature, patches of bald skin overlying specially shaped features also contribute. The prominent black knob on the foreheads of adult mute swans consists mostly of structural adipose tissue, as does the wattle on male turkeys.

Nothing is known of how these structural adipose depots grow. Presumably they take up lipids from the blood during the early stages of their development, and then lose the capacity to do so as they mature. Because it does not expand or shrink once mature, and can form almost anywhere on the body, structural adipose tissue seems to be a promising material to use in the surgical reconstruction of serious injuries. However, as plastic surgeons have found to their cost, although the tissue has no role as an energy store, it is not completely metabolically inert. The transplant must have a blood supply: without it, the adipocytes gradually lose their lipid contents and die. The tissue's consistency slowly changes from soft and rubbery to tough and fibrous, which may be very unsatisfactory for the patient. Establishing a proper blood supply is not easy, because adipose tissue is never perfused from a single major vessel (see Chapter 2). Adipose tissue's potential value in reconstructive surgery is one of the main reasons why biologists should study this much-ignored type of tissue.

Fatty armour?

The fact that certain specialised kinds of adipose tissue are clearly adapted to mechanical roles has led to another explanation for the distribution of adipose tissue in mammals: that it serves as padding, protecting vital organs from the bumps and bangs of everyday life, and sometimes from injuries incurred during fights with members of the same species. Many observers have been impressed by the fact that adipose tissue is found around some delicate vital organs, including the kidney and heart, and have suggested that it protects them.

The hypothesis that mechanical protection is a factor determining adipose tissue distribution can be tested by comparing corresponding tissues from animals of naturally different sizes and habits, in the same way as comparative data were used to evaluate the notion that adipose tissue is adapted to serve as thermal insulation. Larger animals need proportionately more massive skeletons than smaller animals pursuing similar lifestyles. In practice, an elephant's skeleton represents only a slightly greater proportion of its body mass than that of a mouse because elephants use their limbs more cautiously: they do not gallop, climb or leap but mice always travel at a gallop, and have no hesitation in jumping from a great height. In larger animals, soft tissues, especially those such as the liver that are not encased by or attached to the skeleton, also have more skeletal material in the form of a network of collagen. That is why both muscle and organ meat (liver, kidneys, etc.) from the adults of large species (e.g. cattle) are tougher than the corresponding tissues from small animals (e.g. poultry). The skin is also tougher and more massive in rhinoceroses, elephants and other large animals than in smaller species.

The theory can be tested by comparing animals of a wide range of sizes. If the function of adipose tissue were to act as a sort of armour, protecting underlying or enclosed tissues from injury, then the component responsible for most of its mechanical properties, the collagen, should be proportionately more abundant in larger species, as it is in the liver and muscles. Measurements of the collagen content of corresponding adipose depots in mammals of widely different sizes are not consistent with this prediction.[9] The perirenal adipose tissue of larger animals does not have proportionately more collagen, and so it is not mechanically tougher than that of smaller ones. The comparative data show that adipose tissue is not *adapted* to protect vital organs, making such explanations for its anatomical arrangement much less plausible.

So it remains to be explained why there is at least a small quantity of perirenal adipose tissue near the kidneys in almost all mammals. The kidneys and adipose tissue do not share a blood supply, and they are separated by a sheet of stout connective tissue so, unlike the heart or lymph nodes, exchange of metabolites between adjacent tissues is most unlikely. The most probable explanation for the association is thermal but not insulatory:

depending upon circumstances, heat may pass from the adipose tissue to the kidneys and other viscera, or in the reverse direction.

The dorsal wall of the abdomen is one of the warmest places in the body, close to the liver, gut and kidneys all of which produce a lot of heat as a by-product of their metabolism, and need to be warm to function properly. In some neonates, including human babies, the perirenal depot contains brown as well as white adipose tissue, though the former disappears shortly after birth. Lipid reserves have to be not only sufficient in quantity, but also accessible when they are most needed, such as when arousing from hibernation. Several of the perirenal depot's biochemical properties suggest that, except around the time of birth, it is primarily a general energy store for the whole body (see Chapter 4). Its situation in the warmest place in the body may enable it to supply lipid to brown adipose tissue and to the muscles when the reserves in superficial depots are too cool to be mobilised efficiently.

In many harem-forming mammals such as red deer and elephant seals, the males become fatter before the breeding season, and they eat little and lose weight during the rut. Some of the additional adipose tissue may accumulate on visually conspicuous sites, such as the back or neck, making the animal appear more massive and therefore stronger. In a few species that engage in fights, notably elephant seals, prominent superficial adipose depots may sustain many of the injuries inflicted by rival males. Such observations have been invoked to explain some aspects of the distribution of human adipose tissue.

The idea can be tested by exploring how well it accounts for known sex differences and age-related changes in the relative masses of adipose depots. Although there is much argument among anthropologists about the role of hunting and intra-specific conflict, it is generally accepted that only young adult males were involved in most forms of physical combat. But in almost all living races, superficial 'protective' adipose tissue is thicker and more extensive in women than in men, and in children than in young adults. Adipose tissue is not distributed in a way that would protect delicate and vulnerable parts of the body, being minimal over exposed vital organs such as the head and neck.

Final words

Measurements of the arrangement of white adipose tissue in wild animals are not consistent with the long-established theory that fat is adapted to thermal insulation or mechanical protection in terrestrial mammals. But such notions have been useful in establishing the concept of a functional relationship between anatomical location and site-specific properties. Biologists see no problems with accepting a functional interpretation for the differences in the composition of triacylglycerol fatty acids and the abundance and arrangement of collagen in the structural adipose tissue in the paw or around the eye. Metabolically active storage adipose tissue is also partitioned into several depots, but the ideas of local interactions and site-specific properties being matched with the tissue's anatomical relations have been slower to gain acceptance. This topic is discussed further in Chapter 8.

7

Fat people

Wild animals do not become obese without good reason. They accumulate as much fat as they need for the journey, or hibernation, or breeding period, but no more. Most species are fat only briefly, and remain lean for a large proportion of the year. The capacity to fatten evolves quite quickly as and when it is needed, and, so far as we can tell, disappears again equally easily when the trait is no longer useful. There is a strong tendency to assume that our own predisposition towards obesity is a relic of a situation when the trait was useful for one or other of the reasons that animals have evolved to become fat. But is there really any evidence that our species or its ancestors evolved habits for which fatness was essential? If so, when did this state of affairs occur, and for how long? And why has it ended, but the tendency to fatten remained after the reason(s) for its existence disappeared?

You might be surprised to know that, in spite of much discussion, the answers to these questions remain vague and hence controversial. Experts cannot even agree upon whether our ancestors become fat millions of years ago, when sharp stones were first used as tools; hundreds of thousands of years ago, when they first invented cooking and using skins for clothing; or just a few thousand years ago, when people took up agriculture and pastoralism. This chapter tries to bring together comparative biology, anthropology and evolutionary theory to consider the special features of human obesity, and how our habits and habitats may have promoted fattening.

When our ancestors fattened

Fossilised remains of human ancestors are very rare: far more fossils are available to guide our understanding of the course of events leading to the evolution of horses or cats than that of our own species. The few remains we have do not readily form a coherent, unambiguous story, so much is left open to guesswork and opinion, particularly the evolution of soft tissue features, such as fatness and the loss of body hair, which cannot be discerned from fossils anyway. Comparing ourselves to living animals with whom we share common ancestry offers the best chance of understanding how and when such features arose.

Primate lives

Compared with horses, cattle, whales or elephants, primates are anatomically unspecialised mammals: we still have the basic five fingers and five toes, all the major arm and leg bones and almost as many teeth as the earliest mammals. It is not clear when or how primates took to grasping food first with the hand, instead of with the teeth or tongue as most mammals do, or adopted the 'sitting up' posture while feeding, but even the most primitive living species show these characteristics, so they must have appeared very early.

The diet and habits of primitive prosimian primates, tarsiers, bushbabies, lorises and lemurs, resemble those of the first mammals. Most kinds eat large insects and other invertebrates such as worms and snails, and carrion if they are lucky enough to find it. The larger lemurs also eat fruit, flowers and certain leaves. The problem with an insectivorous diet is not chewing or digesting it but finding it, and avoiding becoming prey oneself. Success calls for accurate visual perception and hearing, powerful, well-controlled movement, fast reaction time and a good memory. So while the ancestors of deer and antelopes were perfecting ways of getting as much nourishment as possible from tough, toxic vegetation by rumination, primate evolution concentrated on adaptability and the mental and physical agility needed to reach the most digestible morsels. These qualities are usually associated with relatively slow growth, and prolonged dependency of infants on parents, which in turn favour small litters and elaborate social and sexual behaviour that keeps family groups together.

Until the 1970s, biologists confidently asserted that monkeys and apes other than humans were primarily herbivores, eating mainly leaves and fruit. More thorough observations of animals in the wild, and more sympathetic and enterprising management of them in captivity, show clearly that nearly all the monkeys and apes include some lipid-rich foods in their diet. Most small monkeys from time to time eat oily seeds and nuts, the larvae or pupae of large insects, and birds' or reptiles' eggs, as did their prosimian ancestors. Larger monkeys such as mandrills and baboons take carnivory more seriously. They challenge other scavengers such as vultures, jackals and even hyenas for access to carcases, and occasionally kill and eat mammals as large as a young antelope.

The primatologist, Jane Goodall, and her colleagues were among the first to report that groups of wild chimpanzees at Gombe in western Tanzania were killing and eating mammals, including sometimes members of their own species. Such activities are not readily observed, because they are brief and don't happen very often, but there is now no doubt that from time to time, the common chimpanzee (*Pan troglodytes*) kills large prey. Her descriptions of organised violence as normal chimpanzee behaviour were greeted with scepticism from other scientists: they spoilt cherished notions about the uniqueness of meat-eating in humans, and the ideal of apes as peaceful animals, uncorrupted by aggression and strife.

Chimpanzees seem to prefer monkeys, especially colobus that are less than a fifth of their size, but they also take bushpig and small antelope, and sometimes chase predators as large as leopards off fresh carcases. They almost always hunt in groups consisting only of adult males, led by a senior dominant animal, which usually gets most of the spoils. Female and juvenile members of the troop also try to get a share of the flesh, but where competition is fierce, adult females tend to concentrate on collecting insects. Noisy and spectacular as chimpanzee hunts are, they make only a small contribution to the animals' overall diet. Field observations suggest that adult chimpanzees eat about 10 kg of mammalian meat in a year, or an average of 27 g (about 1 ounce) per chimp per day. For comparison, Bushmen in the

Kalahari desert eat about ten times as much meat, and British people in the 1990s about six times.

Animal foods and nuts are not as easy to obtain as fruit and leaves, but primates are more than willing to put in the extra time and effort required to find and handle them. Primate teeth are strong enough to open small nuts, but larger, tougher ones call for more ingenuity. Chimpanzees use hard stones to crack large nuts, an indisputable case of tool-using by a non-human primate. Chimps clearly find nut cracking to be hard work: the technique is only effective with an easily held 'hammer' used on a suitably shaped 'anvil'. Sometimes chimps gather handfuls of nuts and carry them to a good cracking site, or bring a useful 'hammer' to a source of nuts. Nut cracking may have been the activity that set ancestral hominids on the path of extending and diversifying their food supply by using, then making, tools.

At least in part of their remaining range, chimps also use tools for harvesting ants and termites, another rich source of highly palatable lipids. After locating an ant or termite nest, a chimp chooses and prepares a suitable long, fine stick or grass stem and dips it into a hole leading to the interior of the nest. The angry insects run onto the intrusion, and a few usually remain attached long enough for the chimp to pull it out and lick them off. Chimps obviously relish these lipid-rich foods: they spend hours cracking nuts and removing the kernels from the shell fragments, and in 'fishing' for termites. Such manual skills are not easy to learn: juveniles watch proficient adults attentively and practice for months before their technique becomes efficient.

Many other animals that are basically herbivores share primates' taste for the occasional bite of animal food. Rabbits, guinea-pigs and other grazing animals with very efficient digestive systems can obtain sufficient lipids from leaves and roots, but rats, mice and many other rodents prefer more concentrated sources of such nutrients. That's why baiting mouse traps with hard cheese or bacon rind works so well: wild mice never eat cheese or bacon, and can exist for months on a diet of almost pure starch, but they do like fatty foods. Most kinds of parrots can live on starchy seeds and nuts, but they obviously prefer those rich in lipids, and are willing to spend time and effort opening a hard nutshell to reach a delicious kernel. In captivity, they too like some kinds of cheese and meat fats, and readily drink thick cream. Parrots seem to know that these foods are rich in lipids, even though they are in an unnatural form.

From herbivore to carnivore

I am probably not alone in being perplexed as a child by the paradox that people had to eat a mixed diet of fruit, vegetables, grains and meat to stay healthy, but cats eat only meat and horses only grass. Many kinds of animals, such as badgers, starlings and brown bears are part-time predators, eating a mixed diet much like our own. Only a few groups of vertebrates, notably certain fish, snakes, crocodiles, owls, many seabirds, eagles and hawks, and the cat, seal and toothed whale families are exclusively carnivorous. Changes in metabolism and sensory abilities accompany long-term commitment to such a specialised diet. Cats rely so completely on obtaining all the fatty acids they need from their diet that they have almost lost the capacity to produce some of the enzymes that desaturate and elongate fatty acids. Animal tissues contain very little carbohydrate, so the biochemical pathways that make glucose from proteins, which are weakly active in most animals, are greatly enhanced in obligate carnivores, as are the mechanisms that enable them to excrete the waste products of breaking down proteins.

Cats and their relatives cannot taste sweetness, although dogs, badgers, bears and other members of the mammalian order Carnivora, and indeed most other mammals, can do so.[1] So while bears raid bees' nests for honey, and dogs bark for chocolate, the proverbial cat concentrates on stealing the cream (which contains about the same mixture of triacylglycerols as the adipose tissue of the vertebrate prey of its ancestors). There is no evidence that snakes can taste sweet foods, although omnivorous reptiles such as iguanas and tortoises certainly can (and many are very fond of figs, strawberries and other sweet fruit). As well as having no taste for sugars and starches, cats and snakes cannot digest such materials efficiently. More than a very small quantity in the diet leads to serious digestive disorders. Wolves, dogs and foxes have not completely lost the capacity to recognise and digest plant foods, and sometimes eat fruit or grains when hunting is poor, but they cannot live indefinitely without meat and fat.

Ancestral humans are among the many kinds of animals that made the transition from a mainly herbivorous diet of leaves and fruit, supplemented with a little fish or carrion and the odd insect, to living mainly on meat, with a little fruit or other plant matter from time to time. Dozens of insect families consist mainly of herbivores, with a few predatory species, or the other way round. Parrots are primarily seed eaters: their finely controlled, powerful beaks and grasping feet enable them to crack and dismember very tough nuts. Nonetheless, most of the larger species occasionally eat animal food, and one species, the kea (*Nestor notabilis*) of New Zealand's Southern Alps, lives mainly on carrion and small prey when plant food becomes scarce in winter, and farmers accuse it of killing sheep.

The bear family, Ursidae, are members of the order Carnivora, but all but one living species are omnivores. Their diet includes almost all plant parts except wood and tough stems, plus worms, snails and other large invertebrates, and vertebrate prey from time to time. One kind of bear, the polar bear, is almost exclusively carnivorous. Similarities of bones and teeth point to the conclusion that polar bears (*Ursus maritimus*) evolved from the ancestors of brown (grizzly) bears (*Ursus arctos*); indeed, the two species are genetically so similar that zoo specimens caged together sometimes hybridise successfully, producing khaki-coloured cubs. The giant panda of China, which is probably best defined as a racoon-like bear, has gone to the opposite extreme: the few pandas that remain in the wild are seen eating bamboo stems and little else. But the best way of catching them is to bait the trap with rich, raw meat.

Diet is one of the main differences between brown and polar bears. The change has taken place very recently in evolutionary terms: polar bears appeared about 100 000 years ago, during one of the long cold periods of the Pleistocene Ice Age, when the climate over much of Europe was similar to that of northern Canada today. When really hungry, polar bears scavenge carrion and nibble at berries or sea-weed, but their main food is seals that they kill themselves. Growing juveniles and lactating females eat much or all of the meat as well as the blubber, but mature non-breeding adults often strip off the blubber and leave most of the meat (which is quickly eaten by scavenging birds and arctic foxes – good food does not go to waste in the Arctic). Newly born and orphaned seals often have so little blubber that the bears do not bother to skin them: they eat the head, presumably to get the lipid-rich brain, and leave the body.

But after going to all the trouble of making a kill, why not eat all the meat? Except for animals metabolically equipped to deal with it, protein is inefficient as a source of energy. Several energy-using biochemical transformations are necessary to convert the carbon and hydrogen components of the protein into glucose or lipid, and the nitrogen must be removed and excreted. Birds and most reptiles can excrete nitrogen as insoluble uric acid, that is familiar as the semi-solid white component of bird droppings, but all mammals produce urea, which can only be excreted in solution. Thus getting rid of excess nitrogen after eating too much protein requires a lot of water, and, because it involves some difficult biochemical reactions, is energetically expensive and fairly slow.

If protein intake exceeds the body's capacity to deal with it, free amino acids remain at high levels in the blood, upsetting the chemistry of transmission between nerves and other important processes. In humans, overindulgence leads to a severe headache, nausea and a firm wish not to repeat the experience. Animals other than specialist carnivores such as cats and snakes also prefer to forego nutritious food rather than risk poisoning themselves. Triacylglycerols do not contain nitrogen and so such metabolic problems do not arise. With an adequate supply of low density lipoproteins to transport it and sufficient adipose tissue in which to store it, there seems to be no upper limit to the bears' consumption of triacylglycerols.

People are even worse than bears in this respect: too much protein leads to nitrogen poisoning, and fails to satisfy hunger. Travellers in remote regions and at sea relish oily fish and seek out the fattest animals they can find, rejecting very lean specimens even when they are hungry. Darwin[2] mused upon this effect in 1832 while travelling south-west from Buenos Aires into the pampas grasslands, where the gauchos lived almost entirely on fatty meat. He tried a similar diet himself for a few days, but was glad to reach a small town where he could buy some biscuits, concluding that the preference for fat over lean meat 'appears to me a curious physiological fact'. Modern research indicates that the upper limit for most adults is 50% of the total energy needs derived from protein, less for children, whose higher energy expenditure can only be satisfied by substantial quantities of carbohydrates or lipids, preferably both.

Even Eskimos cannot live indefinitely on lean meat alone. As well as reindeer (caribou) meat, the traditional winter diet of Canadian Eskimos[3] included the meat and blubber of seals and, if they could catch them, whales and arctic char, an oily fish of the salmon family, all eaten raw. Several observers remarked upon the large quantity of water drunk during meals. Indeed, preparing drinking water by melting snow or fragments of icebergs (sea ice is too salty) was the main use for pots and fires fuelled by willow twigs and seal oil. Eskimos relished as delicacies the faeces and gut contents of reindeer (which eat mosses and other plants) and the stomach contents of walruses (that include clams and other invertebrates). In summer, they collected various wild berries and roots, a few of which they stored and ate during the winter. This small quantity of vegetable matter, and certain animal tissues such as the skin of narwhal whales eaten fresh and raw provided sufficient vitamin C. The deficiency disease scurvy, which decimated European expeditions to the Arctic, was unknown among the indigenous people.

The possibility of protein poisoning on a carnivorous diet has led to the suggestion that, as novice predators, humans, like polar bears, were more interested in eating the lipids in their prey than its meat. It is, of course, very difficult to assess the validity of such a suggestion from the fossil remains of the humans themselves, but support for the idea comes from the study of the debris they left behind. Limb bones of large mammals such as femurs and tibias resist weathering and further destruction from scavenging animals, so it is not surprising that they are often found among the relics of human activity. A great many prove to have been smashed, apparently by blows from heavy, blunt rocks. The only material made available by such operations is bone marrow, which in adult animals would be mostly adipose tissue rather than cells engaged in producing blood. Such observations have given rise to the suggestion that early humans broke large bones in order to get at the fatty marrow, presumably to eat it.

Hominids would not have been the only animals on the African savannah to exploit this nutritious source of food: the lammergeier, a kind of vulture native to southern Europe and Africa, breaks similar bones by grasping them with their feet, flying high over outcrops of suitably hard, flat rocks and dropping them with great precision. These birds seem to be among the last to get to a carcase, normally only gaining access after the vultures, marabou storks and jackals have had their fill, and the bone marrow may just be the only edible bits left. Brown hyenas scavenge the west coast of South Africa and Namibia, where cold, nutrient-rich currents from the Southern Ocean support huge breeding colonies of seals. As well as the meat and blubber, they eat whole seal bones, marrow, hard tissues and all. Baboons always eat the brains and bone marrow of small prey such as birds or hares, and make serious attempts to reach both tissues in prey such as antelope, which have much tougher skeletons. They try to suck the marrow out of bones and poke their fingers in skull orifices to scoop out brain tissue.

Should the habit be taken to indicate that early people had already developed an appetite for fat? Why was more accessible adipose tissue, such as that of the large intra-abdominal depots, not sufficient? The answer to the first question is almost certainly yes. Hominids had probably long sought out and eaten lipid-rich nuts and fatty insects, as other apes and baboons still do. The answer to the second question depends upon the definition of 'sufficient'. The supply may have been enough to satisfy minimum nutritional needs, but not enough to satisfy appetite for such palatable foods. Dietary change by itself is unlikely to have led to obesity, though it may be a physiological means of accumulating storage lipid without increasing the capacity for fatty acid synthesis. It could also have influenced the structure of the adipose tissue itself.

Human adipose tissue

Some contrasts between the cellular structure of adipose tissue in mammals and birds of different dietary habits were described in Chapter 2. In general, animals that are mainly or entirely carnivorous as adults have four times as many adipocytes, each about a quarter of the size, as non-ruminant herbivores of similar body size. In all mammals, mature adipocytes form faster under the influence of the high-fat diet during suckling than at any other time of life (see Chapter 5). The low-carbohydrate, high-fat diet of carnivores maintains such conditions after weaning, so pre-adipocytes may have longer in which to divide and mature into adipocytes than in herbivores, for whom a high-fat diet comes to an abrupt end at weaning.

The (admittedly rather crude) estimates of the adipocyte complements of modern western people shown in Figure 9 (page 49) suggest that humans have around ten times as many adipocytes as expected from their body mass. This increase is greater than that observed in any of the naturally obese arctic mammals studied. Is the relatively recent adoption of a carnivorous diet partly to blame? In the sample upon which Figure 9 is based, the carnivores were not, on average, fatter than the herbivores, because their adipocytes were smaller. But having more adipocytes provides the opportunity to become fatter, because the storage cells can be both more numerous and larger. Even alley cats become fat cats when they find an unlimited supply of delicious, nutritious food.

The situation is probably made worse by the peculiar features of primate growth. Most carnivores grow quite fast when young but slow down or stop as soon as they approach sexual maturity, which is at an early age relative to their body size. Humans have inherited the habit of slow juvenile growth and delayed maturity from their primate ancestors, which means a longer period during which new adipocytes can form. Thus the predisposition to obesity could have evolved without having any immediate function.

The predatory habit is obviously most useful to non-hibernating omnivores during the winter, when leaves, fruit, flowers and invertebrates are scarce. The Pleistocene Ice Age increased the need for ancestral humans to kill and eat other animals, just as it promoted the evolution of carnivorous polar bears from omnivorous brown bears. After millions of years of evolution in tropical Africa, the adoption of carnivory extended the geographical range of the human population beyond that of any other primate. Hominids were then and still are the only primates to establish permanent populations at high latitudes and in temperate regions where winters are long and cold.

Food availability more than climate limits the distribution of other primates: most of the large monkeys and apes kept in zoos in Europe and North America can safely go outside every day of the year (although they do like a sheltered place to sleep), and the Japanese macaque monkey lives wild in regions where it snows every winter.

As explained in Chapter 2, obesity among wild animals is more prevalent at high latitudes and in rapidly changing, unpredictable climates. As the glaciations spread, people probably ate more animal fat, and obtained more of their energy from fatty acids because the prey animals were fatter, at least at certain seasons. But (with the possible exception of Eskimos), people did not become as completely or irreversibly committed to carnivory as polar bears did over roughly the same period. As the climate became warmer, humans quickly returned to gathering plant food, and later to growing crops. The bears moved further north, and are now confined to high latitudes. Until very recently, people lived at the highest densities, and had the greatest genetic diversity between the tropics. Cold regions were sparsely populated or occupied only seasonally.

In the 1920s and 1930s, 'man the hunter' was the fashionable image of Pleistocene people, but many students of human evolution now believe that the role of predation in human ecology has been exaggerated. People have probably always engaged in a fair amount of scavenging and gathering of plant material, but few relics of these activities have survived, while stone axes and smashed or scraped bones have. Although the cut-marks of stone tools can be seen on remains as old as 1.5–2 million years, they are nearly as numerous on the bones of paws and hooves as on those such as the back and thighs that were surrounded by meat. This observation has been interpreted as indicating that using the skins or tendons for making clothing or bags was as much a reason for butchering carcasses as obtaining food.

Carnivory and the evolution of intelligence

The transition to meat-eating coincided with rapid enlargement of the cranial cavity of fossil hominids, and hence presumably in the size of the brain. Many anthropologists assume that the association is causal, that carnivory directly promoted improvements in intellectual abilities and hence paved the way to the evolution of modern humans. When viewed in the context of the transition between herbivory and carnivory, this notion appears more controversial.

Carbohydrates are normally obtained by grazing or browsing, both time-consuming activities that do not call for much intellectual ability. Carnivory requires greater skill and agility in tracking and catching the prey, which would promote better sensory acuity, better memory and more skilled movement. Meat is also such a concentrated source of nutrients that carnivores have a great deal of free time, which, as mentioned in Chapter 5, under some circumstances can favour the evolution of complex social relationships.

The social organisation seems to be much more important to the evolution of such habits than the dietary nutrients *per se*. Lions spend much of the day dozing, squabbling and soliciting sexual attention while their prey, zebras and antelopes are busy eating, but carnivory has not done much for the social lives or intellectual development of spiders, snakes or crocodiles. It could be argued that they are solitary 'sit-and-wait' predators that ambush rather than chase their prey. The behavioural repertoires of hedgehogs, shrews and the monotremes (duck-billed platypus and spiny anteater or echidna), all active hunters and survivors from the earliest lineages of mammals, are not obviously more advanced than that of similar-sized herbivores such as rats, mice, squirrels and guinea-pigs in spite of millions of years of carnivory.

The crab-eating macaque monkey (*Macaca fascicularis*) is native to the lowlands and islands of South-East Asia, and, as the cynomolgus monkey, became famous as an experimental animal in the 1950s in the development of oral vaccines against polio. For many years, *M. fascicularis* has been bred in spacious outdoor facilities in North Carolina (USA) for research into the role of high-fat diets in the long-term development of heart disease, one of the best funded of all topics in lipid biology. The monkeys are able to live in more or less natural troops, though from time to time it is necessary to disrupt the established social structure of family groups.

Researchers noticed some unexpected differences in the social behaviour of monkeys that were maintained on different diets.[4] Those that had been fed for more than 2 years on diets enriched with triacylglycerols and cholesterol groomed each other more often, and spent more time within touching distance than those that received a nutritionally adequate, low-fat diet. The fat-fed animals were also better able to tolerate being moved from one group to another, usually a very unsettling experience for monkeys. In brief, a high-fat, high-cholesterol diet seems to make the monkeys more friendly. The observers concluded that a low-fat diet promoted anti-social behaviour, possibly by affecting brain lipids and thereby altering production of the brain messenger molecule, serotonin, the lack of which is known to cause aggression in humans and monkeys.

People with higher blood cholesterol have also been found to be capable of faster mental processing, while those whose levels of cholesterol are unusually low, or have been artificially reduced with drugs, seem to be more prone to suicide and aggressive behaviour. While there are many possible explanations for these associations, a taste for lipids may have led to a more sociable personality and/or quicker wits, and, as an incidental consequence, to obesity.

These traits may have been valuable for the development of co-operative hunting, ever more complex language and cultural activities such as singing and dancing. They would have been even more useful following the agricultural revolution, when the population increased abruptly, and people began to live in denser, more settled communities. Food could be stored in much larger quantities, making the supply more reliable, and, for the first time in human history, dairy products became available. Milk is rich in lipids because it is intended for fast-growing neonates, but it is fattening for adults, especially when concentrated as butter or cheese. Humans are the only species that eats milk when adult.

Thus the agricultural revolution provided easy opportunities for fattening, and possibly also an incentive to become fat, if the feature was associated with personality traits that were better suited to the newly adopted, potentially stressful habit of living in close proximity. Such a scenario could explain why secondary sexual characters enlarge and become more conspicuous with even moderate levels of obesity. Social adaptability is obviously a desirable trait in a marriage partner: as well as promoting family harmony, the union may produce well-adjusted offspring whose quick wits and agreeable personalities could enable them to advance their social status in the group. The association between contentment and obesity may explain why some deities and the Buddha in China and India are often represented as fat, sometimes grotesquely so, although the hard work of subsistence farming keeps most of the indigenous populations lean.

Another aspect of the relationship between diet and intellectual ability relates to heat.[5] A large brain produces a great deal of heat, and is easily damaged by even brief periods at a few degrees above normal temperature. Digestion of plant material produces a great deal of heat: the need to avoid overheating is believed to be why hippos spend hot days almost completely submerged in water, but come out to graze during the cooler nights. Improved thermoregulation is often suggested as a reason for the evolution of hair reduction in a large primate, which, unlike hippos, had better things to do in the daytime than wallow in water-holes.

Hominids may have further reduced the risk of overheating by taking to carnivory. Animal food is much less bulky than plant food supplying the same amount of nutrients, and it can be digested more quickly in a shorter gut. According to this argument, the evolution of a large brain is made possible by the gut becoming smaller, and the digestive processes simpler so less heat is produced, following the switch to a more carnivorous diet.

Brain development may be more directly linked to obesity. As pointed out in Chapter 1, the chemical composition of the nervous system has much in common with that of adipose tissue. The brain needs adequate supplies of particular fatty acids throughout its growth period, though once formed, its preferred fuel is glucose. Shortages of particular fatty acids lead to permanent impairment of function and can be fatal (see Chapter 5), but the brain cannot synthesise or store its own supplies. Growth of the foetal brain depends entirely upon essential raw materials extracted from the mother's blood, which in turn derive from her diet or her adipose tissue.

Contrary to the simplistic assumption that more brain tissue is essential for more thinking, intelligent animals do not invariably have larger brains than those which, by our criteria, are stupider.[6] Except in the case of hominids, the brains of adult primates are not consistently larger than expected from their body size. But the primate brain does develop more rapidly than that of other mammals, reaching a larger size earlier in gestation: the brain of neonatal primates is around 12% of the total body mass, compared with around 6% in other kinds of mammals. Humans have to some extent reversed this trend: at birth, a baby's brain is only about 40% of its final mass, which, although absolutely larger, is proportionately smaller than the brain of a newly born ape.

In higher primates, an adequate supply of the right kinds of long-chain polyunsaturated fatty acids is essential to nourish the foetus's brain during gestation, and, particularly in humans, it remains important into the suckling period. Maternal obesity could help to ensure adequate supplies, though rather inefficiently, since most adipose tissue triacylglycerols contain saturated or monounsaturated fatty acids that are not suitable for making neural phospholipids. There is certainly significant movement of fatty acids from mother to foetus across the placenta: during the first half of pregnancy the lipids go mainly to the brain and other neural tissues, but in the final three months they accumulate in adipose tissue, with the result that human babies are unusually fat at birth. It could be argued that neonatal obesity is an incidental consequence of the evolution of mechanisms for transferring larger quantities of lipids from mother to foetus. Lipids cannot be easily used for energy production during gestation, so as the nervous system's need for them wanes, they have to be deposited in the adipose tissue.

A larger brain uses more energy so, as well as requiring more lipid early in pregnancy for the raw materials of brain growth, more carbohydrate is needed to sustain it in action. Its metabolism accounts for a substantial fraction of the foetus's total heat output, but overheating is potentially dangerous and must be avoided. Hair reduction, and minimising the amount of adipose tissue in and on the abdomen would help the mother to dissipate heat efficiently. The necessary lipid stores accumulate elsewhere in the adipose tissue on the buttocks and thighs.

This anomalous growth pattern may also explain why primate milk is exceptionally low in lipids: brain development in primate infants is so nearly complete shortly after birth that they don't need many more precursors for phospholipids to complete the job. Since the other tissues are growing quite slowly and the babies are carried by their mothers, they don't need energy-dense milk as much as calves and fawns that grow fast and have to run for their lives from birth onwards. These contrasts in the timing of the use of lipids in development, and the amount of strenuous activity that neonates perform also mean, of course, that feeding babies on cows' milk provides them with lipids to spare and so they fatten, but feeding calves on women's milk would leave them abnormally thin. Veal is very lean because the calves are separated from their mothers within a few days of birth and raised on diluted or even synthetic milk.

None of these ideas is mutually exclusive, and all may have a grain of truth. They illustrate the difficulties involved in understanding whether, when and why syndromes of characters – carnivory, large brain, hair loss, obesity – might have evolved together. Compared with other species that were adapting to the rapidly changing Pleistocene climate, people did not advance very far down to the road to carnivory. The trend has proved to be easily reversed: the traditional diets of many native peoples of India, Thailand and much of Central America, and those adopted by vegetarians in meat-eating societies, contain hardly more animal food than that of chimpanzees. Whether Adam was tempted to eat the snake as well as, or even instead of, the fruit of the Tree of Knowledge we do not know, but in the event, he chose the plant food, and his descendants' diet has remained largely vegetarian. Had he chosen otherwise, the evolution of metabolism as well as of metaphysics would have been different.

Why people became fat

The simple explanation for human obesity is that, like bears, camels, penguins, migratory birds and most other naturally obese species, humans fatten so they can fast for long periods or travel long distances between food supplies. Medical texts often offer explanations along these lines for how people fell from pristine leanness to fatty degeneration. On this scenario, fasting should follow fattening, and people should have evolved adaptations to both, but almost no other feature of human ecology, social habits or physiology appears to be so adapted.

Our diet is, and almost certainly always has been, nothing if not varied: if one kind of food becomes scarce, people simply switch to another. Humans can move long distances to places that offer rich pickings, assembling into larger or smaller groups as hunting and gathering strategies require. There is no evidence that people, or any other higher primate, 'hibernated' or otherwise sat out periods of food shortage as specialised predators like polar bears do: they looked for alternative foods or moved elsewhere, eating whatever they could find on the way.

Food sharing within families, and later food storage, are fundamental features of human ecology. As well as, or perhaps even instead of, eating the food where they found it, as apes and monkeys do, people carried food back to camps to distribute among children, the elderly, unsuccessful foragers and anyone else who had no direct access to food. Freeing the hands to carry food and infants has been suggested as an early function of the upright posture. Bags and baskets of quite sophisticated design have been found among Neolithic remains, and may have been made and used much earlier but, being of more perishable materials, have not survived as well as stone axes or lamps. The habit probably preceded, and may have led to, the use of fire and other forms of food preparation. Cleaning and cooking carcases could not easily be performed in the open, where predators threatened and getting enough fuel together could be difficult, but the job could be done properly if the food was carried back to camp.

Like other large animals, ancient people probably experienced famine from time to time, when they were trapped in unproductive habitats by abrupt climate change, or natural disasters such as floods, volcanic eruptions or since at least Neolithic times, each other's hostilities. But there is no evidence that the natural selection imposed by such events was severe enough or consistent enough to promote the evolution of adaptations to prolonged fasting sustained by obesity, as has happened in bears, penguins, whales, etc. Similar calamities must have affected other large, long-lived tropical mammals including chimps and gorillas from time to time, and they are not exceptionally obese in the wild. The occasional famine could not by itself lead to the evolution of continuous obesity.

Modern anthropologists[7] justify the opinion that obesity evolved in response to periodic food shortages by the facts that !Kung hunter-gatherers lose 1–2% of their body mass during the dry season in the Kalahari desert, and some African peasant farmers lose about 4–6.5%. Such changes are small for people whose body mass is 10–35% fat, and compared to migratory birds and bears, people do not fatten very rapidly or very efficiently. In 1988, a group of French scientists[8] monitored the diet and body mass of young Cameroonian men aged 23–35, who regard fatness as desirable and traditionally overeat during the 2 months of the year when food is plentiful. Eating meals of sorghum and cow's milk every 3 hours night and day, these initially lean men (BMI 18–23) gained between 11.7 and 23.2 kg, 64–75% of it as lipid, at an average rate of 0.27 kg per day or less than 0.4% of the body mass per day, a modest achievement compared with that of seal pups or migratory birds (see Chapter 5). In spite of the hot climate, the subjects' heat production increased by over 40%: instead of fattening as efficiently as possible, the men's bodies seemed to be 'trying' to get rid of the excess energy.

Resistance to fasting

Persistent overeating eventually fattens people, and, like bears, penguins and migratory birds, they can be moderately fat for years without their obesity generating serious health risks. But unlike these naturally obese animals, people cannot easily reduce their adipose tissue. Slimmers complain of tiredness, depression and irritability; starving people have poor complexions, scruffy hair and a greatly increased susceptibility to a huge range of ailments from boils and pustules, to tuberculosis and parasitic worms. Loss of body fat lowers resistance to disease so effectively that people rarely die of starvation alone unless they are kept away from sources of infection and treated with modern medicines: the immediate cause of death is usually an infectious disease or pneumonia-like inflammation of the lungs.

These adverse consequences of even brief fasting are so universal that many people would assume that penguins, bears, migratory birds and the many other animals discussed in Chapter 5 that naturally forego food for an extended period experience similar symptoms. It is, of course, very difficult to determine with certainty whether an animal, particularly a wild animal, 'feels quite well', but careful observations fail to find any evidence for distress, discomfort, diminishing mental capacity or poor health in naturally fasting animals, unless the fast continues for much longer than normal.

Fasting impairs mental and physical performance much more severely in humans than in animals that are adapted to perform important activities while very hungry. Few modern people would claim to function better after several days without food: prolonged fasting causes weakness, lethargy, poor concentration and irritability, a far cry from the business-like progress of migrating birds, or the patient, alert tracking of prey by a hungry bear or lion. Australian aboriginals, American Indians and other tribal people use fasting, along with various drugs, to induce trance-like states in which they claim to make contact with the spirit world, but are incapable of the clear thinking and steady hand needed for hunting or travelling.

The effects of food on mood and mental processes are the basis for certain scientific theories proposed to account for overeating, and hence for its (almost) inevitable result, obesity. All mothers know that sweet food, even in very modest quantities, soothes crying babies. The calming action of sugar is particularly obvious in young children, but psychological tests reveal transient but measurable mood changes even in adults who view themselves as being completely in control of their emotions. Fatty foods are thought to modulate other chemical processes in the brain, relieving anxiety and inducing drowsiness, while chocolate craving may owe its origins to the capacity of one of the essential ingredients of chocolate to simulate the action of cannabis-like drugs.[9] According to such theories, food affects our brains like a short-acting drug, and without frequent doses, we feel depressed and become ineffectual. How fasting animals avoid such unpleasant sensations remains to be determined.

Hitherto healthy young adults can survive 1–2 months of total starvation, less if they are lactating, exposed to cold or engaged in strenuous exercise. Obese people last longer than lean, the record for a fat person of sedentary habits living in warm surroundings being as long as 11 months. Fasting people start to break down proteins, especially those from the limb and back muscles, long before their adipose tissue lipids are anywhere near exhaustion. The muscles waste, significantly reducing the capacity for running, jumping, carrying loads and other strenuous activities. After a few weeks, impairment of muscle function can be so severe that sufferers become incapable of foraging for themselves and survive only if others feed them.

Regular, strenuous exercise during fasting helps to retard protein loss from muscles, especially in young people, but it cannot be prevented entirely. Severely obese people do somewhat better than those of average body composition, obtaining about 5% of their energy from the breakdown of protein, compared with almost 20% in the non-obese. But both are much worse than 'professional' fasters such as hibernating mammals, bears or penguins, in which hardly any loss of muscle mass, or impairment of function can be measured. Female polar bears routinely fast for at least 3 months, longer if necessary, while suckling twins in a very cold climate. Compared with animals that naturally alternate obesity and leanness, humans seem to be peculiarly inept both physiologically and psychologically at dealing with feasting, fasting or fattening. Clearly we must seek alternative explanations for the evolution of human obesity.

Did hair loss lead to fat gain?

Humans are the only primates to have large quantities of subcutaneous fat and the only ones in which hair is so sparse as to be almost useless as insulation over almost the whole body. It is easy to assume that the two evolutionary changes were closely linked. The fossil record is silent over when and why our ancestors lost most of their body hair, so the only evidence we have comes from physiological studies of living people and animals.

Hair loss would have facilitated the dissipation of the excess heat generated by running, thus making possible hunting on open plains under the midday sun. The evolution of a larger brain would have exacerbated the problem of overheating arising from vigorous, prolonged muscular activity, because it alone produces around 20% of the body heat in resting adults, as much as 45% in infants. The mammalian brain is very sensitive to high temperature, possibly because of the importance of membrane lipids, and is most susceptible early in gestation. A rise of a 1.5 °C stops foetal brain cells from dividing, and the tissue is killed at 3 °C above its normal temperature of 0.5 °C above that of the mother's abdomen. Even brief episodes of high fever during pregnancy can lead to birth defects in the brain and skeleton.[10]

Erect posture and bipedal walking would reduce the amount of radiant heat absorbed from the sun, especially near the equator, where the sun is directly overhead at midday. The retention of thick hair over the back and top of the head would help to prevent local heating of parts of the body unavoidably exposed to full sunshine. But in spite of these measures, just panting, as dogs and lions do, probably could not dissipate heat fast enough: sweating over bare skin was the only form of cooling that could match the heat generated by strenuous exercise on hot tropical days.

We have to turn to the circumstances promoting the evolution of hair loss in other mammals for clues about what may have happened in our own evolutionary history. Heat dissipation was probably a major factor promoting hair reduction in very large mammals such as rhinoceroses, hippopotamuses and elephants. Although the adults of both living species are essentially hairless, newly born elephants have quite a lot of dark hair, and young rhinos have more hair than their parents. There is no evidence that hair loss promoted higher fatness in these large pachyderms, or in naked mole-rats or any other mammals in which drastic hair reduction has evolved. Living as they do in equable climates with year-round food supplies, these tropical animals do not normally become fat. But when they found themselves in colder conditions with more variable food supply, either because of climate change or following colonisation of higher latitudes, fur *and* fat reappeared in abundance.

Mammoths are close relatives of elephants and were widespread over much of Europe, northern Asia and North America during the Pleistocene. They almost certainly had shaggy black coats, and they were fat, at least for part of the year: a mummified *Mammuthus primigenius* found in a frozen river bank in Siberia about a century ago proved to have subcutaneous fat up to 8 cm thick. Like other arctic animals, they probably needed a lot of storage lipid to supply them through the long winters. Living in herds and being so large, they could probably see off predators with a threat and a snort, and never ran far or fast, any more than modern tropical elephants do. There is nothing incompatible about abundant subcutaneous fat and thick fur, as long as the climate and exercise habits pose no risk of overheating: even very large animals such as mammoths and bears have both.

The relics and images that have survived suggest that humans regained very little hair when they colonised southern Europe and Asia during the Pleistocene glaciations. It is true that modern European races have more body hair as adults than Africans or Chinese, but it is so sparse that its contribution to keeping warm must be very small. Aboriginal Australians, who have lived in a warm climate for at least 40 000 years, have quite a lot of hair. The largest members of the species, adult males, have the most hair, while small children, who are most susceptible to cold, have the least. Although baby elephants and rhinos are noticeably hairy, human infants all over the world are born with less hair than their parents. Human foetuses develop body and facial hair (laguno) during the final months of gestation but it is usually seen only in premature infants, because it disappears by full-term.

It can't have been all that difficult to re-evolve our body hair: we are said to have 99% of our genes in common with those of hairy chimpanzees, and there are several reports of whole families of exceptionally hairy modern people. Modifications of colour, texture, distribution, age changes and other aspects of facial and body hair have evolved very recently and are now among the most conspicuous differences between modern races, and between individuals. The standard explanation for this state of affairs is that more hairy people were not favoured by natural selection because humans took to keeping warm by wearing animal skins, and thus had no need of body hair. The problem with such notions is the lack of evidence for any correlation between the use of clothing and the loss of their own hair.

The 'aquatic ape' hypothesis

Modern western people are much concerned with why they get fat so much more readily than other primates do. In their search for explanations, some theorists have been impressed by the fact that whales and walruses are apparently fat, almost completely hairless – and aquatic. Fanciful thinking along these lines has led to the suggestion that massive adipose tissue evolved because at some stage in their evolutionary history, humans also lived in water. The aquatic ape hypothesis is almost entirely without palaeontological foundation: the fossil record offers no hint of when or where the human species may have adopted an aquatic lifestyle for long enough for such adaptations to have evolved. It is also not supported by comparative and functional studies of adipose tissue.

The notion that human superficial adipose tissue is adapted to act as insulation against heat loss is one of the most firmly established of all theories in biology, but there is very little anatomical or experimental evidence that supports it, and much that is inconsistent with it. The form and function of blubber in marine mammals was described in Chapter 6. In adapting to its new role as a specialised insulator, the chemical composition, metabolism and anatomical arrangement of the adipose tissue have undergone radical changes, at least as fundamental as the adaptations to aquatic life found in the blood, respiratory system, eyes and ears. There is no evidence that a comparable reorganisation of human adipose tissue has ever taken place: as explained in Chapter 2, the distribution of human adipose tissue is almost identical to that of furred, terrestrial primates.

The relative thicknesses of the superficial depots are not consistent with a role as insulation in air or in water: the back, head and neck are much more exposed to the elements during both walking upright and swimming than the ventral trunk or the inner surface of the thigh, but superficial adipose tissue is thickest over the ventral abdominal wall (paunch) and, in women, on the thorax (breast adipose tissue), the upper arms and the thighs. As in most terrestrial mammals, adipose tissue along the back above the waist is minimal, amounting to a thin, fibrous layer in all but the most obese people. If 'subcutaneous' adipose tissue were important for thermal insulation, one would expect it to be conserved during fasting, and the internal depots depleted first. There is no evidence that the superficial depots are selectively spared in starvation: in obese women who live for weeks on a severely restricted diet, the superficial and intra-abdominal depots are depleted at about the same rate.

Because babies are small and much more sedentary than adults, adipose tissue might be expected to be more important as an insulator. But comparisons between babies born with widely different physiques show that their ability to maintain a constant body temperature correlates more closely with their lean body mass than with the thickness of their superficial adipose tissue.[11]

Humans are among the most widespread of all species, and have evolved various minor adaptations (e.g. darker skin in those living in sunny climates) that equip them to live at almost all latitudes and in many different habitats. We might expect, therefore, that people whose ancestors have lived for hundreds of generations in very cold climates would have more and better superficial adipose tissue than those living in the tropics. Canadian biologists measured the skinfold thickness of traditional and recently urbanised Canadian Eskimos and found that they do not consistently have more abundant or differently arranged subcutaneous fat than those native to tropical countries. Eskimos pursuing a traditional lifestyle and diet are short and stocky, but they prove to have less, rather than more, subcutaneous adipose tissue than Canadians of European descent living in similar climates. In young adults of both sexes, 'normal' adipose tissue is mainly internal but 'additional' fat is deposited in superficial depots. The greatest differences between arctic and temperate-zone races are in the proportions of the skeleton and musculature, not in the distribution or abundance of adipose tissue.

There are no living apes, monkeys or other primates that have obviously become aquatic, in the way that otters and polar bears have evolved from ancestors that were similar to living species of terrestrial carnivores, or water voles and coypus from terrestrial ancestors resembling rats or guinea-pigs. Several species of macaque monkeys sometimes feed on or near beaches and in mangrove swamps, and adults may swim or wade some distance from the shore. One or two such species swim far and often between feeding areas, but such habits are not associated with any detectable changes in the distribution or abundance of adipose tissue.

Since at least the Neolithic, people have built boats, first coracles and canoes, later more elaborate ships for travelling across and foraging in water. Our expertise as sailors has enabled us to colonise the entire world, but that does not make us aquatic. There is no evidence that commitment to aquatic habits was ever prolonged or extensive enough to account for the evolution of hair reduction or obesity. As discussed in Chapter 6, large mammals have proportionately more superficial adipose tissue than smaller ones of similar total fatness. Humans are among the largest primates, and certainly the most obese of the living species: our preponderance of superficial adipose tissue may simply be a consequence of our large body mass, as it is in large, obese carnivores such as bears (Figure 14, page 198).

Fat as a heat producer

Mammals and birds are warm-blooded, but only a tiny fraction of the endogenously produced heat arises from dedicated heat-generating tissues like brown adipose tissue. The rest is a by-product of other metabolic process in all the other tissues, including protein synthesis, selective transport of particular molecules, and muscle contraction. Not all the products of these metabolic processes are strictly necessary: sometimes a substance is synthesised, only to be broken down again almost at once by a reverse reaction (that involves different enzymes), together forming a 'substrate cycle'. The rates of such substrate cycles, and hence the amount of heat they produce, are finely controlled by hormones and the sympathetic nervous system.

Several substrate cycles occur in the liver, skeletal muscle and elsewhere. Some, perhaps all, may help to keep the system in readiness for abrupt changes in the rates at which tissues use certain biochemical materials. For example, those in muscle enable the body to switch from lying in bed to running fast in just a second or two. Another important role may be the heat they produce, which can be significant, though it does not generate as much heat per gram of tissue as the uncoupling of mitochondria in brown adipose tissue (see Chapter 2). The main substrate cycle in white adipose tissue is lipolysis and re-esterification of triacylglycerols. The fatty acids released from the adipocytes by lipolysis re-enter the cell and are re-esterified, but these cells cannot take up glycerol efficiently, so a glucose molecule is used to make more glycerol for each round of the cycle. Heat generated by this substrate cycle could contribute more to thermoregulation than any insulating role that adipose tissue may have in conserving the heat generated in other tissues. In dwarf hamsters,[12] the highest rates of substrate cycling occur in the small intermuscular depots, in which the maximum can be nearly three times higher than in the much larger perirenal and inguinal depots.

Until very recently, only hibernators and the newly born were believed to have substantial amounts of thermogenic brown adipose tissue (see Chapter 2), but advances in molecular biology now make it possible to detect the molecular message (mRNA) that conveys instructions from the genes to direct the synthesis of particular proteins far more precisely and in far lower concentrations than older biochemical methods could reveal. Such techniques have shown that the message for synthesising uncoupling protein is present in inguinal 'white' adipose tissue from certain strains of mice, especially after they had been fed continuously on a high-fat diet for 4 months.[13] Similar studies on white adipose tissue excised from adult humans undergoing routine surgery also revealed variable, usually quite small quantities, of the message for uncoupling protein.

These findings were a surprise: typical brown adipocytes have not been seen in most depots, and so we have to conclude either that they are so scarce that there is almost no chance of seeing them in tissue samples collected at random, or that cells that pass for white adipocytes when viewed under the microscope are capable of doing a bit of brown adipose tissue metabolism on the side. We don't know whether ordinary human white adipose tissue uses uncoupling protein to generate significant amounts of heat in the same way as brown adipose tissue, but the molecular biologists have shown us that it has some of the key apparatus for doing so.

Researchers looking for ways of simulating substrate cycles and/or thermogenesis of the brown adipose tissue type were much encouraged by the finding that rats could be induced to increase their heat production by as much as a quarter by placing them in the cold, or by administering drugs that stimulate a certain type of receptor. Increasing heat production looked like a safe, controllable way of burning off excess energy, i.e. a means of slimming. But the results of similar experiments on people were disappointing: their adipose tissue proved to have very few of the appropriate receptors. The dose required to stimulate any measurable heat production also affected the heart, producing uncomfortable, even dangerous palpitations.

People out in the cold

The subjects of this experiment were Europeans who had all since infancy worn clothes, lived and worked in heated buildings and slept under bedclothes. There is much indirect evidence that people who have not been raised in an artificially controlled climate can, like rats, adjust their body heat production in response to environmental conditions, if necessary producing quite large quantities of heat continuously for long periods. Many observers have remarked upon the extraordinary tolerance for cold of children raised from infancy without clothing or shoes. Fishermen, farmers and others whose work involves being outside throughout the year also complain less from being too cold and/or too hot, even when exposed to extremes of climate, than those who spend most of their lives at an almost constant temperature.

Until quite recently, everyone may have had similar capacities. Captain Cook and other early visitors to Australia in the 1770s noted the Aboriginals' tolerance for heat and cold without any kind of protective clothing. Although it can be, by our standards, quite cold at night over much of Australia, the Aboriginals slept in the open or under crude wooden huts in shallow depressions scooped out of the ground. During exceptionally cold weather, they sometimes moved to caves and cliff shelters, and covered themselves with rugs made from kangaroo or opossum skins, but in general they were more concerned with decorating their bodies than with insulating them. The Aboriginals attached great importance to nose rings, necklaces, feathered head-dresses and ritual scarring, with which they often produced elaborate patterns. Modern research[14] shows that blood flow in and near the skin falls when such people are exposed to cold, thereby reducing heat loss, and their sleep is undisturbed by a drop in core body temperature of several degrees.

Both Cook and his eminent ship's botanist, Joseph Banks, noted in their journals that the natives of all ages appeared to be lean and nimble. Yet they ate fatty foods, including emus, black swans, geese and many other birds and birds' eggs, kangaroos, possums and various lizards and snakes. Those living near coasts also harpooned fish and – one of their favourite foods – huge marine turtles, which can be very fat indeed in the weeks before they come ashore to lay their yolky eggs. Aboriginals were also very fond of honey, which could be collected without risk of injury, because Australian honeybees are stingless.

Captain Cook observed that even in the cool, damp climate of Tasmania, the native people wore only capes and belts made from kangaroo hide. The men and boys were often completely naked, although they were slightly built and appeared to have less fat than was usual on Europeans. Disease and persecution killed most of the native Tasmanians during the 1830s, but a handful survived and adopted European dress and habits. Photographs taken of them in middle age show that all the women were obese, with thick waists and prominent bellies.

Shortly before Christmas 1832, Charles Darwin arrived in Tierra del Fuego, off the southern tip of South America, aboard the surveying ship, HMS *Beagle*. Although it was the middle of the austral summer, the weather was dreadful, with gale-force winds and almost continuous rain, sleet and snow. Darwin noted in his journal[2] how the native inhabitants, including tiny infants, appeared completely comfortable in such weather, protected only by a short cape made from the skin of an otter or guanaco (a relative of llamas). Whale and seal blubber was among their favourite foods, and they carried large chunks of it over long distances to supply families living inland. They also caught fish and otters, gathered fungi and berries, and collected the eggs of the many kinds of seabirds that nested on the steep cliffs.

According to Darwin, famine was frequent and severe, and, like the Australian Aboriginals, Fuegians would eat as much as they possibly could when food was plentiful. Yet they were not fat, quite the contrary. A teenage boy named Jemmy Button, who had been 'bought' for a pearl button and taken to Britain on a previous voyage, was described as 'short, thick and fat' after five years of wearing European clothing and eating European food. Returning him and two other Fuegians to their native land was one of the reasons why the *Beagle* went to Tierra del Fuego. Jemmy made contact with his mother as soon as he landed, and immediately resumed his traditional lifestyle.

The *Beagle* stayed in the area for several months, while the crew explored and mapped the islands and collected specimens. When Jemmy Button made a final visit to the ship's company just before they departed, Darwin described him as 'a thin haggard savage, with long disordered hair, and naked except for a bit of a blanket around his waist.' Yet he did not complain of cold or hunger, and insisted upon staying in his native land, despite the ship's captain's earnest pleas for him to return to his British friends. Far from suffering from cold or exposure, he seems to have remained in good health and prospered, achieving high status among his countrymen. More than 25 years after the *Beagle* set sail, he led a raid on a tiny missionary settlement, in which six Europeans were killed.

Anthropologists agree that humans reached Australia overland from New Guinea and South-East Asia, during a glacial period when sea-level was so low that the straits between what are now islands were dry. Aboriginals have been isolated there for at least 40 000 years, possibly very much longer. South America seems to have been colonised from the north by people who crossed what is now the Bering Strait from Siberia, also during a glacial period. Again, dates are much disputed, but it is certain that these native peoples have not shared a common ancestor for scores of thousands of years. The fact that they both tolerated cold conditions with little clothing strongly suggests that all humans had similar capacities, and may still retain them.

For at least two millennia, people have collected shellfish and other seafoods off the coast of Korea by diving from small boats, sometimes working for several hours in the cold waters of the North Pacific. When they wore only light cotton clothes, the divers' resting heat production was significantly higher than that of non-divers, especially during the winter, and they shivered less readily when exposed to cold. Heat loss at the body surface could be greatly reduced by temporary constriction of blood vessels that reduced blood flow through the superficial tissues, muscles, skin and bone as well as adipose tissue. Divers and non-divers were equally lean, and subcutaneous fat up to 3 mm thick contributed very little to thermal insulation.[15]

The divers' physiological adjustments to cold were re-examined a few years after rubber wet-suits replaced cotton shirts as normal working dress in the 1970s. Their raised heat production had disappeared within 3 years of using the artificial insulation, and the capacity for constricting blood flow near the body surface was no better than that of non-divers after 5 years. If people could work in cold water or stay warm on cold nights by turning up their endogenous heat production, they must also have had ways of reducing it, otherwise they would risk become too hot during strenuous exercise or when exposed to the midday sun.

It is important to remember just how recently our ancestors have taken to wearing warm clothing almost all the time. The Roman historian Tacitus[16] describes ordinary Germans in the first century AD as wearing only a simple cloak made from coarse wool or animal skins. The women were naked almost to the waist. Only the most prominent citizens wore more elaborate, close-fitting clothes. Tacitus was impressed by this state of affairs, not so much for its implications for endogenous heat production, as for the absence of promiscuity and sexual jealousy that so much nakedness would have generated among the more hot-headed Romans.

The anatomical distribution of white adipose tissue is much less important to its function if its contribution to temperature regulation is more thermogenesis than insulation. That would explain why the arrangement (as distinct from the abundance) of human adipose tissue is generally so little altered from that of furry tropical primates. Heat generated by internal depots is probably less easily wasted at the body surface than that generated in superficial depots. If the heat production per gram could be adjusted over anywhere near as wide a range as is possible with 'real' brown adipose tissue, quite lean people could obtain enough heat from their modest quantities of adipose tissue to remain in comfortable thermal equilibrium.

Being able to generate enough heat to maintain body temperature in cold conditions does not preclude relishing external heat or artificial insulation. The wolf ancestors of domestic dogs roamed the Russian steppes and Canadian tundra throughout the winter, but given the choice, dogs opt to lie in front of the fire. Polar bears can sleep outside in severe frost, but captive individuals squabble over access to a heat pad, even in a mild climate. Similarly, people readily took to wearing clothes when they learnt how to obtain animal skins and shape them for their own use. Humans should be compared to hippos and elephants, not to walruses or whales, arctic animals that have exchanged fur for blubber as the main insulating tissue. In spite of the probable irrelevance of the anatomical arrangement of human adipose tissue to its physiological function, its distribution has been a fashionable research topic for more than 30 years.

Where people fatten

Although the distribution of adipose tissue in wild animals has hardly been documented at all, every detail of the age changes, sex differences and pathological implications of that of the human body has been intensively studied. This paradox arises from the general tendency to regard obesity as at best artificially created by the way humans maintain themselves and their livestock, and at worst frankly pathological. The result has been the accumulation of a great deal of largely meaningless detail about the arrangement of adipose tissue in our own species, while almost nothing is known about any others. The discrepancy has seriously distorted scientific thinking about the origin and function of the anatomical distribution of human adipose tissue.

As noted in Chapter 2, a curious but universal feature of the arrangement of adipose tissue in primates is the highly expandable 'paunch' depot, which arises from the midline and at its maximum can extend from pubis to neck, covering much of the belly and chest (see Figure 8, page 44). The fragmentary information available suggests that its metabolic properties, as well as its anatomical arrangement are similar in humans and other primates: it has a low proportion of collagen, low blood perfusion and consists of variable, sometimes very large adipocytes. At least in men, the male sex hormone, testosterone[17] promotes selective expansion of the paunch depot, sometimes to enormous size. We do not know when in evolutionary history it acquired these qualities, nor how its appearance related to posture or social habits, but these features seem to have gone hand in hand for a long time. In all primates in which the matter has been thoroughly studied, a prominent paunch is more frequently observed in dominant males of breeding age and senior females, and the feature is minimal in juveniles and in adults of low social rank.

Because the depot enlarges disproportionately, the increase in total body fatness needed to create a moderately impressive paunch is quite small: the total dissectible adipose tissue in adult ring-tailed lemurs living in a large, forested enclosure was found to be only around 8% of the body mass, about average for a well-fed wild animal. Sitting upright displays the ventral surface of the thorax and abdomen, parts that are inconspicuous in the typical mammalian four-footed posture. As in penguins, the sitting posture that primates adopt while feeding, resting, tending infants or grooming each other provides potential sexual partners and other members of the group with ample opportunity to observe this adipose depot from several different angles. In some primates, especially highly social species such as mandrills, the skin and/or fur overlying the underlying paunch depot is of a contrasting colour, making it clearly visible especially when sitting or standing upright (see Figure 8, page 44). In some men, the pectoral hair covers approximately the same area, suggesting that it acts as an indicator of social status.

As in lemurs and monkeys, the abdominal paunch is best developed, and most clearly visible in male humans of reproductive age and moderate affluence. It is minimal in boys, elderly men and those whose incompetence in finding or getting access to food prevents them from getting fatter. The big differences, of course, between our paunch and that of other primates is that it is no longer completely obscured by hair, and is peculiarly well displayed by the upright posture maintained not only during feeding and tending infants, but also for locomotion and many forms of social interaction. There is much disagreement over the timing and the course of events that led to the erect posture that displays the abdominal paunch so well, but analysis of skeletal remains and footprints indicate that by about 3 million years ago, our hominid ancestors stood upright and walked bipedally. The human hip and foot became specialised for bipedal running over plains rather than for climbing trees or quadrupedal galloping, and that their operation in such roles requires erect posture.

Body shape (due mainly to the abundance and distribution of adipose tissue) may have joined the colour and texture of head hair and beards, the position and shape of the hairline as indicators of people's sex, approximate age, state of health and social position. We instinctively expect an experienced, socially important man to have at least an adequate paunch, grey and/or receding hair, possibly approaching baldness. A nubile woman has thick, abundant, strongly coloured hair, and plenty of breast and leg fat but minimal paunch. A well-developed paunch is an indicator of high social rank, which suggests good health, intelligence and social skills, all characters that might be passed on to offspring, which would themselves achieve high social status and become successful parents. Evolutionary theory predicts that women would choose mates with these qualities, thus selecting for the tendency to develop a prominent paunch, whatever its implications for the health of its bearer in middle age.

Sexual selection

Sex differences in the distribution of adipose tissue are among the most familiar and distinctive features of our species, but they are also among the most unique. The association between site-specific expansion of certain adipose depots and sexual attractiveness titillates the introspective interest in all aspects of human sexuality that have prevailed during the final quarter of the twentieth century. Its origins are one of the most intensively studied and controversial aspects of human biology.

As discussed in Chapter 5, many animals transiently become fat at certain stages of reproduction, but because the sexes differ in the nature and timing of their contributions to breeding, they are not necessarily at their fattest at the same time. So although adult males and females may appear to be different when examined simultaneously, they fatten in similar ways. Even when storing large quantities of fat is integral to reproduction in females, but irrelevant for males, the impact on adipose tissue distribution is minimal: thus there are sex differences in the cellular structure of adipose tissue of adult polar bears (see Figure 11, page 73), but no sex differences in its anatomical distribution. An intensive search in macaque monkeys[18] revealed only a few minor sex differences in the distribution of adipose tissue, all of them more conspicuous in older, fatter specimens. We have to explain how and why men and women differ in the distribution of their adipose tissue in a way that is almost independent of its abundance.

The arrangement of adipose tissue in men is similar to that of other adult male primates, but that of women of reproductive age has several peculiar features whose origin and functions are much more difficult to explain. The breasts, which are mostly adipose tissue, are especially puzzling. There is a small amount of adipose tissue associated with the mammary glands in most mammals, but externally they and the nipples are so inconspicuous that, except in lactating females, they are quite hard to find. But in women, the breasts, together with buttock and leg fat, are often more massive than the paunch depot.

The most obvious explanation, that mammary adipose tissue contributes to milk production, is not supported by physiological or comparative studies. Breast adipose tissue has no special physiological relationship with the mammary gland: in lactating women, lipids seem to be more readily mobilised from the thighs than from the adipose tissue in the breast itself. There is no consistent relationship between breast size and capacity for milk production. In fact, quite lean women with small breasts are among the most prolific producers of milk, as long as they can eat as much as they want.[19] Increased energy expenditure during pregnancy and lactation is so low that it is difficult to measure; among women in The Gambia, West Africa who are engaged in farming and other traditional occupations, the correlation between milk production and skinfold thickness is surprisingly weak, even among those eating a meagre diet. The lipid content of the women's milk is not significantly impaired until their body mass index falls below 18.5.[20]

Although the adipose tissue of well-nourished women contains sufficient lipid to support a baby through the whole of pregnancy and much of the lactation period, there is little direct evidence that it normally does so, even in subsistence economies. Such large quantities of adipose tissue may be critical to successful reproduction only in severe famine. There is no evidence that humans, or indeed any other primate, are adapted to suckle their offspring for long periods without feeding, as do animals such as polar bears. If they failed to evolve sex differences in the distribution of adipose tissue, there seems little justification for postulating that people would have done so for reasons connected with the energetics of reproduction.

Another kind of explanation is sexual selection. Comparative studies in species as diverse as birds, deer, frogs and butterflies show that in general, the sex that contributes most of the hard work of producing and raising the young does the choosing, and its preferences determine the form of sexual characters in the other sex. The situation is different in humans (and a few other species), primarily because while most animals have a well-defined breeding season, outside of which the females are not receptive to males, women are in a state of permanent oestrus, i.e. sexual relations continue throughout the year. But human conception is very inefficient: most episodes of intercourse do not result in pregnancy and men have no means of knowing when a woman is most likely to conceive. So in a word, the males who keep trying all year round are likely to father the most offspring.

Sexually selected characters are usually visually conspicuous but anatomically quite minor. In spite of the importance attributed to them, the breasts comprise a relatively minor adipose depot, typically amounting to about 0.5 litre, which is about 4% of the total adipose tissue in young women. The other depots that are markedly different in form or arrangement in women, the calf, thigh and buttock depots, are also quite small in total mass. They are highly localised, which makes them more visually conspicuous, but their thickness correlates very weakly with that of other depots, so their dimensions are poor indicators of total body composition. The reduction of body hair makes the enlargement of superficial adipose depots such as the breast and buttocks much more easily observed in our species than in furred mammals.

In most mammals, the mammary glands do not mature until towards the end of pregnancy, and usually regress between each breeding season. But in girls, growth of the mammary adipose tissue is among the earliest major anatomical changes in puberty, preceding menarche by about 2 years. The breasts may be almost full size before normal fertility and adult sexual and maternal behaviour have developed. Pregnancy affects both breast shape and nipple colour, but, in contrast to other mammals, the changes are noticeable months before the birth, and persist indefinitely afterwards. Regression of mammary adipose tissue occurs sometime after fertility has declined, and is not necessarily accompanied by a change in total body fatness. These facts suggest that mammary adipose tissue in women and girls is not solely, or even primarily, concerned with supporting lactation.

Women's hips and thighs, and with them their style of walking, are visibly different from those of men and children, because the adult female pelvis is wider and more strongly rotated backwards than that of men. Although they make running less efficient, these features are essential adaptations to giving birth to babies with proportionately large heads, and so probably evolved in parallel with the evolution of a larger brain. The shape of the pelvis is an infallible guide to the sex of adult skeletons, but those of children are not so easily distinguished. Girls' hips are similar in overall shape to those of boys and (apart from being smaller) to those of men. The distinctively female qualities of the pelvic skeleton appear gradually during teenage.

The adipose tissue on the hip and thigh begins to enlarge early in adolescence, although maximum growth of the pelvic skeleton occurs after that of the long bones, and the birth canal does not reach adult dimensions until up to 5 years after menarche. Thus, in spite of its importance to successful reproduction, the maturation of the pelvic skeleton is not complete until long after the development of the conspicuous secondary sexual characters. Selective deposition of adipose tissue simulates and exaggerates the forms of the mature pelvis and breasts, generating the appearance of fertility in girls long before they are actually capable of reproducing successfully.

They are further enhanced by a slim waist, a uniquely human feature that is absent from other kinds of apes and monkeys, in which the pelvis is long and narrow. The erect posture and bipedal walking is made possible by a shorter, wider, more rounded pelvis, that forms a gap between it and the ribs and makes the buttocks protrude, as shown in Figure 15. The waist is less pronounced in men and children than in women because their pelves are relatively narrow and less strongly rotated backwards at the sacral region, producing a flatter, stronger back, more like that of apes. Teenage girls develop a pronounced waist as their pelvic skeleton grows wider and rounder and the spine curves. The sex difference in the skeleton is accentuated by selective enlargement of adipose tissue: young women have plenty of buttock and thigh adipose tissue but only small quantities in the superficial paunch and the intra-abdominal depots. Men have little fat on the buttocks or thighs but may have a bulging paunch.

A slender waist over wide hips is a distinctively human, as well as distinctively female, body shape. As just explained, the erect posture and its

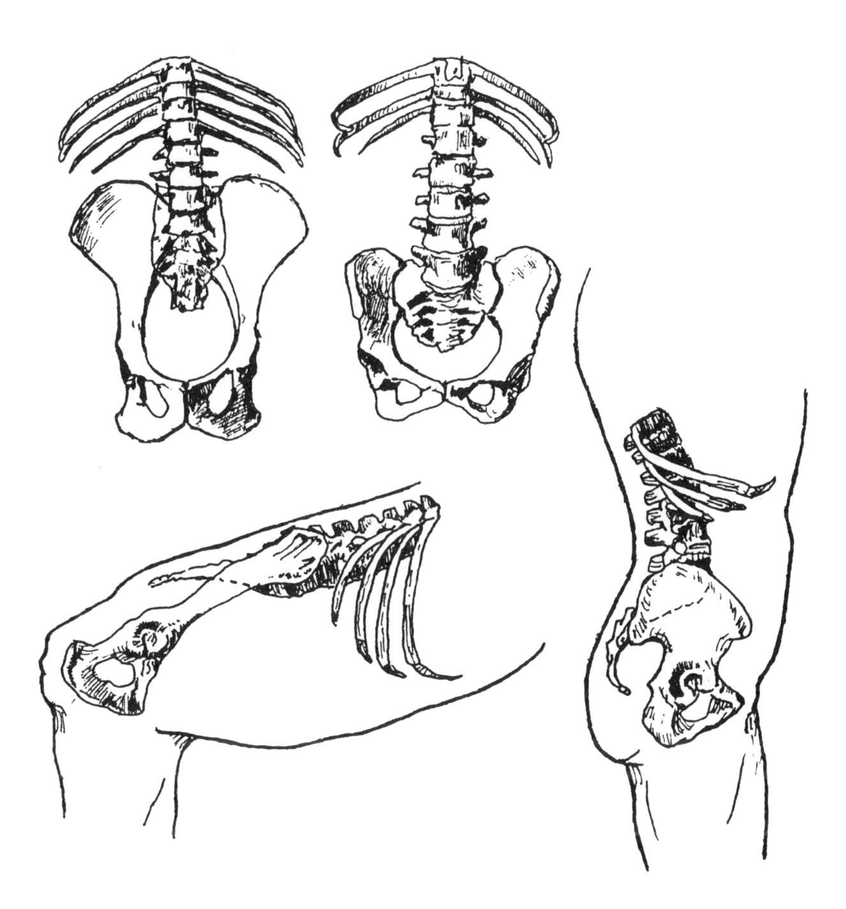

Figure 15. The skeletal origin of the human waist (redrawn and rearranged after A.H. Schultz, 1956 and 1969). Viewed from the front (upper figures), the human pelvic girdle (right) is shorter and rounder than that of apes (left). The resulting gap between the last rib and the anterior crest of the ilium bone forms the waist. The spine also curves backwards at the sacrum where it attaches to the pelvis, so when viewed from the side (below right), the buttocks protrude backwards, forming a waist at the concave lumbar region of the spine. The ape's back is flat in profile (below left), and the buttocks are slight. Accumulation of intra-abdominal and/or paunch adipose tissue erodes the waist, especially when combined with regression of the buttock and breast depots, to form a more ape-like profile in both side view and front view.

associated modifications of the pelvis are believed to be among the earliest of the uniquely human characters to have evolved. So male hominids have had the opportunity to admire their females' waists and hips for at least 3 million years, plenty of time for sexual selection to have led to these features being exaggerated in young women by selective accumulation of adipose tissue on the breasts and buttocks.

All the 'youthful' features of the female body form decline with age, whether or not total fatness changes, as illustrated in Figure 16. With the onset of menopause, women's body shape becomes more like that of men – and that of apes, as shown in Figure 15. The waist becomes thicker and the abdomen more bulging, due partly to weakening abdominal muscles and sagging guts, but also to selective enlargement of the intra-abdominal adipose depots and the superficial paunch. Some social psychologists[21] believe that a low ratio of waist to hip circumference, whether due to an exceptionally small waist or to ample hips, is a key feature by which men recognise women as young, healthy and fertile but not pregnant (i.e. sexually available).

Another aspect of the social organisation of higher primates may also have promoted the premature development of the adult arrangement of adipose tissue in girls. When a male lion or monkey takes over a pride or a troop, ousting his predecessor, his first act is to kill the young. The removal of suckling or otherwise dependent young quickly brings the adult females into breeding condition, and they mate with him and bear his offspring. The mothers resist infanticide, of course, because the practice reduces their lifetime fecundity, but in spite of much screaming, they are rarely successful.

As far as the male is concerned, more breeding females mean more potential descendants for him. So he faces a dilemma: when does a young female cease to be seen as reducing her mother's ability to bear the newcomer's offspring, and qualify as a potential mate? Among monkeys, adolescent females have been observed to avert infanticide by 'acting' as adults. While

Figure 16. (opposite) Age changes and sex differences in the distribution of human adipose tissue. Male and female infants have similar fatness and body shape. Sex differences in the distribution of adipose tissue remain minor in childhood but increase rapidly at the onset of adolescence with the growth of breast and limb depots. They remain pronounced throughout the reproductive period, but gradually disappear after menopause as a result of the expansion of intra-abdominal and paunch depots, combined with regression of breast, buttock and thigh adipose tissue. The body shapes of men and women become similar in old age.

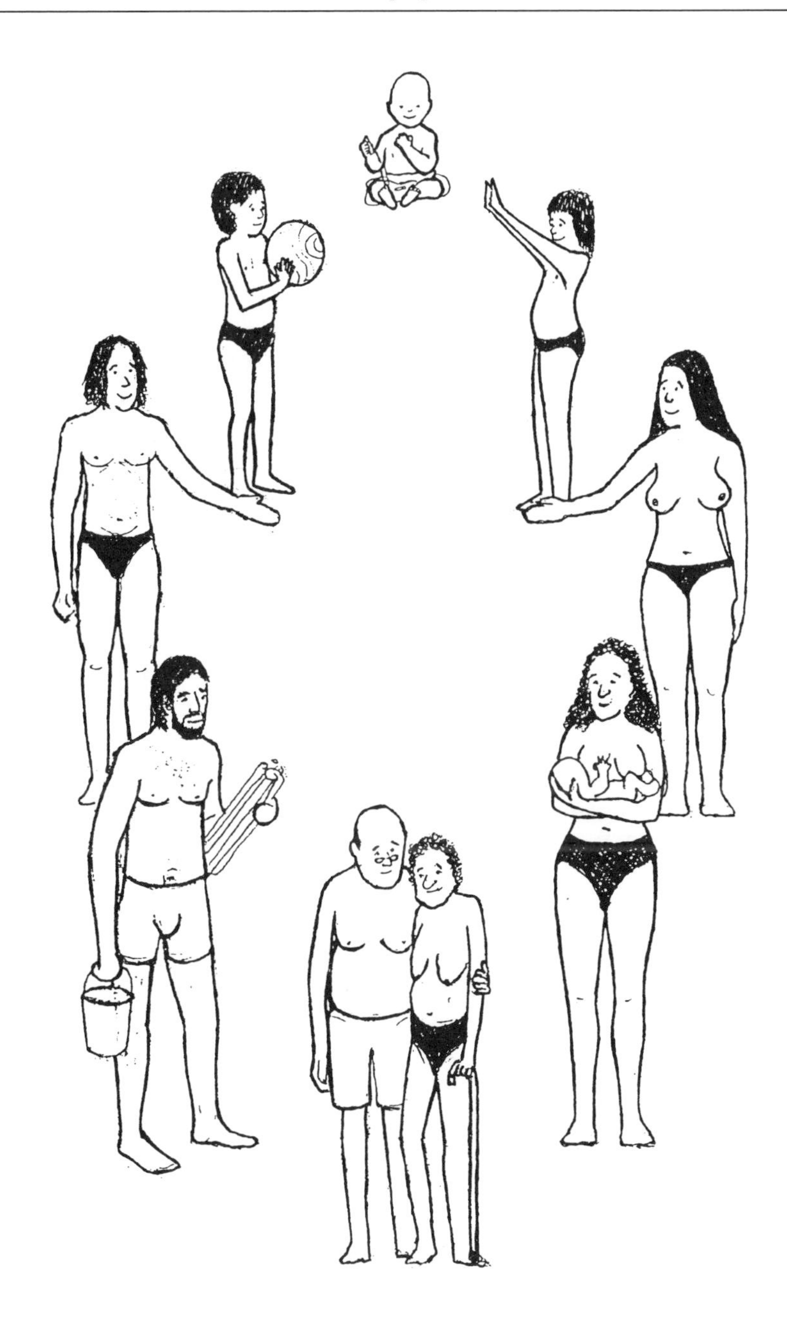

males and infantile females try to save their skin by hiding or running away, nearly mature females adopt sexually submissive postures. Such behaviour is unusual among mammals: in general, adult females attract, and sometimes solicit, the males' attention, but immature ones are not involved in sexual behaviour at all. The fact that the adolescent females are not yet fertile is irrelevant: if the male is deceived and spares them, precocious adult body form and sexual behaviour can evolve. If similar social organisation prevailed at some time in our evolutionary history, girls might have escaped infanticide by appearing to be women. Even if several years elapse before they are actually capable of bearing children, the trick would still mean that their mothers would escape bereavement and eventually become grandmothers.

Little girls resemble boys in having thin, straight thighs, narrow, flat hips, relatively thick waists and protruding bellies. Several years before their ovaries produce mature, potentially fertile eggs, girls develop thicker, more shapely thighs and buttocks, and small but highly visible breasts that together show off the narrow waist, and make them appear obviously different from children as shown in Figure 16. When combined with coquettish behaviour, the appearance of adolescent girls is very effective in persuading men to regard them as potential sexual partners. So much so that it may induce in men sexual behaviour that, when carried too far, we now condemn as criminal. By averting male aggression, it once may have saved the girls' lives.

This long ambiguous period when girls are neither children nor mature women (called adolescent sterility) has no parallel in boys: they produce fertile sperm years before they are either physically fully grown or capable of adult sexual behaviour. It is also rare except in species with the social structure just described. Sexual maturation in boys does not involve adipose tissue as much as it does in girls: puberty is accompanied by growth of muscle and the skeleton, and by changes in distribution and structure of body hair.

Women's fat

As clothing manufacturers know to their cost, the relative masses of the sex-specific adipose depots differ strikingly between women living in similar circumstances and on similar diets. S.G. Shattock, who became chief pathologist at St Thomas' Hospital in London in the 1890s, collected archaeologists' and travellers' accounts of disproportionate growth of minor adipose depots in various inbred tribes. His synthesis of the anthropological, archaeological and biological information about localised enlargement of adipose tissue was published in 1909,[22] and is still one of the most thorough accounts of the topic.

Shattock and contemporary anthropologists in France were particularly impressed by steatopygia, the massive development of adipose tissue over the buttocks. It occurs in slightly different forms in various African tribes and is usually, but not invariably, larger in women than in men, especially young women of marriageable age. The Bushmen, who lived in what is now northern South Africa, are believed to be direct descendants of the earliest groups of *Homo sapiens* to inhabit the region. The breasts, shoulders and abdomen of the young woman shown on the right of Figure 17 are ample, but not enormous, while her thighs and buttocks are greatly enlarged, with the latter forming a highly visible protrusion.

Some very early Egyptian carvings, and stone figurines found at the ancient city of Knossos in Crete dating from around 5000 to 6000 years ago have similar enlargement of the buttocks and thighs. Later Egyptians were well aware of the feature among people of neighbouring countries. The Queen of Punt, shown visiting from Somaliland with her lean husband on a diplomatic or commercial mission (Figure 17, top left), had impressively large buttocks and legs, in her case fat down to the ankle. Her arms are also fairly fat, but her breasts and abdomen are slight relative to the rest of her.

Many modern people regard moderate fatness as a desirable quality in brides, and at least until very recently, young women in some African and Middle Eastern countries were often deliberately 'fattened up for marriage'. But extreme examples of steatopygia and similar disproportionate expansion of certain adipose depots are now rare. Many of the smaller tribes have disappeared or previously segregated populations have intermarried, separating the combinations of genes that promoted their formation. Consequently we know almost nothing about the physiological mechanisms involved in their development.

Animal studies are of little help either. Many mammals have superficial adipose tissue over the hips and around the base of the tail, but in rats, mice and guinea-pigs, the depot is usually too small to provide enough tissue for biochemical analysis, so we know very little about its physiological properties. It is substantial enough in most hoofed mammals for field biologists to get a rough-and-ready estimate of a deer or antelope's fatness by palpating the pelvis around the tail. The depot is also present in most non-human primates, and becomes massive in a few species, notably the fat-tailed dwarf lemur (*Cheirogaleus medius*), which hibernates for 6 months or more during the dry season of its native deciduous forests in Madagascar.

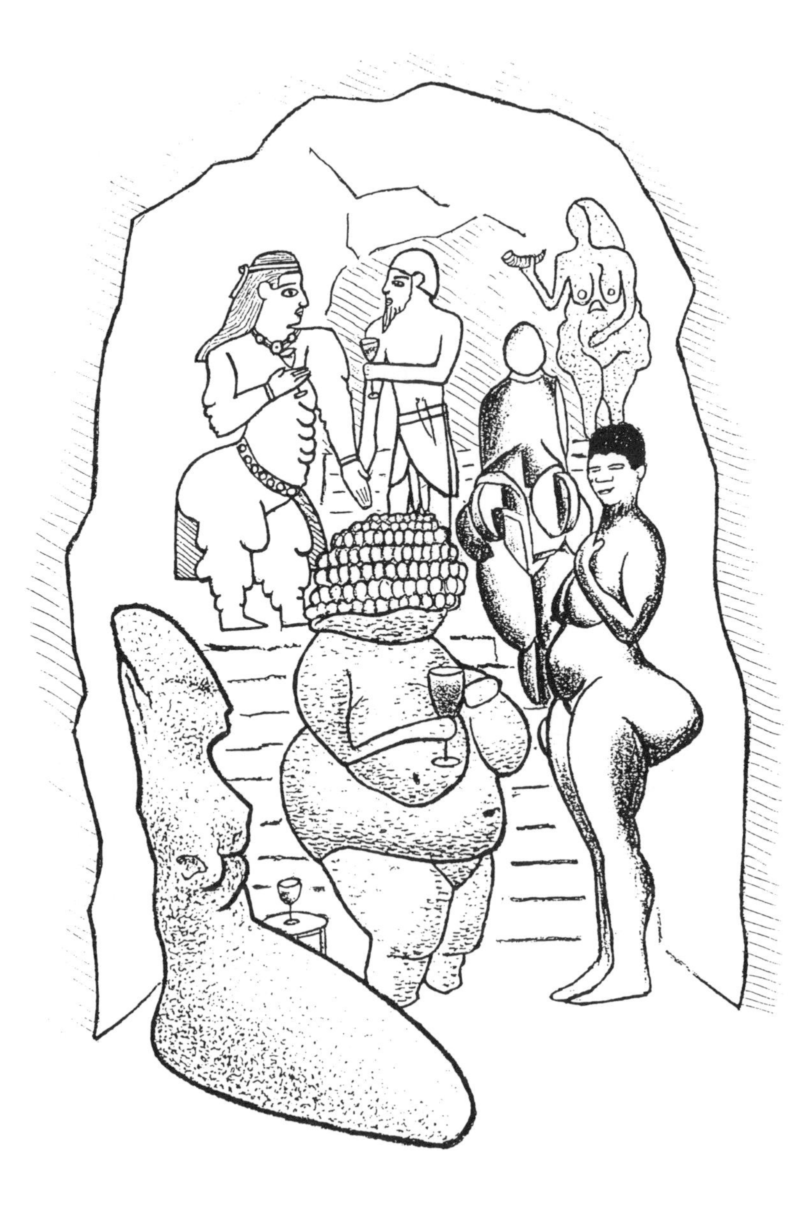

Figure 17. Body shape of young women from the Palaeolithic to the present. Clockwise from the top left: the Queen of Punt (Somaliland) with her husband, about 2000 BC (4000 years ago), after a carving in the temple of Deir el Bahari near the Valley of the Kings at Thebes, Egypt; Vénus à la corne (who is probably pregnant), Palaeolithic relief carving found in the Dordogne valley, western France; Vénus de Lespugue, ivory figurine found in Haute Garonne, France,

As Figure 17 shows, steatopygous women have only moderate amounts of fat in other superficial depots and inside the abdomen. This arrangement may have the same advantages as camels' humps in a hot climate such as Africa: concentrating the adipose depot into one or a few thick depots avoids impeding heat loss by conduction from the rest of the body surface. The relatively minor paunch depot in women of reproductive age not only emphasises the contrasts between their body shape and that of men, it also promotes efficient dissipation of heat from the abdomen, thereby perhaps helping thermoregulation during pregnancy.

There seems little doubt that these body forms evolved under sexual selection exercised by men who, for reasons that neither they nor we would be able to explain, had a preference for certain features. Shattock compiled reports of teenage girls being deliberately overfed to improve their marriage prospects, and of high-status men choosing enormously fat women as wives. The body shape of some of the most highly prized brides bore striking resemblance to the site-specific fatness apparently represented in some Palaeolithic and more recent images (Figure 17), having the fat in the breasts, buttocks or even the ankles greatly enlarged.

We do not know for how long, or how intensively, such selection has been applied, but it is unlikely to have been for longer than about 20 000 years, a much shorter time than people have been living in cold regions. Assuming four to five generations per century, that is less than a thousand generations, about as many as the fruit-fly *Drosophila* has had since it was first bred in laboratories for research into genetics. The ease and speed with which minor but conspicuous differences in the relative masses of adipose depots have evolved in inbred populations, apparently in response to sexual selection, contrast with the failure of any 'adaptations' of the distribution of adipose tissue to a cold climate to appear during the Pleistocene Ice Age.

The changes in adolescent girls' body shape can be startlingly rapid and are due mainly to selective expansion of adipose tissue. In contrast to most mammals, the distribution of adipose tissue in humans is influenced by sex

Figure 17. *cont.*
from the Palaeolithic, about 25 000 years ago; a young Bushwoman from Kalahari region of south-western Africa with pronounced steatopygia, photographed in the mid-nineteenth century; Venus of Willendorf, found in Austria but probably made elsewhere, Palaeolithic oolitic limestone figurine, made about 20 000 years ago; small clay figure found near the west bank of the Nile just north of Thebes, Egypt, from the pre-dynastic period about 5000 years ago.

hormones such as oestrogen that control the enlargement of breast, buttock and leg depots at adolescence, and the changes in adipose tissue distribution around the menopause (see Figure 16, page 251). These secondary sexual characters are enhanced by plumpness, so such selection would also favour genes that promote a tendency to fatten.

Female humans become fatter earlier in their lives than males, remain fat for a greater proportion of the life span, and are more susceptible to massive obesity. But women tolerate obesity better than men (though neither do nearly as well as naturally obese wild animals). They must be substantially fatter than men of similar age and habits before significant biochemical abnormalities are detectable, and health or longevity are measurably impaired.[23] Although moderate obesity is slightly more prevalent in middle-aged men than in women of the same age, extreme obesity in those under 40 years old is much commoner in women and girls. The limited information from animals confined and overfed in captivity suggests that in other species, males are, if anything, more susceptible to obesity than females. Among middle-aged (4–10 year old) cats kept indoors and fed on synthetic pet food, 15% of intact males (28% of neutered toms), but only 3% of intact females were classified as massively obese.[24]

Genes that promote enlargement of adipose tissue turn up from time to time in animals: the single genes that cause spectacular obesity like *ob/ob* in mice have too many harmful effects on the reproductive system to be favoured by natural selection. But the accumulation of genes with less drastic actions was presumably how obesity evolved in arctic foxes, Svalbard reindeer and bears. Long, hard winters were the form of natural selection that promoted their spread throughout the population in these species; sexual selection or some association with social status could have done it in humans. Greater fatness evolves if sexual or other forms of natural selection favour it, unless it is associated with adverse traits so serious, and appearing so early in life, that they impair breeding. Diet and environmental temperature are irrelevant to this theory, as are the physiological mechanisms of fattening, whether by increased appetite, decreased energy expenditure, or greater proliferation of pre-adipocytes.

The idea that it was the tendency to fatten early in adulthood, rather than fatness itself, that was selected also helps to explain the enormous variation between ethnic groups in the incidence of obesity. Chinese civilisation, with its intensive agriculture and animal husbandry, is among the most ancient in the world, dating back thousands of years. Eastern and southern China has had more large cities, where people have lived at higher population density for longer than any other major racial group, certainly for longer than most African tribes. Densely populated cities are prone to famine, siege warfare, epidemic diseases and similar disasters in which obesity might be a useful aid to survival.

Oriental food is as rich and appetising as any, yet obesity is rare throughout China. Ancient drawings suggest it always has been, even among upper class women who could not walk far, or even stand for very long, on feet deformed by foot-binding in childhood. Other people, especially those of Middle Eastern, southern European, Central American or West African descent, are highly susceptible to obesity, even though their traditional diets are not very rich in lipids.

Modern times

We now leave biology and continue the story using evidence from archaeology and cultural history. Art shows us for the first time what people and animals looked like, or more precisely, how artists saw themselves and the world around them. Archaeologists believe that during the Palaeolithic period between about 40 000 and 10 000 years ago, many people of our own species *Homo sapiens*[25] painted cave walls and cliffs, and carved stone, antler, bone, ivory and probably other less durable materials. Many spectacular examples of cave art survive in Europe, especially south-western France and northern Spain, and in Africa and Australia, but are comparatively rare in Asia and the Americas.

The European artists preferred painting large animals, especially mammoths, horses, bison and reindeer, to depicting themselves. Many caves and cliff sites are decorated with dozens of representations of animals but have no human figures at all. Others show 'stick people', so stylised that to our eyes at least, they cannot be identified as male or female. Animals are also the subject of most rock and ivory carvings, but a very few depict human forms that, from their body shape, are obviously women (see Figure 17, page 254).

Archaeologists have argued about the significance of these objects ever since they were first discovered in the middle of the nineteenth century. In a study of 132 Palaeolithic depictions of women,[26] bulging is confined to the abdomen, suggesting that they represent pregnancy, in only about 17%, including Vénus à la corne shown in Figure 17, top right. Some drawings of horses with distended abdomens are similarly interpreted. But the majority of female figures do not obviously represent fecundity: they show enlarged breasts, buttocks, hips or thighs, features that might indicate maternity, or perhaps matriarchy. Certainly, mother goddesses in one form or another are very widespread in primitive religions, but any indications of infants, or even families, are conspicuous by their absence. None of the figures suggests a woman carrying or feeding an infant, and none was found associated with representations or relics of children.

Another obvious interpretation is sexuality, a notion that led the 1860s French amateur archaeologist, the Marquis de Vibraye, to call the carving he found in the Dordogne valley, 'Venus', after the Roman goddess of sexual love. Carvings such as Venuses of Lespugue and Willendorf (Figure 17, centre) may be intended to show 'ideal' female forms. From our modern perspective, we would expect most images with this function to be of young women, but in fact fewer than a quarter apparently represent pubescent girls, with high breasts and flat bellies. Or they may have been charms, serving as sources of sexual arousal or gratification for men. If so, Palaeolithic artists were coy about the outcome: representations of sexually aroused men and intercourse are very rare indeed. Secondary sexual characters such as beards and moustaches are clearly depicted far more often than either female or male genitalia. Men, none of whom are ever shown as obese, are depicted hunting, killing and just standing around. In contrast, many of the deer depicted in caves in France and Spain seem to be in rut, with fully developed antlers and rounded, sassy bodies, actively following females.

A curious property of almost all prehistoric images of women is that facial features, which to us are the mainstay of beauty as well as of personal recognition, are not shown. For example, Venus of Willendorf (Figure 17 centre) has an elaborate hairdo but no face, and her colleague from Lespugue has graceful shoulders, neck and head, but a no hint of eyes, nose or mouth. The careful, accurate representation of the body contrasts with the lack of detail on the face. Was body shape more important than facial features for the cultural purposes of these images? We shall probably never know.

Leaving such speculations aside, these ancient figurines indicate to biologists that at least some women sometimes grew fat long before the abrupt change in diet that followed the development of agriculture and dairy farming. The figures are too anatomically accurate, too similar to modern obese women to be entirely imaginary. Sensible people do not usually put a lot of effort into celebrating things they despise, so we must assume that the sculptors and their patrons regarded female obesity as desirable in some context, whether sexual, maternal or otherwise. What seems to be celebrated is not generalised obesity, but selective enlargement of certain depots, which arises from a relatively modest increase in total lipids stored, and enhanced site-specific properties, so a large proportion ends up where it is desired.

The modern situation is not greatly different: we prize fashion models and film stars for their pleasing site-specific accumulation of adipose tissue – slim waists and thighs but well-developed breasts – not its absolute quantity. Improved measurement techniques show that men engaged in sedentary occupations are almost as often obese as women of the same age and social class, but far fewer men than women feel compelled by cultural pressures to try to slim for cosmetic (as distinct from medical) reasons.

The age of agriculture

Before the historic period, the only source of information about diet is the study of food remnants (i.e. rubbish heaps of bones, shells, discarded seeds, etc.), the tools used to catch, produce or prepare it, and tooth wear. As everyone knows, Palaeolithic people sharpened stone axes, which could be used for cracking nuts and cleaning fish as well as for killing and butchering large mammals. As tool-making became more skilled, and the large mammals became scarce towards the end of the Palaeolithic and increasingly thereafter, people caught fish on bone fish-hooks, collected shellfish and trapped and ate smaller mammals such as rabbits, large birds, and a variety of reptiles including snakes, crocodiles and turtles. We know they did so from the bones and shells left behind, but they probably also ate foods such as insects that required no tools and have left no trace – insect cuticle is much less durable than bone – and a wide variety of nuts, roots and fruit. Diet certainly varied from place to place, and changed with the seasons, in part as a result of people travelling between different localities.

Agriculture and pastoralism introduced major changes in people's diet and habits. All anthropologists who have studied the few remaining people who live by hunting and gathering, such as Australian Aboriginals and Kalahari Bushmen, note that growing crops and tending herds is far harder work than finding wild food. The labour is more arduous, and people have to toil for longer to sustain themselves. Farming does, however, support denser aggregations of people, and enables them to settle in permanent villages where food can be stored, thereby removing the incentive for seasonal migration. Each farming family could produce enough food for several others, so more people were emancipated from the demands of food acquisition and were able to devote themselves to arts and crafts, religion, trade, warfare and, more recently, science. The human population has increased 100-fold during the first 8000 years of agriculture.

Population growth and cultural advance were especially rapid in the natural ranges of species that proved to be suitable for domestication. Horses were first used for food and transport in central Asia. Chickens came from South East Asia. Pigs, ducks, geese and above all the ruminants, camels, sheep, goats and cattle, were domesticated in western Asia and Egypt. People there also tried, but ultimately failed, to domesticate many more species, including oryx, gazelles, cheetahs, hyenas and various large birds including cranes. Domestication favours a tendency to fatten. Farmers protect livestock from their natural predators, making alertness and fast running less important to survival. Those that eat readily grow fastest and produce the largest litters, so are likely to be selected as parents for the next generation of breeding stock.

Ruminant adipose tissue and milk contain a high proportion of saturated fatty acids and very few polyunsaturates (see Chapters 3 and 4), and if they were the only source of dietary lipids, people would have to eat a great deal to obtain enough essential fatty acids. This requirement alone might make people very fat very quickly: the introduction of dairy products into the Japanese diet[27] after the Second World War is widely blamed for an abrupt increase in obesity in a formerly lean population. In fact, the eating habits of traditional pastoralists avoids such problems. For example, the Masai of East Africa take blood as well as milk from their cattle. Blood contains numerous very small cells, each surrounded by a phospholipid membrane. Cell membranes are the main repository of polyunsaturated fatty acids in ruminants, so by eating blood, the Masai obtain adequate supplies of these valuable nutrients.

Jewish dietary laws[28] are among the first written accounts of what people ate. Like many other groups of people living in what we now call the Middle East, Jews around 1000 BC were pastoralists, who tended sheep and goats and, later in their history, cattle. The milk and meat from these ruminants, together with bread made from wheat and barley, formed the major part of the diet. They certainly relished highly concentrated sources of carbohydrates and lipids, describing the 'promised' land as 'flowing with milk (i.e. ewes' or goats' milk) and honey'. The flesh of non-ruminant mammals, including pigs and hyraxes, and that of most kinds of birds, in which the triacylglycerols contain a fair proportion of polyunsaturated fatty acids, was prohibited as 'unclean', but Jews were allowed, indeed encouraged, to eat most kinds of fish, which would be a concentrated source of n–3 fatty acids.

The desert offers few edible oil-rich seeds, and large-scale production of fruit and vegetables is only possible for people living in permanent settlements. However, Jewish law specifically names locusts, grasshoppers and related herbivorous insects as approved foods. These insects are often abundant on the dry grasslands and scrub that covered much of the Middle East before thousands of years of overgrazing by sheep and goats turned much of the land into semi-desert.

Sometimes, as in Egyptian plagues,[29] swarms attack crops, with devastating results. Locusts accumulate large quantities of triacylglycerols just before they set off on migration, which, being derived mainly from plant lipids, contain a high proportion of *n*-6 polyunsaturated fatty acids. Even the occasional meal of fish and locusts would be sufficient to prevent deficiencies of polyunsaturated fatty acids that might arise from a diet of grains and ruminant products. Since the introduction of poultry-keeping, herbivorous birds have largely replaced locusts as sources of *n*–6 fatty acids; chicken fat is still an important ingredient in traditional Jewish cookery.

Ancient Egypt

Written records, paintings and artefacts in tombs and temples provide a more complete picture of everyday life in Ancient Egypt than of any contemporary civilisation. Increasing the supplies of animal fats seems to have been a priority early in the development of animal husbandry. Paintings[30] from the pyramid-building period in Egypt show ibex, oryx, antelope, gazelles as well as bulls and pigs tethered in small pens and feeding from bowls and troughs that probably contained barley or other grain, which is much more fattening than pasture. Farmers are shown forcibly feeding geese and cranes with oval objects, possibly cakes of moistened grain, while the rest of the large flocks stand around preening themselves. Both these species migrate, so have the physiological capacity to fatten, but force-feeding may have produced something like *foie gras* (see Chapter 4). One image even shows a man holding down a hyena while another stuffs food into its mouth. Adult hyenas can break a man's leg with a single bite, so such activities must have been worth the risk.

Many species of the pig family (Suidae) including wild boar, the ancestor of the domestic pig, feed on acorns and the seeds of other trees that 'mast-crop', i.e. cycles of several years of very low seed production followed by a single huge crop. Pigs fatten rapidly during such brief gluts of lipid-rich food, and thus do so readily when well fed in captivity. In other domesticated species, the capacity for fattening was improved by selective breeding. Fat-tailed sheep, a breed in which the adipose depot around the tail enlarges selectively, were established in Egypt (and in much of south-western Asia) by about 4000 years ago. Fifteen hundred years later, the Greek historian, geographer and cultural anthropologist, Herodotus, noted that the tails of mature specimens were so heavy that, to prevent the sheep injuring themselves, Arab shepherds mounted the tails on wheeled trailers. Records of commercial transactions, leases and taxation written on papyrus and preserved in jars and boxes indicate that eggs, cheese and various plant oils, including olive, linseed and sesame, were also produced in large quantities.

While some of the fats harvested from these animals were probably used as fuel, a fair proportion was almost certainly eaten, either as whole adipose tissue or as extracts used in cookery. The Egyptians buried their dead with such care, and the climate is so dry, that many remains are very well preserved. Examination of patterns of wear in their teeth make it clear that barley, wheat and other plants formed a large part of the diet. Meat and fats leave no such traces but the fact that farmers went to such trouble to improve livestock and crop production strongly suggests that these items were highly valued, at least as luxury foods. Early Jewish texts mention feasting on 'fatted calves' as a major feature of celebrations. Other Near Eastern peoples probably had a similar concept of a really good dinner; ordinary Egyptians ate pork, while beef was reserved for priests and royalty.

Ordinary people must have needed the extra energy that fats can provide: building temples and pyramids, or cultivating enough land to produce food for both farmers and builders, requires long hours of hard work, probably more physically demanding than the hunting and gathering of their recent ancestors. But for people such as priests and royalty, who did not perform such arduous work, Egyptian food at the time of the Pharaohs could have been as fattening as our modern diet. Should we really believe that, notwithstanding the wealth of grains, milk, fatty meats and plant oils, all Egyptians were as lean and firm as they appear in their portraits?

Of thousands of carvings and paintings of human figures in Egyptian tombs and temples, none is obviously obese, though other body deformities such as dwarfism are accurately depicted. Though not massive, breasts and slightly protruding abdomens are clearly shown under the women's clothing, and dancing girls have curvaceous thighs and buttocks. The only figures depicted with bulging, sagging abdomens are wizened old men with thin limbs and stooping posture. The Assyrians, who were rivals and sometimes enemies of the Egyptians during the earlier half of the first millennium BC, were more candid. Like the Egyptians, they usually portrayed people (and animals) side-view, an excellent perspective from which to observe obesity. Relief carvings (now in the British Museum) from their palaces near the River Tigris in what is now Iraq, show ambassadors and other senior officials with ample paunches and protruding bellies, though prisoners of war and refugees appear noticeably slimmer.

Ancient Greece

Farming people elsewhere in the eastern Mediterranean region had another reason for fattening their livestock: their gods liked adipose tissue. The burning of sacrificial animals was an integral part of many ancient religions, including those of Assyrians, Persians, Greeks and Jews. Egyptian religious rituals stand out as exceptional: offerings of food, including meat (especially beef), bread, honey, fruit, beer and oil were made, but no materials other than incense were burnt. Ancient Egyptians preferred to mummify, rather than burn, the bodies of the many wild and domesticated animals that they held to be sacred. Perhaps the lack of forests in Egypt made wood too valuable to be used in ritual fires, but the practice was almost universal in the wetter, more densely forested areas further north.

Writing around 2700 years ago, early in the development of Greek civilisation, the poet Hesiod[31] described animal sacrifice in detail and proposed a theological explanation for the procedure. The tradition began after the duplicitous Titan, Prometheus, tricked the gods into allowing people to eat most of the meat of sacrificed animals. He dissected out the tissues of a slaughtered ox and placed the meat in one pile, and the bones covered with fat in another. As the most senior deity present, Zeus was invited to choose which he wanted, and, although he noted that the second pile was smaller, he took a piece of delicious fat, and was annoyed to find that the rest of the offering consisted mostly of inedible bones.

In reprisal, Zeus deprived men of fire, and, in what to many might seem a worse punishment, created women and made living with them a condition of having children. The issue of fire was quickly reversed – Prometheus stole it back, along with the gift of knowledge – though that of women was not so satisfactorily resolved. Ever since then, according to the poets, long bones wrapped in a layer of fat and spiked with small pieces of meat were burnt entirely, while the meat was just cooked and eaten by the worshippers.

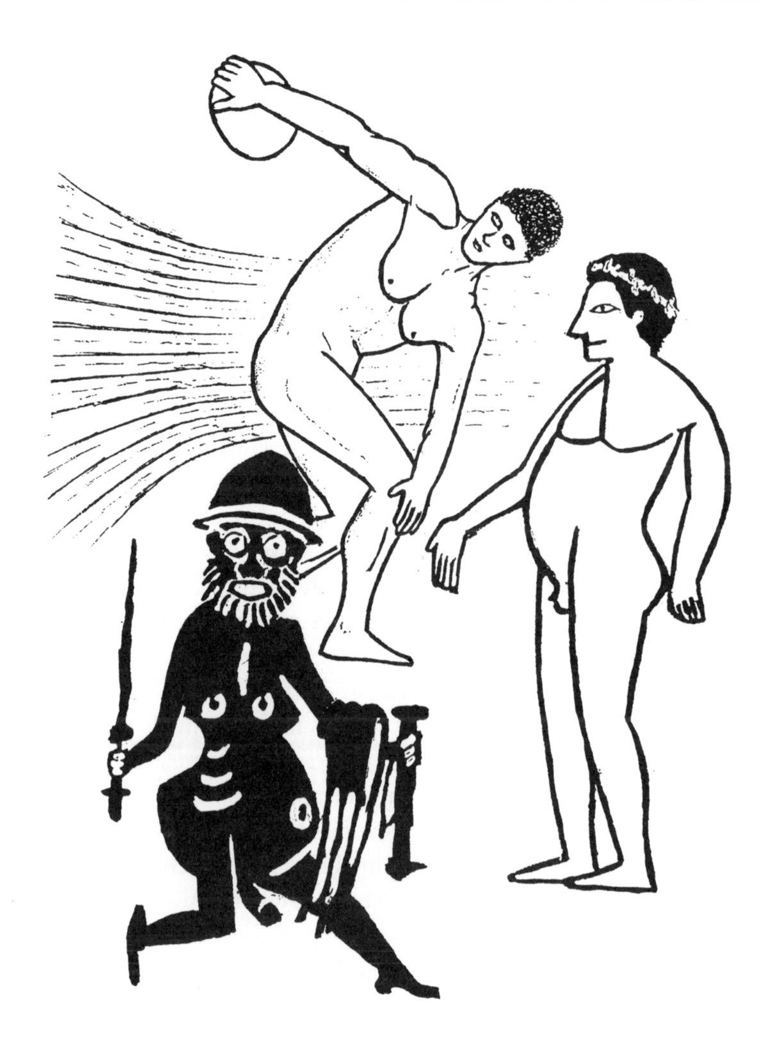

Figure 18. The classical Greek ideal male body form, as represented in formal statues of the fifth century BC (about 2500 years ago) with contemporary images on decorated pottery of an obese young athlete and a middle-aged comic actor playing Odysseus.

In ancient Greece and Rome, eating mammalian meat was firmly associated with religious rituals, and was a welcome addition to the normal diet of grains, fish, cheese, fruit and vegetable oils.

Apart from being the basis of this unfortunate row, adipose tissue has many advantages as a mediator between people and celestial deities. As noted in Chapter 4, lipids burn completely, literally going up to heaven. Burning adipose tissue crackles, sparks, flares and produces a distinctive, pungent odour, all features that may have contributed to the spectacle of religious rituals. Adipose tissue that is depleted of lipid (i.e. contains a high proportion of protein and water) does not burn as impressively as that from well-fed animals.

I have dissected many large mammals in much the same way as Prometheus did, and can thus comment that quite a lot of adipose tissue would be needed to cover a pile of bones, certainly more than is normally found in wild ungulates such as deer or antelope. Artificially fattened domestic animals would make much more impressive sacrifices than wild-caught ones. Overseas trade, wars and the enslavement of defeated peoples ensured that cultural know-how spread from one civilisation to another. Egyptian methods of livestock management may have been adopted elsewhere as a means of improving the quality of sacrificial animals and hence, as a side-effect, the fat content of meat generally.

Free-born male Greeks actively practised sports and games in specially built gymnasiums. A lean, muscular body and success in athletic competitions such as the Olympic games were greatly admired, and celebrated in sculpture, such as those shown in Figure 18. In spite of their high-fat diet, most young men were lean, but the Greeks were well aware that neglecting exercise quickly led to a 'woman-like appearance', in plain language, podginess.

Greek culture valued men and manliness, so attributing womanly qualities to a man was derogatory, even insulting. 'Official' Greek sculpture always represented the lean, idealised male body form, but the sophisticated, sceptical drama of authors such as Euripides, whose long life spanned most of the fifth century BC, spawned more candid images: clay figures whose masks show them to be comic actors have large bottoms and bulging bellies depicted with alarming accuracy (Figure 18, foreground). Goddesses such as Athena always have firm, youthful figures under graceful clothes, but terracotta figures from the second and third centuries BC show obese middle-aged women, and mothers depicted on tombstones and family shrines are sometimes quite ample. The desirability of avoiding obesity may be added to the long list of moral, political, artistic and scientific ideas that originated in fifth century Athens.

Final words

There is no single explanation for why or when people became fat. There is no evidence that the usual reasons for the evolution of obesity in animals – erratic food supply, long-distance migration, hibernation, etc. – ever applied to our ancestors. The climate became cooler several times during the Pleistocene period, and people started living at higher latitudes, but adipose tissue's role in adaptation to cold seems to be supplying fuel and possibly generating heat, not acting as insulation. A direct causal relationship between hair reduction and obesity also seems unlikely, though it may have helped to promote age and sex differences in the distribution of superficial adipose tissue.

Sometime during our evolutionary history, probably after the reduction in body hair, body shape joined the colour and distribution of hair as indicators of social and sexual status. Selective enlargement of certain adipose depots acquired new social significance. Adipose tissue has many advantages for such a role: the relative development of depots can be adjusted by prolonging (or curtailing) pre-adipocyte formation, or by adding (or removing) a few receptors for certain hormones. It does not take an enormous amount of energy to form or maintain, and if properly controlled, is physiologically harmless. Social pressures may have favoured early appearance of womanly body shape in adolescent girls, and selection for this feature may have promoted a tendency towards obesity in female humans of all ages.

Apes, like people, know what they like eating, and are willing to devote considerable energy and ingenuity to obtain desired foods. People have long chosen a diet that is rich in lipids, including animal fats. In one way or another, such preferences may have promoted intellectual development and efficient social organisation. The development of agriculture increased rather than reduced access to animal fats but it, and associated activities such as building, warfare and long-distance travel, also required most people to undertake strenuous exercise throughout their lives. Obesity has so little impact on survival and fertility during the reproductive years that it was probably not seen as a problem – until fashions changed and people wanted to be leaner for cosmetic reasons. With such a variety of body forms, no wonder clothing manufacturers have to make bras in over a dozen different sizes and so many people find that standard-size skirts and trousers do not fit them. Body shape is as much part of our varied genetic inheritance as individual differences in athletic, artistic or intellectual ability.

8

Fat and health

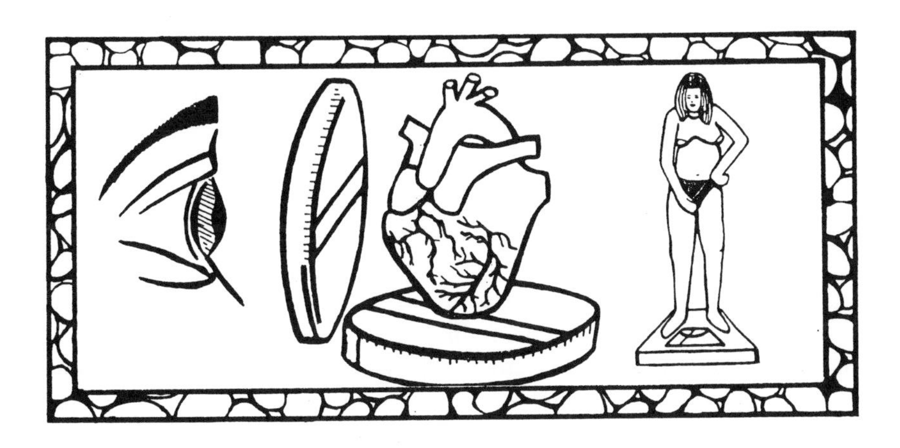

In the late twentieth century, the involvement of adipose tissue and lipids in disease is their most widely publicised role, so much so that many readers would expect this chapter to be by far the most prominent and extensive. In fact, there was some doubt about whether the topic should be included at all, and one reason for doing so was to place it at the end, after the useful functions of fat had been explained. As a further indication of the reluctance to join the chorus of disapproval of all things fatty, this chapter begins with accounts of three quite different ways in which lipids are involved in the prevention and cure of disease and the relief of suffering.

Fats that cure

Lipids have such a central role in the workings of cells and organisms that it would be surprising if their absence or malformation did not cause ill-health.

Most kinds of cells make many of their lipids themselves, so deficiencies arise only from abnormal metabolism, but some fatty acids and vitamins are obtainable only from the diet, and so can be lacking even in physiologically normal individuals. It is often very difficult to determine the exact cause of diseases that arise from dietary deficiencies. Many animals can build up substantial stores of essential nutrients, enough to maintain good health for weeks, months or years, so such disorders almost always develop slowly, becoming noticeable long after the inadequate diet was imposed. Nonetheless, as already hinted in earlier chapters, it is now clear that particular kinds of dietary lipids play pivotal roles in maintaining normal body function, and in resisting invasion by pathogens.

Fatty vitamins

The discovery of vitamins, a miscellaneous collection of substances that are essential components of the diet, albeit in extremely small quantities, was one of the triumphs of biology and medicine in the second half of the nineteenth century and the early twentieth century. At last scientists could explain why debilitating and fatal diseases including scurvy, some forms of anaemia and beri-beri could be prevented, and, unless they were far advanced, cured by regularly including certain foods in the diet. By the mid-1920s, the concept of 'bulk nutrients' – carbohydrates, proteins and lipids – and 'accessory food factors' – vitamins and minerals – was well established.

The first 'accessory food factors' to be identified were believed to be amines (hence the name, vit ('vital')-amine), water-soluble, protein-like molecules. But it soon became clear that some were lipid-soluble and did not fit this definition. Among them was a group of closely related substances now known collectively as vitamin A. They are derived from carotene pigments in plants and are important to vertebrates as the starting point for the synthesis of retinol (see Chapter 3). As well as being essential to vision, vitamin A also helps to maintain healthy skin. Investigations into the causes and treatment of rickets, a disabling disease that reached epidemic proportions after the First World War among poorly fed populations living in crowded cities, led to the discovery of another lipid-soluble vitamin.

At the time, physicians believed that rickets was probably due to an infectious micro-organism, perhaps similar to the bacterium known to cause tuberculosis, a respiratory disease common in the same groups of people. But in 1923, it was discovered that the regular addition of small quantities of cod-liver oil to children's diets could prevent rickets and arrest its progress after symptoms had appeared. At first, vitamin A (which is present in cod-liver oil) was thought to be the agent responsible, but its susceptibility to oxidation proved this assumption wrong. If oxygen is bubbled through boiling cod-liver oil for several hours, the resulting substance (which must smell horrible) retains its ability to prevent rickets, but does not protect from defects of vision. Since the water-soluble vitamins B and C had already been described and named, this second lipid-soluble factor was named vitamin D.

The metabolism of vitamin D turns out to be very similar in all vertebrates in which it has been studied. Avian vitamin D is so like that of humans that many of the early studies into how it supports bone growth were conducted on quail and poultry. Its roles in promoting absorption of calcium from the gut, and in the calcification of bone and teeth are still being investigated. Vitamin D is really more like a steroid hormone than a vitamin, and like other steroids, it readily permeates cell membranes and accumulates in membranes and other fatty biological materials. So it (or very similar substances) is present in small quantities in most of the vertebrate tissues that we, and other carnivores, eat. It is more abundant in egg yolk, and in fish and mammalian livers, with cod-liver oil being by far the most concentrated source known.

In humans and other bare-skinned animals such as elephant seals,[1] it is synthesised from cholesterol in the skin during exposure to bright sunlight. This source is more than sufficient in people who live in sunny climates such as Africa and who wear few clothes. Even if the skin is not exposed regularly for long enough to sunlight (due to social conventions in clothing, and/or to being confined to buildings or narrow streets) for the body to synthesise sufficient vitamin D for itself, human diets rich in animal foods provide plenty of the vitamin. The dual origin of vitamin D in humans was first demonstrated conclusively by the British physiologist, Dame Harriette Chick, who was born in London 14 years before Chevreul died, and, like him, lived to be 102. Thus two of the world's longest-lived scientists made important contributions to the study of animal fats, the consumption of which is now believed to be a major cause of premature mortality.

Vitamins can be toxic in large quantities, but whereas any excess of the water-soluble ones like vitamin C is simply excreted in the urine, getting rid of too much of the lipid-soluble ones is not so straightforward. As early arctic explorers found to their cost, the livers of certain fish, seals and polar bears are toxic, especially when eaten raw, because they contain high concentrations of these vitamins. Abnormal accumulation of calcium salts in elephants' arteries has been attributed to excessive production of vitamin D, caused by overexposure to sunlight after shade trees were cut down for firewood. Most vitamins can now be synthesised artificially and are often added to 'fortified' foods such as breakfast cereals, or sold as capsules. Synthetic vitamins have become so widely available in industrialised countries that overdoses are now as much of a hazard as deficiencies for people, and for captive animals fed on tinned or dried food.

Fatty acid deficiencies

The discovery of vitamins A and D proved beyond doubt that lipids could be the vehicle by which these minor but crucial nutrients were obtained, but other scientists were sceptical when George and Mildred Burr[2] suggested in 1929 that a fatty acid that had been considered to be just a fuel, was also a dietary 'essential'.

Working in the Botany Department of the University of Minnesota, the Burrs observed that rats fed very low fat diets failed to grow normally and developed skin disorders such as scaly feet and tail, and dandruff. The males lost interest in mating, and the few that could be persuaded to do so proved to be infertile. Adding small quantities of triacylglycerols containing linoleic acid (C18:2*n*–6) to the diet corrected these problems. As little as five drops of linseed oil or corn oil (in which about 15% of the fatty acids are linoleic acid) per day for 40 days added to the purified food cured the ailments, but additives such as coconut oil that contain only saturated and monounsaturated fatty acids were not similarly effective.

The Burrs concluded that this particular fatty acid could act either as a co-factor or as a precursor for some vital metabolic process, or, when present in excess, could be used as an energy source like any other fatty acid. They thus proposed the concept of an 'essential fatty acid' (see Chapter 3). Fatty acids of the same *n*–6 family as linoleic acid, with longer chains of carbons and/or more double bonds (e.g. arachidonic acid, C20:4*n*–6) were almost as effective as linoleic acid itself, which suggested that linoleic acid's importance arises chiefly from its role as a precursor for the synthesis of these more highly unsaturated fatty acids.

Further investigations into essential fatty acid deficiency revealed a host of metabolic abnormalities, mostly but not exclusively related to lipid metabolism: the properties of internal cell membranes change, notably those of mitochondria, so they do not use fatty acids as fuels so efficiently, and the synthesis of steroid hormones is impaired. The absence from the diet of lipids containing linolenic acid (C18:3*n*–3) was also found to have specific effects, especially on the eyes, and they too could be corrected by other fatty acids of the *n*–3 family, such as eicosapentaenoic acid (C20:5*n*–3) and docosahexaenoic acid (C22:6*n*–3).

Substantial quantities of linoleic and linolenic acids, and the longer-chain polyunsaturated fatty acids derived from them, can be 'stored' as components of membrane phospholipids, especially those of the liver, or as minor ingredients of triacylglycerols. Such reserves are sufficient to last rats and people for many months (in the cases of some wild animals, possibly years) on a diet in which they are entirely absent. Supplies of these fatty acids first become critical in cells of the kidneys, lungs, heart and other inaccessible organs.

An understanding of their metabolism enables biologists to identify warning signs of impending deficiency, before overt symptoms of organ malfunction appear. The desaturase enzymes work on any monounsaturated fatty acid they find, but in a well-fed mammal, those of the *n*–3 or *n*–6 series are preferred. However, when these fatty acids are scarce, the enzymes desaturate more of the monounsaturated *n*–9 fatty acids, notably oleic acid. Such products accumulate in storage lipids in superficial adipose tissue, where they can be detected in biopsy samples, thus serving as an indicator that reserves and/or intake are so low that essential fatty acid deficiency is imminent.

Until very recently, highly unsaturated fatty acids of the *n*–9 series were believed to be harmless but unimportant by-products of the biochemical pathways that produce useful *n*–3 and *n*–6 polyunsaturates. But now C20:3*n*–9, C22:3*n*–9, C22:4*n*–9, C24:4*n*–9 have been found in significant quantities in the membrane phospholipids of sperm from healthy, fertile rats. What they are doing there is not clear. One idea is that their presence relates to the fact that mammalian sperm usually mature only at temperatures below those of the internal organs.

Essential fatty acid deficiency is a symptom of cystic fibrosis, one of the commonest inherited diseases among people of European descent, but the causes are now known to be indirect. The primary defect is in the gene that makes part of the mechanism that assists small water-soluble molecules across cell membranes. The pancreas is among the organs affected: it fails to secrete enough pancreatic lipase, so digestion of lipids becomes much less efficient, and many remain in the gut contents as whole triacylglycerols. As explained in Chapter 4, lipids are not absorbed unless they are properly digested. Chronically inefficient fat digestion eventually leads to deficiencies in essential fatty acids, even if they are plentiful in the diet.

The nervous system is rich in membrane phospholipids (see Chapter 1) containing highly unsaturated fatty acids. Docosahexaenoic acid (C22:6*n*–3) seems to be particularly concentrated at synapses, the connecting points between neurons, and in the light-sensitive retina of the eye, which in all vertebrates including ourselves, develops as a protuberance of the brain and consists of numerous very small cells, each containing layers of internal membranes. Up to 60% of the fatty acids in the phospholipids of the membranes in the retinal cells are docosahexaenoic acid, a far higher proportion than in the phospholipids of other kinds of cells. Their exact role in vision is unclear, but it probably relates to the formation or transmission of the nerve signals from the light-absorbing pigments to the brain.

In all vertebrates, a large proportion of the body's stock of this polyunsaturated fatty acid are located in the eye, especially in ruminant animals, which have little of it in triacylglycerols. The body cannot make this vital component of the eye for itself, but relies upon dietary supplies. Docosahexaenoic acid appears to be conserved efficiently and directed towards the appropriate cells of the eye. Noticeable defects in vision were detected only in the third or fourth generation of guinea-pigs raised on diets deficient in this fatty acid, suggesting that supplies of this valuable ingredient of retinal membranes may be passed from the mother to the foetus. It is not clear where such reserves are held: all mammals, including ourselves, lose neurons at a rate of thousands per day throughout much of the life span, but the fate of the debris from such destruction has not been followed. Mothers whose reserves of docosahexaenoic acid are almost exhausted would appear to be infertile: foetuses in which development of the nervous system is severely impaired would be aborted or reabsorbed early in gestation.

As pointed out in Chapter 3, polyunsaturated fatty acids such as linoleic acid (in linseed oil) readily oxidise when exposed to air. On the whole, they don't come to much harm when in the membranes of living cells, in which oxidising conditions are actively controlled. So there are polyunsaturates in the membrane phospholipids of most kinds of fresh, raw leaves and flowers, and freshly killed fish and invertebrates (even if not specifically 'oily' species). But in dead or dying cells, such as are found in stale, dried or overcooked vegetables, or rancid fish, the polyunsaturated fatty acids may be oxidised or otherwise degraded to the point that they are of little nutritional value to the mammal eating them. Diets that contain few fresh vegetables or fish may not provide sufficient essential fatty acids and it may be necessary to eat more concentrated sources, such as fish oils or the triacylglycerols of certain kinds of seeds (see Chapter 3).

It is impossible to prescribe exactly how much of what foods would provide sufficient essential fatty acids. The concept of minimum dietary requirement is based mainly upon rigorously controlled experiments on rats that are fed on highly purified synthetic diets. People greatly prefer 'real food' made from natural ingredients. As pointed out in Chapter 3, the exact fatty acid composition of plant tissues depends on the weather conditions under which they were growing, and how they are stored and cooked, as well as upon the species.[3]

When western people are abruptly transferred to a diet that contains plenty of carbohydrates and proteins, and apparently more than enough vitamins, minerals and essential fatty acids, as happened in prisoner of war camps in the 1940s and more recently in artificial situations such as the BioSphere experiment, they function well initially. But after a few months on what is for them a very low fat diet, they become weak, depressed, irritable and develop an extreme craving for fatty foods, even though they are not emaciated and the fat content of their food is not much less than that which would be adequate for some Asian people raised on low-fat vegetarian diets. Such observations suggest that, in some way not readily demonstrable in animals, people need to eat fats: mobilising lipids from their adipose tissue apparently cannot or does not meet the demand and prolonged deprivation produces symptoms that can only be relieved by eating fat.

For reasons explained in later in this chapter, it is becoming clear that long-chain n–3 fatty acids promote health and longevity in doses far above the minimum necessary to avoid overt symptoms of fatty acid deficiency. Nutrition experts currently recommend eating 2 g of n–3 fatty acids per day, but only maritime nations such as Portugal with a strong fishing tradition even approach this intake. With no sign that people would willingly eat more fish, and wild stocks of edible species declining almost everywhere, food technologists are investigating ways of extracting n–3 fatty acids from low-grade fish or even from certain plant oils, and adding them to manufactured foods such as cakes, ice-cream and mayonnaise.

Insects

Of more than a million living species of insects, nutrition has been studied in only a handful of economically important species, so generalisation is impossible. But it has been known since 1946 that the larvae of the common flour moth (a pest in stores of grains and nuts) need small quantities of both linoleic and linolenic acid in their diet. In experiments in which colonies were fed on synthetic diets without these nutrients, the larvae grew slowly and, although many formed pupae, viable adults did not emerge. More recently, caterpillars of the common cabbage white butterfly have been found to have similar requirements.

Experiments similar to those performed on rats showed that only the basic polyunsaturates, linoleic and linolenic acid, restored normal growth: lipids with oleic acid or fish oils were only partially effective. In the natural situation, such requirements would not present a problem: these insect larvae, like those of the great majority of moths and butterflies, eat leaves or seeds (flour is just ground wheat seeds), which would provide more than enough of these fatty acids. Insect eggs are rich in lipids, including essential fatty acids, which meet the larva's requirements for some time. The effects of dietary shortage become obvious only when this endowment is exhausted, which in some species may not be until the formation of the adult stage, or even until the mature females are provisioning the next generation of eggs.

It is interesting that polyunsaturated fatty acids seem to be as 'essential' to insects as they are to vertebrates. The fact that they are required in quite small quantities is consistent with the idea that, as in vertebrates, they are used for the synthesis of some crucial structural or messenger molecules. Some species have the enzymes that convert arachidonic acid to lipid messenger molecules that may control the assembly and actions of roving immune cells, in much the same way as they do in vertebrates. After years of neglect, invertebrate immunology is now being studied in detail, revealing many complex mechanisms that turn out to be strikingly, some would say extraordinarily, similar to those of rats and ourselves.

Insect desaturase enzymes and those of vertebrates insert additional double bonds at slightly different places, so insect long-chain polyunsaturates in phospholipids are distinct from those of vertebrates. Endogenous supplies of such fatty acids cannot be supplemented by feeding insects on fish oils, as can be done for mammals, which is not surprising, since very few insects live in or on the sea, and almost none naturally eats marine fish. In view of the many fundamental contrasts in structure and life history between vertebrates and insects, it is perhaps more remarkable that the biochemical differences are so slight.

Some other insects, such as *Dermestes*, familiar as the common carpet beetle, have no demonstrable dietary need for particular fatty acids: emancipation from such dietary requirements certainly contributes to its ability to thrive and breed on apparently unnutritious foods, such as carpets, fur coats and decaying skeletons. But of course, the lipids and proteins in dirt derived from people's sweat or spilt food make a welcome contribution to the nutrition of *Dermestes* and that of the larvae of the clothes' moth, so they favour dirty materials.

Certain large caterpillars, beetle larvae, and adult insects including locusts and giant water bugs were, and in some parts of Africa and Asia still are, much valued as food for people. Their triacylglycerols are rich in polyunsaturated fatty acids obtained directly from plants (or in case of *Lethoceros* and similar water bugs, from freshwater fish), insects are nutritious, more so in many ways than mammals or birds. A diet of locusts and wild honey, that sustained St John the Baptist[4] while he was living in the wilderness, would have been much less spartan than it now seems to us, especially if the locusts were fattening up before migration, and there was sufficient honey to prevent undue dependence upon oxidation of proteins and fatty acids.

Lipids and pain relief

Infusions of the bark of the common willow (*Salix alba*) have been used in western Europe to relieve acute fever and pain since the middle of the eighteenth century, but the physicians and apothecaries who prescribed and prepared the remedies had no idea how or why they worked. Starting in the 1820s, German and French chemists sought to identify the pharmacologically active ingredient of willow bark. By the middle of the century, several different forms of salicylic acid (named after *Salix*) had been extracted from fever-relieving herbal remedies including oil of wintergreen and meadowsweet, as well as from willows. But the extraction processes were expensive in both human labour and natural resources: several kilograms of bark were needed to extract enough for a few days' treatment of a single patient.

At the time, infectious fevers, including measles, mumps, scarlet fever, chickenpox and typhoid were very common, especially among poorer people in densely populated cities and, for children and the elderly, were often fatal. Millions of adults suffered chronic pain arising from long hours of repetitive or strenuous manual work, inept or non-existent dentistry, untreated internal disorders or imperfectly healed wounds. Many could find relief only by drinking large quantities of alcoholic liquor, with all its attendant complications.

Industrial chemists realised that the market for a cheap, reliable form of salicylic acid was enormous, and set about finding ways of synthesising it in a form that could be easily and safely administered. In 1898, after nearly 40 years of research and product testing, the German company Bayer perfected aspirin, a chemically modified form of salicylic acid that was even more potent than the naturally occurring drug, and was less inclined to cause indigestion (though stomach ulcers and other digestive problems are still common complications of prolonged use). This simple pill, so stable that it keeps for years at room temperature, quickly became, and still remains, a best-seller for the relief of fever, headache, toothache, arthritis, rheumatism, sprains and similar ailments. Synthetic salicylates are among the cheapest, safest and most effective drugs known: unlike pain-killers such as morphine (and the related drugs, codeine and heroin), they are not addictive and do not harm the nervous system.

For a long time, the mechanism of action of aspirin and related drugs sold under names like tylenol, paracetamol and ibuprofen, remained a mystery, but in the early 1970s, research into lipid-based messenger molecules suggested an answer. The British biochemist, Sir John Vane and his colleagues demonstrated that aspirin binds to, and thereby inactivates, an enzyme essential to the synthesis of certain prostaglandins. In doing so, it suppresses pain and fever, and, at higher doses, the mechanisms of tissue inflammation and swelling. Paracetamol and ibuprofen bind to other enzymes that catalyse different stages of prostaglandin synthesis.

Vane's concept also explained many of the side-effects of prolonged use of aspirin: the fact that it seems to protect regular users from strokes and heart attacks, and weakens the ability of blood to clot. These phenomena involve the aggregation of certain blood cells into clumps or clots, a process that is mediated by prostaglandins and other lipid-based messenger molecules, whose synthesis or action is attenuated by aspirin. More recent research has demonstrated other effects of aspirin and aspirin-like drugs, and Vane's hypothesis has been modified and extended, but the principle that they can act on lipid-based messenger molecules still stands.

Thus aspirin and related drugs do not 'heal' the wound, remove the infection (e.g. a disease-causing virus or bacterium) or correct the imbalance (e.g. a hangover after drinking alcohol). They just relieve the unpleasant symptoms associated with the physiological response to injury or infection. Sometimes curtailing these responses promotes healing by, for example, reducing swelling that interferes with blood flow, or permitting free movement of a joint otherwise too painful to move, but the contribution to 'curing' the original causes of pain is indirect.

The biochemistry of menstruation has much in common with that of tissue inflammation: both are controlled by a balance between pro-inflammatory and anti-inflammatory prostaglandins. Most of the former are derived from *n*–6 fatty acids, while the latter are based mainly on *n*–3 fatty acids. Biologists have become interested in whether the production of one or other type of prostaglandins could be enhanced, or curtailed, by controlling the dietary supply of the fatty acids from which they are made. Three major enzymatic steps are required for the synthesis of arachidonic acid (C20:4*n*–6) from linoleic acid (C18:2*n*–6): desaturation to introduce a third double bond, elongation of the chain by two carbon atoms and further desaturation (requiring a different desaturase enzyme) to introduce the fourth double bond. Taking evening primrose oil for a week or two before a period helps to prevent first-day menstrual pains because it contains an exceptionally high proportion, 5–10% of the total fatty acids, of the first intermediate in this pathway. Bypassing the first desaturation step tips the balance in favour of prostaglandins of the *n*–6 family and allows menstruation to proceed freely.

The balm of hurt minds

The search for the mechanism(s) that induce natural sleep in mammals such as ourselves has a long and venerable history. Numerous different proteins found in brain cells and the surrounding fluids have been suggested as the 'sleep signal'. But the current favourite is a fatty acid, oleamide,[5] which is simply oleic acid (C18:1) in which the $O^- H^+$ component at the COOH end of the molecule is replaced by an $—NH_2$ (amide) group. Oleamide can be isolated from samples of the fluid that bathes the spinal cord and brain of cats, and when it is injected into them, they go to sleep, for many hours with a large dose.

The membranes of certain nerves of the central nervous system contain an enzyme, oleamide hydrolase, which hydrolyses oleamide to ammonia (NH_3) and oleic acid, but it is not clear which of these three substances actually induces sleep. As in the case of leptin (see Chapter 2), the molecular biologists have produced the complete amino acid sequence for the hydrolase enzyme before physiologists have had time to answer this question. Like so many enzymes that act on fatty acids, it turns out not to be specific for oleamide, but works just as well on similar amides formed from other common fatty acids.

When the key to peaceful, refreshing sleep is discovered, drug companies will synthesise it by the ton and sell it to insomniacs. Lipid biologists are much attracted to the idea that humble oleic acid may, with minor modifications, turn out to be, in two quite different ways, 'chief nourisher in life's feast'.[6]

Adipose tissue and the immune system

The presence in vertebrate blood of 'white blood cells', so named because they lack the bright red haemoglobin pigment of the much more numerous red blood cells, has been known since the middle of the nineteenth century. Further research, based almost entirely upon observations with a simple light microscope, showed that white blood cells occur in many distinct types, and assemble in large numbers at septic wounds, boils and pustules. They have long been recognised as part of the immune system that defends the body against disease, though it is only in the past few decades that biologists have become aware of their omnipresence and their immense range of capacities. As pointed out in Chapter 1, their outer membranes are much extended by ruffling, and consist of phospholipids whose fatty acid composition is crucial to the cells' function.

Although first identified in the blood, 'white blood cells' actually proved to be far more numerous in their main home, the lymphatic system, and a major class of such cells is now called 'lymphocytes' in recognition of this fact. The lymphatic system consists of fine ducts, though which a watery fluid called lymph moves slowly (much more slowly than blood in all but the smallest blood vessels), propelled mainly by gravity and by squeezing from adjacent structures such as muscles. In mammals, the most widely known role of lymph in lipid metabolism is taking up chylomicrons formed from newly absorbed lipids from the small intestine (where lipids are absorbed). The lymph from the intestine pours into the blood at junctions at the base of the neck, thereby bypassing the liver, which would otherwise take up much of the newly absorbed lipids. Nothing resembling this arrangement is found in birds, in spite of the fact that many have similar diets to mammals. Lipids from their gut enter the blood stream and go to the liver, from whence they continue to the peripheral tissues.

Lower vertebrates, and many invertebrates, have immune cells which look and behave much like those of mammals. In spite of the similarity at the cellular level, the location of centres of lymphocyte proliferation are fundamentally different in the major classes of vertebrates. These anatomical facts have been known for a long time, but scientists' attention has been focused so strongly on the molecular and cellular aspects of immunology that the functional implications of different arrangements of the lymphatics and associated tissues have hardly been discussed.

In mammals, but not in other kinds of vertebrates, the lymphatic system has nodes that act as 'garrisons' for the proliferation, maturation and dispatch of lymphocytes and other kinds of immune cells to a particular part of the body. The arrangement of nodes and vessels in the neck, chest, abdomen and limbs follows the same basic plan in related kinds of mammals, and in general, lymph nodes of larger species are both larger and more numerous in any one location, than those of smaller species. Thus nearly all adult mammals have just one popliteal lymph node (Figure 4, page 29) behind each knee, which weighs about 1 mg in mice, 5–10 times as much in rats, and around 1 g in ourselves.

Lymph nodes have an elaborate internal structure, and are provisioned by their own blood supply as well as inflowing and outflowing lymph ducts. When activated either by a general disease state in the whole body, or by a local infection in their 'patch', lymph nodes swell, sometimes to more than ten times their quiescent size, and blood flow to them increases, so they become firm, pink and warm. In ourselves, the lymph nodes in the groin, in the axilla under the armpit and in the popliteal groove between the major tendons at the back of the thigh are close enough to the skin to be easily palpated, especially when they are swollen. The increased size accommodates additional immune cells that assemble from elsewhere in the body, or form inside the node by division of existing lymphocytes, and usually regresses after a few days if the invading organisms are successfully exterminated.

That much is made clear in all textbooks of immunology. Less widely discussed are the facts that most lymph nodes are surrounded by adipose tissue, and most mammalian adipose depots enclose at least one lymph node, while some, such as the mesentery, contain dozens. The adipose tissue associated with many minor nodes represents such a small fraction of the total that its presence usually goes unrecorded in anatomical drawings and texts. The detailed structure of lymph nodes differs slightly between sites and species, but in all cases, the lymph ducts run through the adipose tissue and divide into numerous fine branches as they approach each node, thereby generating points of entry over much of its surface, and coming into close contact with a large proportion of the adipocytes that immediately surround it.

The need to swell when fighting infection was, until recently presented as the main, if not the sole, reason for the anatomical association between adipose tissue and lymph nodes. However, as pointed out in Chapter 6, adipocytes embedded in their network of collagen are not very compressible.

It is difficult to see why adipose tissue should be preferred as a container for expandable nodes over a mainly extracellular, genuinely extensible material such as connective tissue. The lymphoid tissue of birds and lower vertebrates also expands when activated, but it is not closely associated with adipose tissue. In many species it could not be, because adipose tissue is confined to a few centrally located fat bodies, instead of, as in mammals, being partitioned into numerous small depots, where it can be associated with lymph nodes.

Immune responses

There are many different kinds of lymphocytes, some identifiable by their microscopic appearance and others distinguishable only by their response to certain kinds of messenger molecules or particular antigens. Different types play leading roles in dealing with different kinds of infections, and in many cases, the co-ordinated activities of several classes of cells are needed to combat the invading organisms. The mechanisms by which immune cells communicate with each other and with cells of other tissues have been intensively studied, with a view to manipulating their activities with drugs and other interventions. A great many molecules prove to have 'messenger' properties, in that they are released from one kind of cell and modulate the activities of others of the same cohort, and/or one or more different kinds of cells. At least a dozen different protein messenger molecules have been identified, and are collectively known as cytokines. They operate alongside, and possibly interact with, the lipid-derived messenger molecules, prostaglandins, leukotrienes and thromboxanes (see Chapter 4).

Like all proteins, cytokines are made from amino acids that are readily available in the blood, or from within the producing cells themselves. 'Old' proteins are broken down all the time and their constituent amino acids recycled, so protein synthesis is curtailed for lack of precursors only in severe starvation. The same cannot be said of the precursors for the synthesis of lipid messenger molecules: they are derived from essential fatty acids that are always a minority in adipose tissue triacylglycerols.

As well as requiring messenger molecules, major immune responses involve rapid division of lymphocytes. The newly formed cells need phospholipids of appropriate fatty acid composition to build their membranes, creating further demand for polyunsaturated fatty acids. There is not much hope that the diet could supply the necessary fatty acids. Animals as well as people lose their appetite and become sleepy when ill with a systemic or major local infection. They do not go out foraging and often refuse to eat even when food is put in front of them.

Partly as a consequence of stopping eating, and partly as a direct effect of cytokines produced from activated immune cells, the concentration of fatty acids in the blood usually rises during infectious disease (and in many other kinds of illness, including severe wounds and burns). The fatty acids are presumed to arise from increased lipolysis in general storage adipose tissue, as happens in fasting or after prolonged exercise. Once in the blood, a nutrient is 'up for grabs' by any tissue that can get hold of it, for use in any role for which it is suitable. Most of the blood-borne fatty acids, whether saturated or unsaturated, are taken up by tissues such as muscle and the liver and oxidised to produce energy. Consumption of lipids by such tissues is even higher if the immune response stimulates fever, i.e. the body's rate of energy utilisation is increased until it produces so much heat that its temperature rises. Fever accelerates the rate at which rare and, at least for the time being, irreplaceable, fatty acids are released into the general bloodstream from where they may be lost to oxidation.

Most kinds of immune cells, including lymphocytes, are very robust, and there are well-established ways of persuading them to proliferate, synthesise certain kinds of proteins, stick to each other and to other cell types and perform various other interesting antics in tissue culture. It is very much easier to obtain samples of blood from veins than it is to sample the contents of lymph vessels, so lymphocytes are usually isolated from blood or, if the animal is killed, from the largest lymph nodes or the spleen. Information about how and when lymphocytes synthesise messenger molecules comes almost entirely from the study of such cells in culture. Much less is known about their activities in small lymph nodes, and almost nothing about whether other tissues contribute to these processes. If provided with more than enough suitable precursors, lymphocytes certainly can make a range of cytokines and lipid messenger molecules on their own, but that does not mean that they always do so in the intact animal.

Dietary lipids and immune function

Rheumatoid arthritis, multiple sclerosis and several other long-term, non-infectious diseases are now believed to be caused by the immune system attacking the very tissues it is meant to protect from foreign invasion. Sufferers from these painful and debilitating 'auto-immune' diseases have long claimed to obtain some relief, sometimes even remission, by replacing butter, red meat and other sources of animal fats in their diet with fish or plant oils. Although cod-liver oil was used in the Scottish highlands and elsewhere in the treatment of rheumatism centuries before its role in preventing rickets was described, in the late twentieth century, such treatments were regarded as cranky, harmless enough but without scientifically proven therapeutic value. More thorough studies in both laboratory animals and people have shown clearly that lipids can alter immune responses.

In one kind of experiment,[7] mixtures of lymphocytes extracted from rat lymph nodes were cultured for several days with different kinds of pure fatty acids and their responses to various stimulants were compared. Another kind of investigation[8] involves examining the properties of immune cells taken from rats or people who have spent several weeks on diets containing different natural oils. Whether fatty acids were applied directly to isolated lymphocytes, or allowed to make their own way to the immune system via the gut, blood, etc., the conclusions are similar: polyunsaturated fatty acids of the *n*–3 family (most abundant in fish oils) suppress immune responses most strongly, followed by those of the *n*–6 family (common in plant lipids), then monounsaturates, with saturated lipids having the smallest effect. The exact mechanisms for these actions remain to be determined. Possibilities include changes in membrane fluidity that alter the cells' mobility or adhesiveness or their capacity to support receptors for messenger molecules, synthesis or release of lipid-based messenger molecules or direct action of fatty acids on the genes themselves.

While curtailing immune function may offer relief from auto-immune diseases, it might also impair resistance to invading micro-organisms. For example, mice were fed for a month on chow of different lipid composition and their capacity to resist experimental infection with *Listeria* bacteria was compared. Eight out of ten of those fed on fish oil died, while all of those that had been eating lard survived,[9] because their immune systems rounded up and killed the bacteria before they had time to proliferate out of control. Some (but not all) similar experiments reveal less efficient resistance to other kinds of disease-causing bacteria and slower healing of wounds in animals fed on fish oils. Infectious micro-organisms flourish when their hosts have frequent, intimate contact with each other. There may have been good physiological reasons why Neolithic farmers who started living at high density in villages favoured fatty meat and milk, and the mice that built up huge populations feasting in their granaries came to traps baited with cheese or suet.

This research, based largely upon the study of isolated cells, analysis of samples of blood, and feeding-and-infection experiments has shown that diet can influence immune function and the course of diseases, but it reveals nothing of whether that specialist in lipid management, adipose tissue, is manipulating the relationship between food and the activities of immune cells behind the scenes.

Adipose tissue around lymph nodes

The association between minor depots of adipose tissue and lymph nodes seemed too common to be coincidental, so Christine Mattacks and I[10] decided to ask the lymphoid cells themselves to point out which bits of adipose tissue they interacted with most strongly. The experiment was simple: take a rough and ready mixture of lymphoid cells from some large, accessible lymph nodes. Incubate identical samples of it for several days with little cubes of adipose tissue taken from near to and away from nodes of various depots of the same animal, and measure how many new lymphocytes have formed, and how much lipolysis has taken place. The experiment can be made more exciting by adding a mitogen, which activates immune cells and prompts them to proliferate as though they were under attack from bacteria.

The presence of adipose tissue always curtailed both spontaneous and mitogen-stimulated proliferation of lymphocytes, but the extent of inhibition depended greatly upon the source of the sample. In all the eight depots studied, but especially the mesentery, omentum, forearm, popliteal and cervical depots that contain one or more lymph nodes, the samples taken from near a lymph node suppressed the formation of new lymphocytes more strongly than those taken from elsewhere in the same depot. The least effective sample was that from the perirenal depot.

The same experiments revealed that lymphoid cells consistently induced more lipolysis in adipose tissue from near to nodes in samples from elsewhere in the same depot, especially in the cases of the small intermuscular popliteal and cervical depots, and the omentum and mesentery. Lipolysis rises by more than threefold in samples from some of these depots, a greater increase than is observed when adipocytes isolated from similar sources are stimulated with large doses of noradrenalin. The adipose tissue around the nodes seems to be more responsive to lymphoid cells than neighbouring tissue from just a centimetre away.

The gross anatomy of these nodes and their surrounding adipose tissue suggests an explanation for the strong local interactions. The mesenteric nodes, being the first to come into contact with material absorbed through the gut, are in the 'front line' of defence against pathogens invading through the intestine. The omentum also contains a great deal of lymphoid tissue and is believed to remove debris from the abdominal cavity. The popliteal lymph node is the most distal of the lower limb nodes, and protects the whole of the hindlimb below the knee. The cubital lymph node (in the 'fore-arm' adipose depot) is also located at 'the end of the line', and performs similar functions for the lower part of the forelimb.

Hands and feet, paws and hooves are continually exposed to abrasion and assaults from parasites and pathogens, so the nodes that serve them are nearer 'the front line' in dealing with local, minor injuries, infections and inflammations than the more centrally located inguinal and axillary ('behind arm') nodes. The popliteal depots are small, representing less than 5% of the total adipose mass in guinea-pigs and most other mammals, but they contain relatively large nodes. Indeed, enclosing these important lymph nodes may be their main role. As pointed out in Chapter 2, they do not enlarge as much as the superficial and large intra-abdominal depots as animals fatten, and seem to be conserved in starvation.

In guinea-pigs (and many other mammals), the perirenal adipose tissue (part of the dorsal wall of abdomen depot) contains no lymph nodes and when samples from it are incubated with lymphoid cells, its lipolysis rises by less than 5%, a negligible increase compared with that of the node-containing depots. The perirenal depot responds satisfactorily to all other known local and blood-borne stimulants of lipolysis, and indeed is often taken as a representative adipose depot. Evidently, it is atypical as far as interactions with the immune system are concerned. In guinea-pigs and many other mammals, the perirenal is among the largest of all depots and undergoes extensive changes in size as total fatness changes. Its lack of interaction with lymphoid cells may simply be a necessary corollary of its role as an energy store for the body as a whole. The other, smaller depots expand and shrink less readily because part of their adipose tissue is conserved for special, local functions.

To find out more about what lymph node lymphoid cells might be getting by stimulating lipolysis in the adipose tissue around them, we compared the fatty acid composition of triacylglycerols in adipose tissue from different parts of depots that contain lymph nodes.[11] In all those examined, but especially in the popliteal, cervical, omental and mesenteric depots, there were fewer saturated fatty acids, and more polyunsaturates in the triacylglycerols found in the adipose tissue 1–2 mm around the nodes than elsewhere in the depot. The adipose tissue from around lymph nodes that in tissue culture interacts most strongly with lymphoid cells also contains a greater proportion of the very fatty acids that these cells need for their proliferation and correct functioning, and cannot make for themselves. Selective release of polyunsaturated fatty acids when lipolysis is stimulated (see Chapter 4), combined with their higher concentration near the nodes, would maximise supplies to the activated lymphoid cells.

Such site-specific differences in the composition of the storage lipids came as a surprise: previous investigators had assumed that continuous lipolysis and re-esterification of triacylglycerols would eventually homogenise the entire store. The only other examples of site-specific differences in fatty acid composition of triacylglycerols hitherto described were the extremities and superficial adipose tissue of some arctic mammals (see Chapter 6), which, although similar in principle, differ in some important details. The adaptations of adipose tissue triacylglycerols to cooler conditions mainly involved substituting saturated fatty acids with monounsaturates. Since it is found in internal depots, the pattern of site-specific properties around lymph nodes probably has nothing to do with adaptation to temperature, and the saturates decreased as the relative abundance of the polyunsaturates increased, with the proportions of monounsaturates remaining constant. Almost nothing is known about the mechanisms that produce selective accumulation or exclusion of particular fatty acids.

Similar site-specific differences in the composition of triacylglycerol fatty acids were also found in pigs, hyraxes and other non-ruminants, but a thorough search for them in ruminants such as reindeer, sheep and antelopes produced only a vague indication of a pattern. For the reasons explained in Chapters 3 and 4, ruminants are chronically short of polyunsaturated fatty acids. Highly specific enzymes direct the few that are available into phospholipids in cell membranes, leaving almost none for adipocytes to take up and incorporate into triacylglycerols.

The bone marrow is another part of the body where adipocytes and cells of the immune system are found close together (see Chapter 2). The association between them has rarely been studied, because very little marrow can be extracted from the bones of laboratory rats. Bone marrow adipocytes are usually smaller than those elsewhere in the body and, presumably because they are never subjected to mechanical deformation while enclosed inside the bone, they have much less collagen, in fact so little that the 'tissue' falls apart when removed. Their triacylglycerols usually contain a greater proportion of polyunsaturated fatty acids than those in adipose tissue elsewhere in the body, even in marrow cavities (such as those of the pelvis and upper segments of the limb bones) that are continuously at core body temperature.

In rabbits, bone marrow adipocytes retain their capacity to take up fatty acids, and do not release any of their lipid content even after 3 weeks of starvation, but in deer, antelopes, muskoxen and other wild ruminants, the bone marrow becomes depleted of its lipid when the rest of the adipose tissue has shrunk to below about 5% of the body mass. This effect is so reliable that the lipid content of bone marrow can be used to estimate total fatness in these species. But the similar methods do not work well in wolves and other large carnivores[12] because, as in the rabbits, the marrow lipid is not depleted even when adipose tissue elsewhere in the body almost empty. Perhaps bone marrow adipocytes interact specifically with adjacent immune cells in non-ruminants, but have lost this capacity in ruminants: their lipid thus becomes available for general use by all tissues, and is released into the blood when stores elsewhere run low. Non-ruminants save theirs to help the immune system.

The mammalian immune system seems to have organised its own private, local supplies of the polyunsaturated fatty acids they need, thereby removing competition with other tissues, and, what is perhaps even more important, avoiding transporting lipids through the general circulation. As explained in the next section, lipids cause more trouble while being transported around in the blood than when sequestered in adipose tissue.

Fats that kill

High-fat diets, we are told, can do far more than just make you fat: they can be lethal, often killing embarrassingly suddenly. When infectious diseases were major causes of premature death, only a few people lived long enough to die in this way, so the association between obesity and sudden death was less noticeable, but it was still strong enough in ancient Greece for Hippocrates to observe that obese people were prone to sudden death. Long after the correlation became inescapable, the mechanism remained mysterious: active, mentally sound people who ate well, were without fever, pus, diarrhoea, pallid colouring, swellings, body wasting, fits or any other symptoms of serious illness, and whose only complaint was transient, often quite minor pains, suddenly dropped down dead. No wonder many people sought explanations in witchcraft and curses. Intensive research on many fronts, from epidemiology to molecular biology, now offer an explanation for how seemingly innocuous habits can lead to sudden death or disablement.

Fat in arteries

Most kinds of mammalian cells can make all the cholesterol they need for themselves, but the human diet usually includes more than enough to supply our cells. Only the liver can degrade and eliminate cholesterol, so any excess has to be transported there from other tissues. The small, high density lipoproteins take it up from the surfaces of cells where it is esterified to a fatty acid to form cholesteryl ester, and transferred, with the help of specific proteins, to low density, or very low density, lipoproteins that return it to the liver. All these processes take place in the blood, which can, of course, be easily sampled. Cholesterol and its ester are easily shaken out of their lipoprotein packaging and can be measured as 'blood cholesterol' level.

Most low density lipoproteins remain in the blood for a few hours until they are taken up by liver cells for dismantling and reassembly. A few seep through the thin, delicate lining of major blood vessels where they become trapped under the endothelial (lining) cells. Some escape back into the blood but others remain, attracting macrophages, roving immune cells that engulf any debris they find, including bacteria, dead cells and dust particles. Normally, macrophages take up low density lipoproteins via specific receptors on their surface, and degrade them in their cytoplasm. The cholesterol released by the breakdown of the lipoproteins inhibits the formation of more receptors, so the macrophages do not take up more than a manageable load. Problems begin when the lipoproteins stuck in the lining of blood vessels are around for long enough to react with other components of the blood, especially oxygen. Vitamin C and other anti-oxidants in fresh fruit and vegetables and certain ingredients of red wine (not the alcohol itself) may protect people from heart disease by preventing the oxidation of lipoproteins at this stage.

Such slightly modified lipoproteins seem to be able to enter macrophages in unlimited quantity, causing the cells to swell to several times their normal size. Hundreds of pale, greasy particles trapped inside cells resemble foam when viewed under the microscope, so (long before the mechanism of disease was elucidated) macrophages replete with lipoproteins were named foam cells. Some die, releasing their lipid contents into the intercellular space. Together with surviving foam cells, the decomposing lipoproteins accumulate to form fatty streaks that are conspicuous enough to be visible to the naked eye when arteries are opened for post-mortem inspection. Fatty streaks seem to do little harm, and are common in the arteries of people whose cardiovascular system functions normally, including young children.

The disease process can remain at this stage for years, perhaps decades, until further changes lead to atherosclerosis, which can cause serious illness or death within weeks.

Something, perhaps a local shortage of oxygen, kills some of the endothelial cells over the fatty streaks. Tough extracellular proteins take their place and the arteries become 'hardened' with fibrous plaques overlying bloated foam cells and a wad of cholesterol. This dangerous mixture narrows the 'bore' of the vessel, so higher pressures are needed to get even a reduced flow of blood through it. The functioning of the delicate sense organs that monitor blood pressure and guide adjustments to it is also impaired. Such degeneration can occur in any blood vessel, but its consequences are most serious in arteries, which are narrower than veins and have thick, muscular walls through which the blood flows under pressure, bringing more oxygen to the tissues. Narrowed arteries bring less blood even if the heart pumps harder, and stiff walls are unable to accommodate changes in pressure. Tissues downstream from the deformity may weaken and atrophy from lack of oxygen and nourishment. They may be the focus of intermittent pain, called angina when it affects the heart, which becomes especially evident during exercise or emotional stress, when the tissues' demand for oxygen increases.

Since the 1960s, this stage of atherosclerosis has been treatable by inserting a healthy length of blood vessel (usually a vein taken from the patient's own leg) that 'bypasses' the narrowed section. For reasons that are not entirely clear, the coronary arteries that bring blood to the heart itself are particularly susceptible to atherosclerosis, and because of their crucial role in supplying the heart muscle, they are usually the subject of such bypass operations. The treatment just re-routes the blood flow, it does not remove the cause of arterial narrowing. Only diets that minimise the abundance of low density lipoproteins in the circulation and drugs that lower cholesterol can prevent recurrence.

In the final stages of the disease, the stiff, narrow arteries ulcerate, with a high risk of forming a large blood clot, or thrombosis, which acts like a plug, cutting off blood flow completely. The tissues downstream from the blockage soon die from lack of oxygen but how much dies, and what the consequences are, depends upon where the blood clot settles: if in the brain, the result is called a stroke; in the heart, a heart attack; in the legs, ischaemic gangrene. Post-mortem studies of people who have died from heart attacks, show that an average of 2.7 of their four major coronary arteries are narrowed to half the normal diameter (flow area reduced by 75%). Perhaps we should be surprised that they lived long enough for atherosclerosis to become so advanced.

This distressing form of heart disease is associated with obesity because obese people often choose a high fat diet, and more adipose tissue leads to more lipids being on the move between one depot and another. Both these factors mean higher concentrations of low density lipoproteins in the blood for a greater proportion of the time, thus favouring infiltration of lipoproteins into the artery linings. The blood plasma[13] of an obese person often contains so much lipid that it appears pale and cloudy, while that of a lean animal or person is clear and yellowish-brown in colour. In humans, the adipose tissue in the abdomen seems to release its lipid contents more readily than that of the peripheral depots, so weight for weight, intra-abdominal adipose tissue contributes more to the lipoprotein problem. What role, if any, the adipose tissue on and around the heart (see Chapter 4) has in coronary atherosclerosis remains to be explored.

High blood pressure also assists the process of infiltration and increases the risk of damage to endothelial cells. Some components of cigarette smoke hinder oxygen transport in the blood, thereby exacerbating oxygen shortages, and others may trigger abnormalities in lipoproteins or cells. A stressful lifestyle produces more frequent and probably greater stimulation of the sympathetic nervous system, leading to more lipolysis from adipocytes. Unless the fatty acids are quickly burnt off by exercise, they remain in the circulation until they are taken up by other adipocytes. A sudden emotional or physical shock may trigger an abrupt rise in lipolysis, that, especially in obese people, floods the blood with lipids, making clot formation more likely.

Heart attacks

By the 1950s, it was clear that atherosclerosis led to heart attacks, and that this cause of sudden death was becoming alarmingly more frequent, especially in middle-aged men. So biologists set about investigating it in earnest. One of the first contributions was a collaboration between medical pathologists[14] and vets at the London Zoo to study the course of the disease in animals. Fatty streaks, fibrous plaques and blood clots remain visible for many hours after an animal dies, so the stages of the disease could be easily identified in animals that died at the Zoo.

Examination of over 2000 such specimens revealed that, in the majority of mammals, all signs of arterial disease were conspicuous by their absence, even in elderly, long-term residents of the Zoo that had had very little exercise for many years. The incipient stages of atherosclerosis, fatty streaks, were observed in around a quarter of the birds and mammals, though in fewer of the reptiles, in spite of the greater longevity of such animals. Fibrous plaques were rarely found, though they were noticeably more common in elderly apes and monkeys and some hoofed animals than in the other species examined. They also occurred in certain kinds of birds, especially birds of prey such as vultures, falcons and eagles, and carnivorous cranes and storks, but also some herbivorous families including geese and parrots. The most surprising conclusion was that advanced atherosclerosis appeared to have been the immediate cause of death in very few of the animals: actual blockage of vessels by blood clots was very rare, even among the large primates that had a fair number of fibrous plaques.

This research was largely forgotten in the search for suitable experimental animals in which to study the causes and progression of heart disease. Attempts to induce atherosclerosis by manipulating the diet have revealed some interesting differences in the susceptibility of animals to the disease. Pigs readily develop atherosclerosis if allowed to live to porcine middle age (8 years +),[15] and then they die of heart attacks and strokes nearly as often as humans do. Adding 0.5% cholesterol and 5% butter fat to the diet of rabbits, which naturally eat little except grass and herbs, increases the cholesterol levels in their blood by 7- to 15-fold in 2 weeks, and by 30- to 50-fold in 8 weeks. Not surprisingly, rabbits on such a highly abnormal diet quickly develop all the symptoms of atherosclerosis, as do other mainly herbivorous animals including most kinds of monkeys and pigeons.

Dogs prove to be resistant to atherosclerosis: the condition is rare even in elderly, pampered pets, who are as overfed and underexercised as their owners. The disease appears sometimes as a complication of diabetes, but it cannot be induced in healthy dogs just by increasing the fat content of the diet. Special drugs and/or surgical intervention plus a high-cholesterol diet must be applied. Once produced, the atherosclerotic lesions are similar to those found in humans. The basic mechanisms that lead to obstruction of the arteries are there, but these 'professional' carnivores seem to have evolved ways of preventing the disease from getting started.

The ancestors of dogs and cats have been active predators on other vertebrates for at least 50 million years, but those of modern humans have been eating large quantities of meat and animal fats for less than a tenth of that time. Susceptibility to atherosclerosis seems to be another consequence of our being such novices to carnivory that we cannot deal efficiently with large quantities of animal fats. People have suffered from atherosclerosis for at least 5300 years, probably much longer. The condition was well advanced in some of the arteries of the body known as the Iceman, that was found frozen in ice high in the Alps on the border of Austria and Italy in 1991. He was about 45 years old when he died, but he had not lived a life of ease: his bones and joints showed evidence of frequent, strenuous exercise since childhood.

Although fatty streaks in arteries are equally common in men and women, young and old, and occur at about the same frequency all over the world, advanced atherosclerosis is much more common in certain populations. In Europe, Britain, Ireland and Denmark are far ahead, with around three times as many deaths from this cause as France, Greece or Spain, although the French, Greeks and Spaniards and are just as obese, and smoke and drink just as much. Within Britain, the rates are highest in Scotland and northern England. People in all these places eat fried foods, dairy products and fatty meat that contain both saturated fat and cholesterol. Analysis of diet questionnaires and blood samples from thousands of people throughout the western world showed that the concentration of cholesterol in the blood correlated much more closely with how much saturated fatty acids they ate (as triacylglycerols) than with intake of cholesterol *per se*. The main reason is that only a fraction (in humans, between a third and a fifth) of the 'new' cholesterol comes from the diet, the rest is made by the tissues themselves. So the important question is what determines how much cholesterol is retained in the blood.

Ancel Keys, a Minnesota physician famous for his studies of the physiology of starvation in the 1950s, is among a growing number of dieticians who advocate the 'Mediterranean diet'[16] with its emphasis on bread, pasta, fruit, fish, vegetables and plant oils, especially olive oil, and low meat content, as protection from atherosclerosis and obesity. Such views were slow to attract interest because no-one could envisage a possible mechanism for the association, but the epidemiological studies and experiments in which people who had already suffered a minor heart attack were put on controlled diets showed that triacylglycerols containing a high proportion of n–3 polyunsaturates reduced mortality from atherosclerosis. It is now generally agreed that lipoprotein composition and metabolism are the links between dietary lipid and heart disease.

Linoleic acid (C18:2n–6), and perhaps also oleic acid (C18:1), may substitute for saturated fatty acids in macrophage membranes, thereby decreasing the number of receptors for low density lipoproteins that they can support, and curtailing their development into foam cells. Linolenic acid (C18:3*n*–3) may reduce the average size of very low density lipoproteins and perhaps inhibit fat synthesis in the liver. Replacing butter with margarines that contain more polyunsaturated fatty acids thus helps to reduce the risk of atherosclerosis, though by itself this exchange contributes nothing towards slimming.

Fish oils may also be what saves polar bears from heart disease: although their diet is among the world's fattiest (see Chapters 2 and 5), and their blood lipids reach concentrations that would kill dogs or rabbits, the large quantities of *n*–3 fatty acids, especially eicosapentaenoic acid (C20:5*n*–3) obtained from the seals they eat seem to prevent atherosclerosis. Polar bears apparently don't care for reindeer flesh (which contains mostly saturated lipids): although the two species come into contact quite frequently, bears scavenge dead reindeer only when very hungry, and hardly ever attack them.

Saturated fatty acids, especially myristic (C14:0) and palmitic (C16:0) acids, exacerbate atherosclerotic processes, helping to accumulate cholesterol even when the diet contains only a little of it. The mechanisms are complicated and not fully understood. They probably depend upon the fact that lipoprotein lipase and the enzymes that esterify cholesterol act slightly more efficiently on certain fatty acids than on others. Such differences are so small that they are difficult to measure in laboratory experiments that last only a few hours, but over months and years, they can have important consequences for the accumulation of cholesterol.

Other kinds of food can affect cholesterol metabolism in humans. The complex carbohydrates in oats bind cholesterol in the gut, preventing it from being absorbed. The cholesterol in the bile salts is thus excreted with the faeces, instead of being recycled from the gut contents back into the body, and the liver has to use more to make bile acids to digest the next meal. As Dr Samuel Johnson noted in his dictionary, oats are fed to horses in most of the civilised world, but in Scotland, they sustain the population. Thus until very recently, the Scottish diet reduced the risk of atherosclerosis, but since turning to fried food and fatty meat, mortality from heart attacks and strokes in Scotland is among the highest in the world (although their average blood cholesterol is not enormous). Certain other natural and synthetic carbohydrates bind to cholesterol and other lipids in the same way as oats do. Large doses taken just before a meal help to curtail cholesterol absorption. However, these drugs do not 'cure' the tendency towards high blood cholesterol, but simply counteract it, so to be effective, they have to be taken continually.

The putative role of cholesterol in heart attacks has made it famous, and drug companies rich, as its expulsion from body and diet has become a health fashion. But such enthusiasm should be kept in perspective: exceptionally low blood cholesterol affords no protection from heart disease relative to people with average values, and, as pointed out in Chapter 7, this condition has its own, quite different, problems. An alternative view of the relationship between diet and heart disease was championed by the physician, biochemist and explorer, Hugh Sinclair, who lived with Eskimos on the islands off the north coast of Canada for several years in the early 1940s.

Sinclair noted[17] that Eskimos rarely suffered from heart disease or strokes in spite of a very high-fat diet that included reindeer meat. The Masai people of Kenya eat large quantities of ruminant milk and meat, and Jamaicans eat saturated fats in coconut oil, but few of them die from heart attacks. He suggested that, as in the zoo animals, what mattered was not the changes in the lining of the arteries, but the final event in arterial disease, the formation of blood clots. This process involves short-lived, lipid-based messenger molecules interacting with blood cell membranes, both of which are affected by availability of polyunsaturated fatty acids on a time-scale of days rather than years. He argued that the large proportion of polyunsaturates in the fish and seals of Eskimos' diet made their blood clot less readily, thus preventing thrombosis. Certainly, Eskimos are notorious for their tendency to bleed profusely from quite minor wounds because their blood clots unusually slowly.

Heart attacks are thus seen as arising from a deficiency of polyunsaturated fatty acids rather than from an excess of saturates or cholesterol. Describing arachidonic acid as a 'vitamin', Sinclair attributed various diseases, including atherosclerosis and heart attacks, to the frequent consumption of artificially hydrogenated marine and vegetable fats (see Chapter 3), without simultaneously eating a suitable source of polyunsaturated fatty acids. His ideas were not well received when first put forward, but 30 years on, they may prove helpful if reducing blood cholesterol turns out to be neither feasible nor useful in preventing heart and vascular disease.

Obesity

Many people in western countries, by some estimates as many as a third of the population, are obese. In some fat people, half the body mass may be adipose tissue. Its expansion is due to enlargement of adipocytes and, at least in many cases, to the presence of more adipocytes than would be expected from the comparisons with wild animals (see Figure 9 page 49). It is difficult to estimate the number of adipocytes accurately, but people who have more than ten times as many as expected do not seem to be rare. As pointed out in Chapter 2, wild animals that naturally become obese do not have such enormously increased numbers of adipocytes. Although very numerous, human adipocytes are relatively small except in very fat people, barely half the size of those in corresponding depots of sheep, Svalbard reindeer or brown bears, which are about the same body mass. It seems that even in lean people, something has promoted proliferation of pre-adipocytes and/or their maturation into adipocytes. Their owners have eaten more energy-rich foods than they need for long enough for the excess lipid to fill them up.

Another reason for regarding human obesity as abnormal, and in the evolutionary sense of the term, non-adaptive, is that the wrong depots enlarge selectively. In naturally obese mammals and birds, additional storage lipid is accommodated in the superficial depots which expand disproportionately, but in people, especially middle-aged men, the opposite happens: the intra-abdominal depots expand, while the 'safer' superficial depots remain constant or even shrink.

What's wrong with being fat?

Obese people complain of being ugly, unhappy and unable to fit conveniently into chairs, turnstiles and other equipment designed for average people, but in general, they do not feel ill. Perhaps that is why fat people are often reluctant to put a lot of effort into losing weight, no matter how much physicians regale them with the horrors that await them from obesity-related illnesses. Some sources of such ill-health are fairly obvious: bulky adipose tissue can compress veins and lymph vessels, leading to poor blood perfusion which causes local swelling and allows skin infections such as boils and gangrene to become established more efficiently.

Being very heavy and not fitting into standard furniture make very obese people more prone to accidents, but in old age, moderate fatness actually reduces the risk of serious injury, mainly because the bones of thin old women (and, at an older age, men) very often lose mineral and become fragile. Substantial adipose tissue helps by padding delicate joints, especially the hip, and the extra weight promotes mineral retention and therefore stronger bones. Lipid-soluble hormones especially oestrogen, produced by or stored in the adipose tissue are also thought to be the main reason why some forms of cancer, especially breast cancer, more often develop in fat people, and obese young women tend to be infertile.

The commonest metabolic defect associated with obesity, known as impaired glucose tolerance or insulin resistance, can only be demonstrated by biochemical analysis of several blood samples taken at intervals over several hours. To function efficiently, the chemical composition of blood must be controlled within narrow limits. Glucose and fatty acids are no exception to this generalisation, but there is a problem: how to cope with the abrupt influx that follows a sugary or fatty meal. As explained in Chapter 4, insulin secreted from the pancreas is the main hormone that co-ordinates 'glucose disposal'. Its concentration in the blood rises as soon as food is eaten – sometimes just the sight or smell of appetising food is enough to prompt secretion – and stimulates the muscles and liver to take up glucose from the blood. At the same time, insulin suppresses lipolysis in adipose tissue, so fewer non-esterified fatty acids are released into the circulation. In normal people, even a large dose of glucose is cleared in about one and a half to two hours, and the concentration in the blood returns to its baseline value.

Almost all obese people have higher concentrations of both fatty acids and glucose in their blood, and because their muscles and liver 'resist' the action of normal amounts of insulin, the concentration of fuels in the blood takes much longer to return to fasting levels after a rich meal. The pancreas usually responds to the persistence of glucose by producing more insulin, which raises the level for longer, and produces a higher peak concentration. This condition is very common in middle-aged and older people, and usually causes nothing more alarming than feeling a bit light-headed after eating a lot of sugary foods. But in some people, the insulin-secreting cells of the pancreas cannot sustain the extra workload indefinitely, and, usually after many years of obesity, they gradually fail to produce enough insulin to stimulate the 'resistant' tissues. The glucose concentration in the blood remains higher for longer than it should, producing a disease known as 'non-insulin dependent diabetes',[18] or, because it is almost confined to the middle-aged and elderly, 'maturity onset diabetes'.

The low concentration of insulin in such people's blood makes their bodies 'think' they are starving, so lipolysis in adipose tissue continues to release non-esterified fatty acids. Insulin also activates the enzyme lipoprotein lipase, which enables the adipose tissue to take up circulating lipids from the blood. When insulin is low, the lipoprotein lipase is less active, and lipolysis is increased, so lipids taken up from the gut remain longer in the blood, accompanied by those released from adipose tissue in greater quantities. In summary, adipose tissue no longer does its essential job of taking up, holding, and controlling the release of potentially harmful lipids as efficiently as it should.

Although first identified as a mediator of adipocyte depletion and death, tumour necrosis factor has recently been found to have a role in obesity. Samples of adipocytes from obese people produce up to three times more of it than those from lean people, and it interferes with their receptors for insulin, thereby inducing insulin resistance. It might, in fact, be the adipocytes' cry of 'Enough! We can't cope with any more lipids!' But if such protests go unheeded, and their owners continue to eat fatty foods, the additional lipids just hang around in the liver and the blood, getting up to mischief.

Non-insulin dependent diabetes is not immediately life-threatening – many people remain active and moderately healthy for years with the condition – but the disturbances that result in more lipoproteins and non-esterified fatty acids in the blood greatly increase the risk of metabolic disorders, including atherosclerosis. Obese people's extra bulk has to be perfused with blood, which puts extra work on the heart. Much of the increased mortality associated with obesity is due to heart diseases, but several other insidious changes impair the quality of life. The chronically elevated concentration of glucose in the blood gradually damages small blood vessels, leading to defects in the kidneys, peripheral nerves and eyes.

This form of diabetes is fundamentally different from insulin-dependent diabetes, which arises from irreparable damage to the insulin-secreting cells, and is neither caused by, nor usually leads to, obesity (the two diseases are so different in origin and course that they should really have different names). Maturity onset diabetes is rare is children and young people, but becomes much more common as people get fatter in middle age. In western countries, this form of diabetes can be recognised in as many as 10% of people over 70, though in many it causes only slight inconvenience. Becoming too fat, eating a rich diet and taking too little exercise undoubtedly contribute, but the tendency to develop the condition is also quite strongly inherited in families. The only permanent cure is to slim to normal body mass, but most such diabetics are much improved on a diet that avoids triggering sharp rises in the amount of glucose absorbed through the gut. These days, there are also drugs that promote insulin secretion from the tired pancreas, and/or achieve modest improvements in insulin sensitivity in some people.

It has been known for many years that maturity onset diabetes is associated with the disproportionate enlargement of the intra-abdominal adipose depots, particularly the omentum and mesentery, and the superficial depots on the upper half of the body, including the ventral abdominal paunch. Men and older women are most susceptible: as Figure 17 (page 254) shows, the abdomen is not greatly enlarged in girls and young women, even when the adipose tissue on the buttocks, thighs or breasts is massive. But for reasons that remain obscure, men's buttock and thigh depots are reluctant to expand until the total fat content of the body is high. Their mesenteric and omental adipocytes enlarge at quite modest levels of fatness, and with them, the increased risk of the metabolic complications of obesity. Slim young men have only about 3 kg of adipose tissue inside the abdomen, but it can grow to as much as 70 kg, creating the 'pot-belly' appearance that is now common among middle-aged men, and many post-menopausal women.

Smoking, drinking alcohol to excess, and susceptibility to depression and stress, as well as overeating and lack of exercise, promote selective expansion of the abdominal depots, raising the ratio of waist to hip circumference. In people with a high waist/hip ratio, the muscles utilise lipids less efficiently, so they remain in the blood and eventually pass into the adipocytes. Enlarged omental and mesenteric adipocytes become very sensitive to signals that stimulate lipolysis, so the fatty acids are soon re-released. Blood from these depots flows directly to the liver, which is thus frequently, perhaps continuously, exposed to an influx of fatty acids, which seems to upset its ability to regulate its glucose and lipoprotein metabolism. The hormones and nerve pathways that link personality and lifestyle to adipocyte and liver metabolism are being studied in great detail, in the hope of finding a safe, effective treatment that reverses, or at least curtails, excessive enlargement of the intra-abdominal and paunch depots.

Naturally obese animals have succeeded in doing so: the mesenteric and omental depots are very rarely massive in wild animals and are almost absent in seals, dolphins and moles. Even in large, very fat polar bears, only about 15% of the adipose tissue is intra-abdominal.

How people fatten

Sociological surveys show that people's average food intake has actually declined during the past 40 years, while the incidence of obesity has increased greatly. Comparisons between different countries suggest that the rise in obesity corresponds quite closely to the use of motorised transport, especially private cars, and improvements in central heating in homes and workplaces. As pointed out in Chapter 7, our tolerance of cold is vastly inferior to that of our very recent ancestors. So it is not that we eat more these days, but we eat more than is needed to sustain ourselves doing less physical work and producing less heat, and the excess is stored in the adipose tissue. Only very slight imbalance can, over many years, lead to obesity.

Overeating by less than 1% for a year would lead to gaining 1 kg, which might not sound a lot, but even half that amount over 30 years, say between the ages of 20 and 50, could mean a 25% increase in body mass, from 60 to 75 kg. Perhaps we should be more impressed by the number of people who remain at constant size and shape for years without apparent effort, than by the fact that by middle age, many people are slightly or moderately obese.

In the 1950s and 1960s, the cause of obesity was believed to lie in abnormalities of adipose tissue, particularly its development during the suckling period (see Chapter 5). Such notions stimulated the research that forms the basis of our understanding of the formation, maturation and metabolism of adipocytes (see Chapter 2). Mothers were advised to avoid overfeeding their babies, in the belief that adipocytes are formed mainly, some investigators thought only, during infancy and too many would predispose the child to obesity in later life. While plump children tend to become obese adults, we now know that additional adipocytes can form at any time of life.

Brown adipose tissue was rediscovered in the late 1960s, and the mechanisms by which it generates heat by breaking down lipids were intensively studied. The circumstances that promoted or suppressed thermogenesis were also investigated, and for a while in the 1970s, making the body burn fat faster by stimulating the metabolism of brown adipose tissue was the great hope of obesity research. However, as pointed out in Chapter 7, humans have very little brown adipose tissue, and it contains very few of the receptors that mediate the activation of heat production in rats. It has proved impossible to devise a safe means of stimulating people's heat production sufficiently to make any worthwhile impact on obesity.

The control of appetite

By the 1980s, it was clear that neither controlling white adipocyte formation nor burning off excess lipid in brown adipocytes was likely to offer a foolproof means of preventing or curing obesity. The only way to stop people from becoming fat, and to treat the obese, was to curtail their food intake. Adipose tissue was seen as the passive recipient of excess lipid, the quantity of which was determined only by the balance between energy intake and energy expenditure. At that time, so-called eating disorders, anorexia nervosa and bulimia, were increasing almost as rapidly as obesity, especially among otherwise healthy girls and young women. Drug companies saw the possibility of killing two highly lucrative birds with one stone by developing drugs that would safely and reversibly suppress or enhance appetite. This objective has fuelled an enormous amount of research into the neural mechanisms that determine appetite.

Our understanding of the molecular mechanisms that control feeding advanced very rapidly during the 1990s, due to the exploitation of genetically obese rats and mice (see Chapter 2) and the development of biochemical procedures that identify genes and gene products, selectively 'knock out' particular genes and offer quick, cheap means of producing particular proteins to order. Genetically obese *ob/ob* mice were first described in 1950 but it was not until the early 1990s that the protein produced from this gene was identified as leptin, the messenger molecule that informs the brain that the adipose tissue is adequately replete. The gene has been artificially introduced into bacteria, which then synthesise leptin in sufficient quantities for it to be extracted and purified. Such 'recombinant' leptin provides a powerful tool for the study of appetite control. If *ob/ob* mice are given daily doses of leptin, their appetite declines at once, and over a period of about a week, their energy expenditure increases by up to 20%, causing them to shed as much as 30% of their body mass in less than a month.

Biochemists know of at least a dozen different small molecules that stimulate or suppress appetite when injected into rats or mice. Some are produced by the gut itself, others by adipose tissue or the brain. One of the most potent, at least in rats, is a small protein known by the unromantic name of neuropeptide Y. When injected into the blood, or directly into the brain, it is a powerful stimulator of feeding. Injecting leptin counteracts this action, and suppresses endogenous production of neuropeptide Y, suggesting that controlling its synthesis is one of the main ways in which leptin regulates appetite and hence fatness. Neuropeptide Y is the most abundant peptide in the rodent brain, so it almost certainly has a variety of roles in addition to appetite stimulation. There may also be supplementary or alternative mechanisms that control appetite and fulfil some of neuropeptide Y's other roles: mice that lack the gene for producing it do not lose weight and appear to be surprisingly normal.

This research is progressing very rapidly[19] so this account will probably be out of date by the time you read it. So far, it has told us a lot about what makes mice in laboratory cages eat mouse chow, but its application to understanding what makes people in cafés eat cream cakes is less clear. Although leptin that is almost identical to that of mice can be isolated from human blood, very few of the many different forms of severe obesity can be attributed to mutations in the gene that codes for it, though in some cases, there may be minor defects in an adjacent controller gene. Measurements of the blood concentration of leptin show that fat people tend to produce more leptin than normal, although most kinds of genetically obese mice have too little.

One possible explanation for this situation is that fat people may become 'leptin resistant', if the mechanisms that escort the messenger molecule to its destination in the brain are impaired. In women, adipocytes from the superficial paunch depot contain more than five times as much of the message (mRNA) for leptin synthesis as the omentum (the men studied had only about twice as much).[20] Omental adipocytes may thus have less 'say' in determining appetite, perhaps because they are specially equipped to interact locally with lymphoid cells, while the perirenal and paunch depots do not contain lymph nodes and probably serve as 'whole-body' lipid stores.

These studies may throw some light on how human adipose tissue tries to report its condition to the brain, but they contribute nothing to understanding what features of food composition tempt people to overeat in spite of satiety signals. Psychological studies show quite clearly that food taste, texture, 'mouth feel', and stomach fullness all contribute to what people eat and how much. People eat because they enjoy it, but food preferences depend enormously upon upbringing and habit. Many studies have shown that monkeys, apes, deer, bears and many other animals learn what to eat and what to avoid, as well as where to find food and how to handle it, from their mothers. The formation of human eating habits and attitudes to food is well underway before babies are a year old. Most readers probably concluded that they would not find the traditional Eskimo diet (Chapter 7) appetising, yet it was the norm among such people until less than 50 years ago.

Fatty food

As pointed out in Chapter 4, the human body synthesises very few fatty acids from glucose. The liver can produce the necessary enzymes, but, although rats convert excess carbohydrates into fat as soon as glycogen stores reach normal levels, people have to overeat persistently for several consecutive days before significant amounts are turned into fat. This apparent paradox comes about because our 'normal' level of glycogen storage is far below the maximum possible. Extra glycogen can be stored in the liver and skeletal muscles, which together with its attendant water molecules, may cause these tissues to swell enough to produce noticeable enlargement of the waist and thigh muscles. Glycogen starts to be oxidised and its water excreted after several hours of fasting: you may have noticed the loss of large quantities of water, a reduction in body mass of up to a kilogram and an improvement in the waistline, following a return to a more meagre diet after a day or two of indulgence on a high carbohydrate diet. Even more rapid water loss may follow a bout of strenuous swimming, running or skiing.

Although carbohydrates do not normally get converted into fat, overconsumption suppresses the utilisation of lipids, so any lipids eaten with the carbohydrates – and most highly palatable foods such as chocolate, cakes, biscuits and puddings contain both – are deposited in the adipose tissue. Utilisation of the lipids is delayed until a period of fasting that lasts long enough for the glycogen stores to be depleted to below the level at which mobilising triacylglycerols from adipose tissue is stimulated. Our adipose tissue is full of fatty acids obtained directly from the diet that accumulate there because physiological conditions that favour their release and oxidation occur too briefly or too infrequently for them to be used up. Frequent eating – snacking – curtails lipid utilisation in rats very well, and they slowly get fatter even though the actual food intake over a long period may not be excessive. However, it is less easy to demonstrate a clear-cut association between meal frequency and fatness in people.

Drinking alcohol in any quantity has a similar effect: as well as being usable a source of energy, alcohol is poisonous to the brain (impairing memory, judgement, vision and much else), and must be got rid of as fast as possible. It readily diffuses into cells and takes priority over glucose for oxidation, so delaying the utilisation of both carbohydrates and lipids. Many alcoholic drinks, especially beer, also contain large quantities of dissolved sugars and starches. Drinkers often choose fatty foods such as cheese or peanuts because the presence of lipids delays uptake of alcohol through the wall of the stomach, thereby extending the period of intoxication.

Some people, especially young men, manage to adjust energy expenditure and food intake so that hardly any lipid is accumulated in their adipose tissue. Fatter people burn more lipid (and less carbohydrate) in the course of normal activities,[21] possibly because cellular processes 'get used' to having higher concentrations of lipid circulating in the blood. But even with this adjustment of metabolism to body composition, many people gradually get fatter in middle age. Recently discovered contrasts in the way people's appetite responds to intake of different kinds of macronutrients complicates this simple picture of food selection and fuel utilisation.

An unlimited liking for lipids

Concern about obesity in developed countries and the rise in consumption of packaged meals during the last quarter of the twentieth century has promoted intensive study of people's ability to detect the presence of carbohydrate and fat in composite foods such as pies, ice-cream and milk-shakes, and their capacity to adjust their total food intake accordingly. In a typical experiment, volunteers spent a week eating nothing but synthetic diets of equal energy content in which the proportions of the major classes of nutrients were altered. It was found that eating extra carbohydrate or protein led to oxidation of more of these nutrients. People also adjusted their appetite according to their recent intake of nutrients: if they overate at one meal, they ate less at the next, so their total carbohydrate intake over 24–48 hours was surprisingly constant.

Physiologists are still very vague about what mechanisms are involved in regulating appetite for carbohydrates. Suggestions include insulin, whose production increases within minutes of a small rise in blood glucose, or fast-acting hormones released from the gut itself, but there is little firm evidence for any of them. Whatever the mechanism, it matures early in life and is very efficient: even very young children can adjust their appetite to the carbohydrate content of a meal or snack. Most of us remember being told that eating sweets near a meal time may 'spoil your dinner'.

In contrast to this accurate control of carbohydrate intake, adding more fat to the food without obviously altering its taste had only a minimal effect on the quantity that the subjects ate. It's not that people can't taste lipids: other experiments show that even untrained tasters notice impairment of flavour when the fat content of prepared foods such as pastry or cakes is reduced. People can also readily distinguish a lipid-rich snack, which they may describe as 'satisfying' or 'nourishing', from one that consists only of sugars and starches, or unabsorbable materials such as bran. They are just very bad at quantifying their fat intake and knowing when to stop eating and so they compensate incompletely or not at all for extra lipid 'hidden' in the test meals. It was as though the human body just did not notice that the fat was there.

This failure horrifies and perplexes biochemists and physiologists, but the few evolutionary biologists who have studied the data are less scandalised by their conclusions. As described in Chapter 7, most primates are, and for millions of years have been, mainly herbivorous. Leaves, flowers and fruits are mostly sugars and starches, so primates have had plenty of time in which to hone the regulation of appetite to carbohydrate intake. We are apparently much less capable than other high-fat consumers of adjusting total energy intake to the lipid content of a meal. Other animals seem to do a much better job, which to a biologist implies that they have evolved features that equip them to perform these functions, but humans have not.

For 99% of their history, people ate lipids in a fairly pure form – as adipose tissue, fish roes or nuts – so taste was more than adequate as an indicator of consumption. Cookery, in the sense of complex mixtures of many ingredients produced by multi-stage processes, only became possible when food storage was well advanced, and cooking pots and ovens replaced roasting over open fires, both very recent technological developments on an evolutionary time scale. Relative to their experience in adjusting carbohydrate intake to need, primates are novices in the business of eating large quantities of fat mixed with other ingredients. So perhaps it is not surprising that subjects of the food manipulation experiments, in which the flavour of the lipids was deliberately disguised, were so easily fooled. The synthetic foods fed to intensively farmed livestock may induce overeating, and hence obesity, by similar means: on an artificial diet, the animals fail to adjust appetite to energy intake.

Eating activates many biochemical processes, from production of more digestive enzymes to taking up glucose from the blood and converting it into glycogen for storage, that generate extra heat, called diet-induced thermogenesis. Everyone must have noticed feeling warmer within a few minutes of starting a meal. The increase in heat production contributes to the sense of well-being that accompanies eating and promotes relaxation and convivial conversation during a meal.

Protein is the most effective of the macronutrients, increasing basal metabolic rate by up to 20% (that is a major reason why 'proper' meals are more satisfying than snacks), followed by carbohydrate, but fat alone prompts a meagre 6% or so increase in heat production. Like many studies of whole-body metabolism, it is difficult to identify the tissue(s) or biochemical mechanism(s) underlying these phenomena, but there is no doubt they are real, and they have important implications for the contribution of diet composition to fattening.

Slimming in the past

Since ancient times, physicians have been aware that the safest and most reliable means of becoming leaner is to eat less than is needed to meet average energy expenditure, but unfortunately, it is very difficult to suppress appetite by will-power, the control mechanisms are too efficient. Much medical ingenuity has been devoted to devising short-cuts to slimming. If old medical reports are to be believed, some obese patients submitted to some pretty drastic treatment, including surgical excision of large quantities of adipose tissue, but for obvious reasons, more congenial treatments proved to be more popular.

As mentioned in Chapter 1, the skin and its secretions include a variety of lipids. Sweat seemed the most obvious way by which fats were extruded from the body, so Hippocrates recommended hot baths and exercise to induce perspiration. And of course, it's true: exercise leads to sweating and, if continued for long enough, to weight loss. But thanks to Lavoisier's demonstration that energy is produced by the oxidation of carbon in biological systems, we now know that metabolism contributes far more to the removal of fat from the body than sweating ever can, but the notion that sweating depleted subcutaneous fat simply by expelling its lipids persisted for millennia.

The idea was still generally accepted when in 1757, the Scottish physician, Malcolm Flemyng addressed himself to the problem of 'Corpulency, when in an extraordinary degree, may be reckoned a disease, as it in some measure obstructs the free exercise of the animal functions, and hath a tendency to shorten life by paving the way to dangerous distempers'. He was well aware of the prime cause: 'a luxurious table, a keen appetite and good company (that are) temptations to exceed often too strong for human nature to resist'. Flemyng believed that in obesity, blood and fat were 'imperfectly blended', so fat accumulated when it should be eliminated. He noted that, in the British climate, 'insensible perspiration carried off little oil'. Not much was lost with the faeces either, so he extended the idea of 'dissolving' fat away in body fluids by considering how it might be expelled in the urine.

Flemyng reasoned that soap made lipids mix with water in the laundry tub, so why not in the blood, whence they would be painlessly discharged in the urine. So he prescribed soap, preferably Spanish soap from Alicant, to be taken at bedtime as pills, dissolved in water or in a syrup, at the same time recommending only one meal of 'food plain and lean' a day and suggesting that patients should 'rise from the table with appetite'. He described[22] how one middle-aged friend, in whom 'fatness stole upon him, and kept increasing' until he weighed nearly 21 stone (132 kg), lost 2 stone (13 kg) by taking soap for 3 months. In fact, soap, which consists of fatty acids linked to sodium ions, is at least partially digested, so a good many of the fatty acids would be absorbed, exactly as though the patient had eaten chocolate, cheese or chips.

Other agents, including vinegar and soda-water, were used as slimming aids in the belief that they promoted elimination of fats in the urine or sweat. One devotee of such treatment was the poet and political activist Lord Byron (1788–1824), who, notwithstanding his romantic image and numerous female conquests, was noticeably stout for much of his life.

Perhaps the most ingenious application of this notion was the marketing, in 1898, of 'Amiral' soap as a 'treatment of obesity by a method without change of diet or regimen, and without medicine'. Not only did it claim to relieve the patient of the hazards and inconvenience of 'constitutional treatment', it promised local action, to remove fat that offends, but keep that which makes a glamorous figure. Amiral soap was made from the bile of slaughtered animals, and its manufacturers claimed that, although raw bile does not penetrate the skin, it does so when made into a soap. You lather the soap on, sponge it off, and unwanted fat dissolves away! Two cakes cost 8 shillings, a working man's weekly wage at the time, and treatment was to be continued for 2 months, but Amiral soap nonetheless sold extremely well for many years.

We laugh at such quackery now, but decades of sophisticated research have so far not produced a really effective permanent 'cure' for obesity. There is not even general agreement upon what aspect of the balance between food intake, energy expenditure and adipose tissue expansion is most likely to yield to manipulation. Nor should we be surprised at the apparent success of these potions: sham treatment of obesity with 'placebo' doses of inert materials achieves weight loss alarmingly often, even when compared with scientifically developed drugs.

Slimming now

The search for short-cuts to losing weight and alternatives to dieting has continued at an ever accelerating pace throughout the twentieth century. One obvious route is to increase energy expenditure by taking more exercise. Physical activity and living in cool climates probably help to prevent the development of obesity, especially in children and young adults, but experiments show that exercise is rarely of much benefit to people who are already too fat: they just eat a little more to make up for the extra energy expended.

To make matters worse, overall energy expenditure declines when average energy intake decreases, permanently altering the relationship between food and physical activity. Rats become leaner when given, say, only 75% of the food that they would eat voluntarily for a few weeks, but when again allowed to eat more, they invariably regain, and maintain, their normal body weight on less food than they ate before the reduced rations were imposed upon them.

People respond to dieting in the same way: in sedentary healthy adults (i.e. those not suffering from fever or wasting diseases such as cancer), adipose tissue lipids are not significantly depleted until the fifth day of fasting. The rate of weight loss gradually declines from about 0.5 kg per day for young men (proportionately more for children, less for women and older people) at the end of the first week, to only 0.1 kg per day (about 0.15% of body mass) after 3 weeks, accompanied by lower physical activity and heat production. Energy utilisation increases as soon as dieters return to normal eating, but rarely reaches pre-dieting levels.

Both exposure to cold and eating are less effective in stimulating thermogenesis in obese people than in lean ones, and those on a weight-reducing diet respond worst of all. These effects further impair the satisfaction of eating, and often make exercise unpleasant. Furthermore, weak diet-induced thermogenesis and poor tolerance of cold persist in people who were once obese, even after they have reduced their body mass to normal.

Various natural and synthetic substances such as caffeine (present in coffee and tea) stimulate whole-body energy expenditure a little, but in large doses they, like those that activate brown adipose tissue, raise the heartbeat rate to unpleasant, possibly dangerous levels. Nicotine, the main active ingredient of tobacco, is unfortunately one of the safest and most effective drugs for raising energy utilisation without active exercise. As well as curtailing appetite and replacing eating sweets and snacks, smoking stimulates the sympathetic nervous system, which produces a sense of alertness and slightly increases heat production. Many smokers are elegantly lean, and get fatter as soon as they quit the habit. Fear of gaining weight is one of the commonest reasons for not trying to give up smoking.

When emperor penguins rebuild their body reserves after the breeding season (see Chapter 5), protein is regained more quickly than it was lost during the fast, reaching normal levels in a few weeks, while the fat stores are replenished more slowly. Careful measurements on young male volunteers undergoing experimental starvation and refeeding reveal that humans do the opposite: fat accumulates faster than protein, and the men's appetite remains higher than normal for some time after their adipose tissue has reached the dimensions it was before the experiment began. This 'overshoot' in the relationship between body composition and appetite may explain why famine survivors and undernourished children often become obese in middle age. All these effects are bad news for slimmers: repeated cycles of intensive dieting followed by refattening, whether or not accompanied by exercise, increase susceptibility to obesity by enabling former slimmers to gain more weight on less food, and by undermining the contribution of the adipose tissue to the control of appetite.

By the 1990s, physicians trying to treat obesity had switched from promoting energy expenditure to limiting food intake. Advances in knowledge about digestion, absorption and the control of appetite were applied to the development of several kinds of drugs that might aid slimming. Progress has been very rapid, though most of the drugs are so new they are still under development and testing. Fenfluramine and its close relative, dexfenfluramine, help obese people to eat less by curtailing appetite where it originates, in the brain. The drugs stimulate the release of serotonin, a neural signal molecule that, among many other roles, mediates satiety: more serotonin deludes the brain that food intake meets energy demand even when in fact, adipose tissue is being depleted. Dexfenfluramine also enhances diet-induced thermogenesis, improves insulin sensitivity and counteracts lipolysis from adipose tissue stimulated by the sympathetic nervous system. These effects help to reduce the high levels of glucose and lipids in the blood that usually accompany obesity and hence reduce the risks of diabetes and heart disease.

Dexfenfluramine is sometimes prescribed with amphetamines, stimulants that boost energy expenditure and promote a sense of well-being, making it easier to disregard hunger. Unfortunately, these drugs affect other neural control systems, leading to unpleasant side-effects including dry mouth (due to inhibition of saliva production), and a general change of mood affecting sleep patterns, anxiety and sexual desire. Studies of brain structure and function in experimental baboons[23] have suggested that large doses of fenfluramine administered for several months may permanently damage some neurons that respond to serotonin, especially if combined with an amphetamine to suppress drowsiness. These drugs can also cause permanent damage to the blood circulation in the lungs. Their side-effects are potentially so dangerous that it is unlikely they could ever be taken except under medical supervision, and some experts believe they should not be used at all.

Leptin and neuropeptide Y present obvious possibilities for artificially controlling the relationship between fatness and hunger, and several drug companies are looking for ways of inactivating neuropeptide Y and increasing leptin levels in the blood. Being proteins, leptin and neuropeptide Y would be digested if swallowed as a pill, so they would have to be injected, perhaps as often as four times a day. Unfortunately, the wide distribution of neuropeptide Y in the brain suggests that appetite control may be just one of many roles, and that interfering with it may have as many unavoidable side effects as serotonin stimulators and amphetamines.

The multiple physiological roles of leptin may also limit its use as an anti-obesity drug. At least six specific receptors for leptin have been identified and shown to occur, in various combinations, in a wide range of tissues, including the heart, muscle, ovary and testis as well as adipose tissue and the brain. The effects of leptin on the reproductive organs cause particular concern: mice given leptin mature abnormally and may be infertile, and the drug may act similarly in people. So pharmacologists are now exploring the possibility of developing drugs that suppress appetite by interfering with the signals sent from the gut to the brain that co-ordinate appetite and digestion.

Another way to reduce energy intake is to impair the digestion of fatty foods, so people could eat as much as they like but the triacylglycerols would pass through the body unabsorbed. Drugs that inhibit the major lipase produced by the pancreas, and olestra, the artificial non-digestible substitute for fats, were mentioned in Chapter 4. These products and non-nutritious bulky materials such as bran that fill the stomach and/or delay its emptying, often facilitate dieting for a short time. But careful studies show that even if the taste buds fail to detect the deception, the mechanisms of appetite and satiety are not so easily fooled, and people eventually compensate by eating more. Long-term interference with digestion can be hazardous, especially when combined with changes in the diet. The gut may fail to absorb enough essential fatty acids, vitamins or minerals, leading to the classic diseases of nutritional deficiency.

Slimming aids are likely to be taken for a long period because losing weight is depressingly slow. After the first few days (when most of the loss is glycogen and water, not fat), the maximum sustainable rate for most adults is about 1% of the body mass per week. Only the most drastic (and unpleasant) treatments achieve a total reduction of more than 8–10 kg, or about 10% of the body mass. Even assuming that it is all triacylglycerols, this mass is less than half the total amount in the adipose tissue of a moderately obese adult, a pathetically small reduction compared with that seen in bears, reindeer, penguins, seals or migratory birds (see Chapter 5).

When people lose weight by whatever means, the abdominal depots are the first to be depleted, so for men, the loss of one kilogram translates into an average reduction in waistline of 7 mm, with no consistent change in hip circumference. For pre-menopausal women, it works out at 8 mm off the waist and 8.3 mm off the hips. Measurements from computerised imaging systems[24] show that men can lose as much as 40% of the volume of their intra-abdominal adipose tissue by slimming for several months, but women only manage 33% at best.

Cutting it out

All 'slimming' aids are aimed only towards limiting total abundance of adipose tissue, not with determining its distribution. Their main application is in weight reduction for medical reasons such as reducing the risk of diabetes or heart disease. Cosmetic improvements are, at best, modest and may not correspond to the patient's concept of a desirable body shape. Drugs for a glamorous figure are a very remote prospect because scientific understanding of what causes selective accumulation, or depletion, of certain depots is minimal. Although site-specific differences in metabolism can be demonstrated in animals (e.g. Figure 13, page 135), we know almost nothing about the physiological bases for the extreme conformations illustrated in Figure 17 (page 254). With good luck and will-power as the only known routes to a desirable figure, it is perhaps not surprising that many people are tempted to take a short cut to what they perceive as the ideal body shape, and request plastic surgery to remove adipose tissue.

Liposuction involves sucking out adipose tissue through small incisions, often at considerable pressure. The procedure can cause excessive bleeding as the adipose tissue is literally ripped apart by the suction. Only about a kilogram of tissue can safely be removed in a single operation, limiting the use of liposuction to the remodelling of small but conspicuous depots that fail to respond to diet or exercise. Producing a neat and symmetrical outcome is a skilled job: nobody wants dimples on thighs, or buttocks of different sizes. Cutting out adipose tissue through a large incision is really only worth doing for the breasts and the superficial abdominal paunch. The skin has to be reorganised to cover the reduced volume, itself a delicate piece of plastic surgery, and it is difficult to remove a lot of superficial adipose tissue without impairing the blood supply to the skin, prejudicing prospects for a perfect repair.

The benefit of surgical removal of adipose tissue is entirely cosmetic: by itself, it does nothing to improve insulin resistance or any of the other metabolic abnormalities associated with obesity. To add to the disappointment, adipose tissue often regrows in the site from which it was removed, or other depots expand, so within a few months of treatment, many people are as heavy as they were before surgery.

Another kind of surgical treatment for obesity is reducing the effective size of the stomach by stapling off part of it. The outcome is, of course, that appetite is curtailed and people eat less, They withdraw the lipids from their adipose tissue, from which all the usual benefits follow. Any surgery that involves opening the abdomen is a major procedure, and interfering with the early stages of digestion in the stomach risks producing problems further down the gut. But as a treatment for severe obesity, gastroplasty is effective. It leads to greater, and what is even more important, longer-lasting weight loss than any drug treatment now available, as well as avoiding the side-effects of prolonged use of drugs.

A simple-minded conclusion from this situation is that people's stomachs are too large, or, if the real benefit arises from the disruption to digestion, their guts are too efficient. Or, putting it another way, modern food is too concentrated and too easily digested. In view of the rapid changes in food production (i.e. agriculture) and food processing (i.e. cookery) during the past 10 000 years, blaming obesity on food quality does not seem unreasonable.

Although we can now replace diseased hearts, reattach severed fingers and control many forms of serious mental illness, medical treatment of obesity is notorious for its poor results and high rate of recidivism. Many obese people who enter treatment lose less weight than they had hoped, and the majority regain what little they have shed within months of stopping taking appetite suppressants or reversal of stomach-reducing surgery, even after as long as four years of maintaining a reduced body mass. The failure of either will-power or fatness to suppress the human appetite contrasts with the ability of migratory birds to eat enough to fill their adipose tissue to the required level, accurately adjusting energy stores to requirements.

Final words

Lipids are essential components of diet in humans as well as other animals. Both the fatty acids themselves and lipid-soluble vitamins affect diverse aspects of physiological function, from the way the immune system responds to infection to how fatty acids are removed from the blood after a meal. People at risk from atherosclerosis or auto-immune diseases are especially concerned about which kinds of fats they should eat, but so many factors make it difficult to propose a universally applicable prescription, except to avoid large quantities of any fat.

Adipose tissue is much more than simply an energy store. Lipids get up to mischief, causing atherosclerosis, when they are milling about in the blood with no homes to go to. Adipose tissue mops up the excess and tidies them away. Loss of appetite is among the earliest and most consistent symptom of a huge range of systemic illnesses. Fasting puts adipose tissue in charge of managing lipid supplies. Its site-specific properties direct scarce lipids to the tissues that really need them.

The causes of obesity in humans are much more diverse than in any laboratory animal so far studied. People eat fat, and grow obese because they enjoy it. That has probably been true in one way or another for at least the past 10 000 years. Hedonism has made huge contributions to music, dance, art, literature and technology, especially that related to keeping warm and high-speed travel, but its importance as an imperative to eat has got ahead of the body's capacity to deal with the metabolic consequences. Obesity may be the price we pay for letting pleasure be our guide.

Epilogue

This book has covered a wide range of topics, literally talking of shoes and ships and (sealing) wax, of cabbages and kings – as well as walruses and oysters,[1] because fats come from such a variety of sources and contribute to so many different aspects of biology. All life needs lipids, and larger, more complex organisms like vertebrates have a special tissue devoted to their storage and management. Large body size and long life span would be impossible, except in an unnaturally constant environment, without the capacity for feasting during periods of plenty, and fasting when there is no time or opportunity for feeding.

The emphasis in protein and nucleic acid chemistry has long been on specificity: only one enzyme or receptor will do, and just one wrongly placed amino acid or base pair out of thousands is enough to spoil the whole molecule. Control and determinism is favoured over plasticity and adaptability: the very name 'genetic engineering' implies that the scientist is in charge, trying to control organisms' careers through their genes. In contrast, the impression that emerges from the study of the biology of lipids and adipose tissue is of plasticity, variability and adaptability. If one kind of fatty acid is in short supply, then another might do instead, certainly as a fuel, and quite possibly as a component of a membrane. Even if it isn't perfect, it is a lot better than nothing, though small differences in its melting point or affinity for other lipids may affect where its owner can live and what it can do.

If lipid biochemistry strikes protein chemists as biologically amateurish, anatomists regard adipose tissue as beyond the pale. All adipocytes look similar under the microscope, so for many years, adipose tissue was believed to 'have no anatomy'. Nonetheless, as the preceding pages have shown, the use of biochemical methods combined with careful attention to the anatomical origin of the samples chosen for study has revealed that similar-looking adipocytes may have contrasting properties, some of which can be explained as adaptations to local interactions with adjacent tissue. Most accounts emphasise the temporal controls on adipocyte activity – short-acting hormones and nerves – but, as we have seen, mammalian adipose tissue is organised in space as well as in time: recent research shows that the contrasts between samples from different sites can be as great or greater than those that can be induced in just one site by applying hormones.

Although the superficial appearance of adipose tissue in individuals of different body composition may seem to have little in common, the tissue's basic arrangement and structure are similar in related species. Now that the organisation of adipose tissue has been shown to be as explainable in terms of adaptation to function or inheritance from ancestors as other vertebrate tissues, further research may lead to a synthetic theory that explains why adipose tissue occurs where it does, and so enable us to predict the consequences of artificial alterations to the normal arrangement.

The distribution of adipose tissue in humans is the most variable of all mammals so far studied, perhaps because sexual selection has distorted its conformation in several parallel but different ways. We are thus the least suitable species upon which to concentrate the search for simple rules governing the anatomical organisation of adipose tissue. The aim of most human studies is not basic science, but the identification of key defects in metabolic disorders and the development of drugs to correct them. Lack of understanding of the normal situation and its natural variation makes detecting anomalies and pathologies very much more difficult.

This essay on fats and fatness illustrates how scientists and physicians approach a problem and, influenced by moralists and health campaigners, establish priorities in what to study. Since the middle of the twentieth century, the prevention and correction of heart disease and obesity have dominated thinking about biological lipids so completely that their other roles, in people as well as wild animals and plants, have been sidelined. Physicians are so preoccupied with how adipose tissue *causes* disease, they don't consider how it might help to prevent it, by, for example, supplying and regulating immune cells in blood and in lymph nodes.

The roles of adipose tissue and how food intake is regulated are unlikely to be explained fully from the study of rats and mice alone. Many naturally obese wild animals have a much wider range of fatness, and spontaneously undergo more abrupt and extensive changes in appetite than should ever be induced in laboratory animals. Research into adipose tissue function and the mechanisms of obesity calls for collaboration between laboratory and field biologists. Physiological adaptations to obesity have evolved in many different lineages of wild animals, so why not in humans? In spite of the much trumpeted health hazards of obesity, we are the longest lived of all mammals, as well as being among the fattest.

Notes and references

Prologue

1 Orbach, S. (1978) *Fat is a feminist issue*. London: Arrow. Ogden, J. (1992) *Fat chance! The myth of dieting explained*. London: Routledge. Klein, R. (1996) *Eat fat*. New York: Pantheon.

1 Introduction to fats

1 The term 'essential oils' is applied to dozens of different complex organic chemicals, mainly terpenes.

2 The measures were drastic by our standards. For example, families were not exempt from tax until they had ten or more children.

3 In 1783 the Swedish chemist Carl Wilhelm Scheele (1742–1786) also extracted a sweet substance from animal fats by heating them with calx of lead (lead oxide), but he did not take the analysis any further.

4 Many organic chemicals were named after the organism in which they were first found, or for some feature of their appearance or properties. By the 1950s the number and variety of such names were causing much confusion, so by international agreement the nomenclature was simplified and rationalised. All names for alcohols were to end in -ol, so glycerine became glycerol.

5 'Chemical research on fatty materials of animal origin.'

6 Hydrolysis of the ester bonds with glycerol accounts for less than 0.3% of the total energy released by metabolism of a whole triacylglycerol molecule.

7 In spite of its culinary importance, common sugar is in fact quite rare as a storage material in nature, occurring in large quantities only in sugar cane (a member of the grass family) and sugar beet (related to beetroot, chard and mangold) and a few others. It is transported in sap in leaves, stems and other plant tissues, from which it can be harvested as, for example, maple syrup, and is an ingredient of nectar.

8 Other ingredients include long-chain alkenes and free fatty acids.

9 Water is anomalous in this respect. The solid form of almost all other substances sinks in its liquid phase, as in the case of lipids.

10 In many technologically advanced countries, including the USA, milk is always homogenised before being sold, so only older readers, and those with the opportunity to observe milk straight from the cow, may be familiar with the spontaneous separation of whey from cream.

11 Materials with a density of less than 1.0 float in water, those of density greater than 1 sink. The average density of a whole fish is determined by the densities and relative abundances of its component tissues.

12 For comparison, the human liver accounts for about 4.5% of the body mass at birth, decreasing steadily to about 2% in old age.

13 Within the past 20 years, scientists have developed special solutions and freezing protocols that permit storage of eggs, sperm and even early embyros at low temperatures, including those of humans. Even with the best procedures, only a fraction of the material is fully viable when thawed to normal body temperature.

2 Introduction to fatty tissues

1 In fact, Lavoisier's wife, Marie Anne (née Paulze), collaborated in the experimental work conducted in their private laboratory, and distributed collections of her husband's papers to scientific institutions after his death, but in the eighteenth century, such assistance was not acknowledged publicly.

2 Hoggan, G. & Hoggan, F.E. (1879) XIX. On the development and retrogression of the fat cell. *Journal of the Royal Microscopical Society* **2**: 353–80.

3 nl, nanolitres is a measure of volume; 1 nl = 10^{-9} litres or one millionth of a millilitre. For cells whose density is exactly 1.0, the values for size expressed as mass or volume are equal. The density of adipocytes changes with the size and composition of the lipid droplet relative to the other cell components and can be below 0.9, so it is important to distinguish between estimates of size based upon mass or upon volume.

4 Martin, A.D., Daniel, M.Z., Drinkwater, D.T. & Clarys, J.P. (1994) Adipose tissue density, estimated adipose tissue lipid fraction and whole body adiposity in male cadavers. *International Journal of Obesity* **18**: 79–83.

5 Pond, C.M., Mattacks, C.A. & Prestrud, P. (1995) Variability in the distribution and composition of adipose tissue in arctic foxes (*Alopex lagopus*) on Svalbard. *Journal of Zoology, London* **236**: 593–610.

6 Such swelling is due to the accumulation of fluid between, rather than within, cells and to an influx of cells of the immune system, but the main point still applies: large changes in volume are unusual for vertebrate tissues.

7 Ailhaud, G., Grimaldi, P.-A. & Négrel, R. (1992) Cellular and molecular aspects of adipose tissue development. *Annual Review of Nutrition* **12**: 207–33. Amri, E.-Z., Ailhaud, G. & Grimaldi, P.-A. (1994) Fatty acids as signal transducing molecules: involvement in the differentiation of preadipose to adipose cells. *Journal of Lipid Research* **35**: 930–7.

8 Young, C., Jarrell, B.E., Hoying, J.B. & Williams, S.K. (1992) A porcine model of adipose tissue derived endothelial cell transplantation. *Cell Transplantation* **1**: 293–8.

9 Pond, C.M. (1992) An evolutionary and functional view of mammalian adipose tissue. *Proceedings of the Nutrition Society* **51**: 367–77.

10 Flier, J.S., Cook, K.S., Usher, P. & Spiegelman, B.M. (1987) Severely impaired adipsin expression in genetic and acquired obesity. *Science* 237: 405–408. Rosen, B. S., Cook, K.S., Yaglom, J., Groves, D.L., Volanakis, J.E., Damn, D., White, T & Spiegelman, B.M. (1989) Adipsin and complement factor D activity: an immune-related defect in obesity. *Science* **244**: 1483–7.

11 Pelleymounter, M.A., Cullen, M.J., Baker, M.B., Hecht, R., Winters, D. Boone, T. & Collins, F. (1995) Effects of the obese gene product on body weight regulation in *ob/ob* mice. *Science* **269**: 540–3.

12 The number of papers published on leptin increased fifty-fold between 1995 and 1996, a rapid rise even by the standards of molecular biology, which is a notoriously fickle follower of fashions. Examples include: Ahima, R. S. *et al.* (1996) Role of leptin in the neuroendocrine response to fasting. *Nature, London* **382**: 250–2. Caro, J. F. *et al.* (1996). Leptin, the tale of an obesity gene. *Diabetes* **45**: 1455–62.

13 Pond, C.M. & Mattacks, C.A. (1984) Anatomical organization of adipose tissue in chelonians. *British Journal of Herpetology* **6**: 402–5.

14 Pond, C.M. (1986) The natural history of adipocytes. *Science Progress, Oxford* **70**: 45–71.

15 Pond, C.M. & Mattacks, C.A. (1985) Body mass and natural diet as determinants of the number and volume of adipocytes in eutherian mammals. *Journal of Morphology* **185**: 183–93.

16 Pond, C.M. & Mattacks, C.A. (1985) Cellular organization of adipose tissue in birds. *Journal of Morphology* **185**: 194–202.

17 Data were calculated from measurements by: Sjöström, L. & Björntorp, P. (1974) Body composition and adipose tissue cellularity in human obesity. *Acta Medica Scandinavica* **195**: 201–11.

18 Prins, J. B. & O'Rahilly, S. (1997) Regulation of adipose cell number in man. *Clinical Science* **92**: 3–11.

19 ATP, adenosine triphosphate. ATP is a universal intermediate that conveys chemical energy released from glucose and fatty acids to power cellular processes such as molecular synthesis and muscle contraction. It is so important and widespread in living cells that, like DNA, it is now often referred to only by its initials.

20 Quetelet, A. (1870) *Anthropométrie ou mesure des différentes facultés de l'homme.* C. Muquardt Brussels. *Sur l'homme et le développement de ses facultés, ou essai de physique sociale* (1835, 2nd edition 1869).

21 Deuterium, ^{2}H, or heavy oxygen, ^{18}O, or sometimes both isotopes in 'doubly labelled water'.

22 The Latin phrases mean 'in the life' (i.e. in the living animal or plant) and 'after death' (i.e. when the organism can be dissected or analysed chemically).

23 Colby, R.H., Mattacks, C.A. & Pond, C.M. (1993) The gross anatomy, cellular structure and fatty acid composition of adipose tissue in captive polar bears (*Ursus maritimus*). *Zoo Biology* **12**: 267–75.

24 Pond, C.M. & Mattacks, C.A. (1987) The anatomy of adipose tissue in captive *Macaca* monkeys and its implications for human biology. *Folia Primatologica* **48**: 164–85. Pereira, M.E. & Pond, C.M. (1995) Organization of white adipose tissue in lemuridae. *American Journal of Primatology* **35**: 1–13.

25 Friedl, K. E., Moore, R.J., Martinez-Lopez, L.E., Vogel, J.A., Askew, E.W., Marchitelli, L.J., Hoyt, R.W., Gordon, C.C. (1994) Lower limit of body fat in healthy active men. *Journal of Applied Physiology* **77**: 933–40.

26 Pond, C.M., Mattacks, C.A., Thompson, M.C. & Sadler, D. (1986) The effects of age, dietary restriction, exercise and maternity on the abundance and volume of adipocytes in twelve adipose depots of adult guinea-pigs. *British Journal of Nutrition* **56**: 29–48.

27 Those who inherited the mutant gene from both parents, and so have maximum abnormality.

28 Phillips, N.J. & Pond, C.M. (1986) Adipose tissue cellularity and site-specific differences in adipocyte volume in genetically obese (*ob/ob*) mice. *Proceedings of the Nutrition Society* **45**: 101A.

29 Thomas, D. W., Bosque, C. & Arends, A. (1993) Development of thermoregulation and the energetics of nestling oil birds (*Steatornis caripensis*). *Physiological Zoology* **66**: 322–48.

30 Gosler, A. G., Greenwood, J.J.D. & Perrins, C. (1995) Predation risk and the cost of being fat. *Nature* **377**: 621–3.

31 Bears do not hibernate in the strict sense of the term, probably because they are too big. The body temperature drops by only a few degrees, while that of true hibernators (e.g. dormice, hedgehogs, ground squirrels) falls by more than 30 °C.

32 Pond, C.M., Mattacks, C.A., Colby, R.H. & Tyler, N.J. C. (1993) The anatomy, chemical composition and maximum glycolytic capacity of adipose tissue in wild Svalbard reindeer (*Rangifer tarandus platyrhynchus*) in winter. *Journal of Zoology, London* **229**: 17–40.

33 Wolverines, also called gluttons, are fierce predators and scavengers related to weasels, pine martens, mink and otters, but much larger, up to 30 kg. They used to occur over much of northern Europe and North America but are now confined to arctic regions of Canada and Russia.

34 Ramsay, M.A., Mattacks, C.A. & Pond, C.M. (1992) Seasonal and sex differences in the cellular structure and chemical composition of adipose tissue in wild polar bears (*Ursus maritimus*). *Journal of Zoology, London* **228**: 533–44.

3 Diverse fatty acids

1 Jensen, R.G. (1996) The lipids in human milk. *Progress in Lipid Research* **35**: 53–92.

2 Many textbooks use the equivalent name, monoenoic, which, being derived from two Greek words is etymologically more correct than monounsaturated, which combines a Latin with a Greek root. This scheme substitutes 'polyenoic' for 'polyunsaturated' but since the latter term is widely used in food labelling, it is used throughout this book, and so, for consistency, is the term monounsaturated.

3 Suet is ground adipose tissue of cattle (usually from the dorsal wall of abdomen depot) and therefore contains collagen and other proteins as well as lipid. The pure triacylglycerols extracted from it are called tallow.

4 To make matters worse, when counting carbon atoms from the methyl end of a fatty acid was first developed, ω (omega) was used instead of n. This book adopts the n system throughout. Under the old system, Δ (delta) was used to indicate numbering from the acid end. The designation 'br' means that the carbon chain is branched instead of straight.

5 The name 'margaric acid' is now applied to C17:0, which is a minor component of mammalian triacylglycerols, especially those of ruminants (sheep, cattle, goats, deer). Half a century later, the derived term 'margarine' was given, somewhat incorrectly, to the semi-crystalline mixture of artificially modified triacylglycerols.

6 In fact, two distinct fatty acids are called linolenic acid, α-linolenic, C18:3n–3, which belongs to the n–3 family, and γ-linolenic, C18:3n–6, which belongs to the n–6 family. Both have metabolic roles in mammals but the former is much more abundant and widespread.

7 This kind of reaction with oxygen, which involves the addition of one or more oxygen atoms into the chain of carbon atoms, should not be confused with the oxidation in mitochondria that yields energy in a biologically usable form, and breaks down the fatty acids into carbon dioxide and water, described in greater detail in Chapter 4.

8 Simopoulos, A.P. (Ed.) (1995) *Plants in human nutrition*. World Review of Nutrition and Dietetics, vol. 77. Basel: Karger.

9 These cooking fats are almost pure triacylglycerols extracted from adipose tissue, lard from (non-ruminant) pigs, and tallow from (ruminant) cattle or sheep.

10 Geiser, F., Firth, B.T. & Seymour, R.S. (1992) Polyunsaturated dietary lipids lower the selected body temperature of a lizard. *Journal of Comparative Physiolology* **162B**: 1–4.

11 See note 32 for Chapter 2.

12 Kashiwagi, T., Meyer-Rochow, V.B., Nishimura, K., Eguchi, E. (1997) Fatty acid composition and ultrastructure of photoreceptive membranes in the crayfish *Procambarus clarkii* under conditions of thermal and photic stress. *Journal of Comparative Physiolology* **167B**: 1–8.

13 Ito, M.K. & Simpson, K.L. (1996) The biosynthesis of ω3 fatty acids from 18:2ω6 in *Artemia* spp. *Comparative Biochemistry and Physiology* **115A**: 69–76.

14 Sometimes abbreviated to *c* and *t*. Some organic chemistry texts favour the equivalent terms *Z* and *E*.

15 Käkelä, R., Hyvärinen, H. & Vainiotalo, P. (1996) Unusual fatty acids in the depot fat of the Canadian beaver (*Castor canadensis*). *Comparative Biochemistry and Physiology* **113B**: 625–9.

16 The name is derived from the Greek word for an olive tree, presumably to reflect its chemical similarity to oleic acid but it never occurs in significant quantities in olives.

17 Gurr, M.I. (1996) Dietary fatty acids with trans unsaturation. *Nutrition Research Reviews* **9**: 259–79.

18 Larsson, K. (1994) *Lipids – molecular organization, physical functions and technical applications.* Dundee: The Oily Press.

19 Cod-liver oil as sold in pharmacies has long been made from the livers of dogfish, salmon and other oily fish as well as cod.

20 Gurr, M.I. (1992) *Role of fats in food and nutrition* (2nd edition). London: Elsevier Applied Science.

21 Erucic acid has also been tried as a therapy for a rare genetic disorder of lipid metabolism called adrenomyeloneuropathy, which destroys neurons in the brain and peripheral nerves, causing paralysis and dementia. Its use was celebrated in the film *Lorenzo's Oil* (1992), about a child suffering from the disease, but controlled clinical trials in which patients took synthetic triacylglycerols containing only oleic and erucic acids for 2 years revealed no benefit from the therapy.

22 The muscles of the so-called castor oil fish (*Ruvettus pretiosus*), a large, mackerel-like fish that can weigh up to 50 kg, contain some unusual esters of long-chain alcohols and fatty acids, which also have purgative properties, but they are chemically quite different from real castor oil. The fish occurs at depths of 600 m or more in almost all warm temperate oceans (including those around Britain). Although it lives too deep, and is too scarce, to be harvested on an industrial scale, when caught by chance, distillates of its flesh were used medicinally, especially by people unable to import alternative remedies such as castor oil and senna, that are derived from tropical plants.

23 The melting point of pure ricinoleic acid is 7.7 °C, compared with 16.3 °C for pure oleic acid.

4 Lipids in action

1 The drugs were undergoing trials in 1996, but are not yet generally available. Suggested trade names include Xenical and Orlistat.

2 Christie, W.W. (ed.) (1981) *Lipid metabolism in ruminant animals.* Oxford: Pergamon Press.

3 Red blood cells contain the oxygen-binding pigment haemoglobin and are by far the most abundant kind of cell in the blood of almost all vertebrates.

4 Halliwell, K. J., Fielding, B.A., Samra, J.S., Humphrys, S.M. & Frayn, K.A. (1996) Release of individual fatty acids from human adipose tissue in vivo after an overnight fast. *Journal of Lipid Research* **37**: 1842–8.

5 Pond, C.M., Mattacks, C.A. & Sadler, D. (1992) The effects of exercise and feeding on the activity of lipoprotein lipase in nine different adipose depots of guinea-pigs. *International Journal of Biochemistry* **24**: 1825–31.

6 Pennington, J. E., Nussenzveig, R.H. & Van Heusden, M.C. (1996) Lipid transfer from insect fat body to lipophorin: comparison between a mosquito triacylglcyerol-rich liphorin and a sphinx moth diacylglycerol-rich liphorin. *Journal of Lipid Research* **37**: 1144–52.

7 The basic structure and composition of snake, lizard, tortoise and crocodile eggs are almost identical to those of birds.

8 Napolitano, G.E., Ackman, R.G. & Parrish, C.C. (1992) Lipids and lipophilic pollutants in three species of migratory shorebirds and their food in Shepody Bay (Bay of Fundy, New Brunswick). *Lipids* **27**: 785–90.

9 Pond, C.M., Mattacks, C.A., Gilmour, I., Johnston, M.A., Pillinger, C. T. & Prestrud, P. (1995) Chemical and carbon isotopic composition of fatty acids in adipose tissue as indicators of dietary history in wild arctic foxes (*Alopex lagopus*) on Svalbard. *Journal of Zoology, London* **236**: 611–23.

10 Most modern candles are made from paraffin wax, produced from refining mineral oil (petroleum), though candles used for certain religious rituals are still made from beeswax, with its mild but distinctive smell.

11 Noradrenalin and adrenalin are known in American English as norepinephrine and epinephrine respectively. The names are derived from the fact that they were first isolated from the adrenal gland which is located near the kidney.

12 Raclot, T. & Groscolas, R. (1993) Differential mobilization of white adipose tissue fatty acids according to chain length, unsaturation, and positional isomerism. *Journal of Lipid Research* **34**: 1515–26. Raclot, T., Mioskowski, E., Bach, A. C., Groscolas, R. (1995) Selectivity of fatty acid mobilization: a general metabolic feature of adipose tissue. *American Journal of Physiology* **269**: R1060-7. Connor, W. E., Lin, D. S. & Colvis, C. (1996) Differential mobilization of fatty acids from adipose tissue. *Journal of Lipid Research* **37**: 290–8.

13 Pond, C. M. & Mattacks, C. A. (1991) The effects of noradrenaline and insulin on lipolysis in adipocytes isolated from nine different adipose depots of guinea-pigs. *International Journal of Obesity* **15**: 609–18.

14 See note 9 for Chapter 2.

15 Marchington, J.M., Mattacks, C.A. & Pond, C.M. (1989) Adipose tissue in the mammalian heart and pericardium: structure, foetal development and biochemical properties. *Comparative Biochemistry and Physiology* **94B**: 225–32.

16 See Chapter 2 notes 5, 23 & 32 and: Pond, C.M., Mattacks, C.A., Colby, R.H. & Ramsay, M.A. (1992) The anatomy, chemical composition and metabolism of adipose tissue in wild polar bears (*Ursus maritimus*). *Canadian Journal of Zoology* **70**: 326–41.

17 Marchington, J.M. & Pond, C.M. (1990) Site-specific properties of pericardial and epicardial adipose tissue: the effects of insulin and high-fat feeding on lipogenesis and the incorporation of fatty acids *in vitro*. *International Journal of Obesity* **14**: 1013–22.

18 Newsholme, E., Leech, T. & Duester, G. (1994) *Keep on running.* Chichester: Wiley.

19 Chloroplasts are tiny cell components, most famous for their role in photosynthesis.

20 Acheson, K.J., Schutz, Y., Bessard, T., Anantharaman, K., Flatt, J.P. & Jéquier, E. (1988) Glycogen storage capacity and *de novo* lipogenesis during massive carbohydrate overfeeding in man. *American Journal of Clinical Nutrition* **48**: 240–7.

21 Colby, R.H. & Pond, C.M. (1993) Site-specific differences in the responses of guinea-pig adipose tissue to changes in the fatty acid composition of the diet. *Nutrition Research* **13**: 1203–12.

22 Daniels, C.B., Orgeig, S., Smits, A.W. (1995) The evolution of the vertebrate pulmonary surfactant system. *Physiological Zoology* **68**: 539–66.

23 McIlhinney, R.A.J. (1995) Mechanisms, biochemical regulation and functional consequences of protein acylation. *Biochemical Society Transactions* **23**: 549–53.

24 Porter, J. A., Young, K.A. & Beachy, P.A. (1996) Cholesterol modification of hedgehog signaling proteins in animal development. *Science* **274**: 255–9.

25 Shakespeare, W. (1603) *Hamlet, Prince of Denmark.*

5 The functions of fattening

1 Pennycuick, C.J. (1992) *Newton rules biology: a physical approach to biological problems.* Oxford: Oxford University Press. Schmidt-Nielsen, K. (1984) *Scaling: why is animal size so important?* Cambridge: Cambridge University Press.

2 Pennycuick, C.J. (1996) Stress and strain in the flight muscles as constraints on the evolution of flying animals. *Journal of Biomechanics* **29**: 577–81.

3 Alerstam, T. (1990) *Bird migration.* Cambridge: Cambridge University Press.

4 In the dialect of the Eskimo language that is spoken in Canada and Greenland, the people's name for themselves is Inuit, as the German word for a German is Deutsche. The use of the term 'Inuit' to mean 'Eskimo' in English has recently become fashionable, but is not used in this book.

5 Hamilton, W.D. & Zuk, M. (1982) Heritable true fitness and bright birds: a role for parasites? *Science* **218**: 384–8.

6 The allergies to cows' milk that some babies develop, necessitating feeding goats' milk or even synthetic 'milk' made from soya beans, is due to features of the proteins.

7 Stirling, I. (1988) *Polar bears* (with photographs by D. Guravich) Ann Arbor: University of Michigan Press.

8 Oftedal, O.T., Bowen, W.D., Widdowson, E.M. & Boness, D.J. (1989) Effects of suckling and the postsuckling fast on weights of the body and internal organs of harp and hooded seal pups. *Biology of the Neonate* **56**: 283–300.

9 Girard, J., Ferré, P., Pégorier, J.-P. & Duée, P.-H. (1992) Adaptations of glucose and fatty acid metabolism during perinatal and suckling-weaning transition. *Physiological Reviews* **72**: 507–62.

10 Ramsay, M. A. & Dunbrack, R.L. (1986) Physiological constraints on life history phenomena: the example of small bear cubs at birth. *American Naturalist* **127**: 735–43.

11 Noble, R.C., Speake, B.K., McCartrey, R., Foggin, C.M. & Deeming, D.C. (1996) Yolk lipids and their fatty acids in the wild and captive ostrich (*Struthio camelus*). *Comparative Biochemistry and Physiology* **113B**: 753–6.

12 A few groups of birds, including Galliformes and ducks, lead their precocious young to food and supervise them, rather than bringing food to them. The eggs of brush turkeys (megapodes) are incubated on a mound of rotting vegetation, which the father tends diligently, but the chicks are large and mature at hatching, and do not associate with their parents.

13 Groscolas, R. (1986) Changes in body mass, body temperature and plasma fuel levels during the natural breeding fast in male and female emperor penguins *Aptenodytes forsteri*. *Journal of Comparative Physiology* **156B**: 521–7. Groscolas, R. (1990) Metabolic adaptations to fasting in emperor and king penguins. In L. S. Davis & Darby J.T. (eds.), *Penguin biology* (pp. 269–96.). San Diego: Academic Press.

14 Secor, S.M. & Diamond, J. (1995) Adaptive responses to feeding in burmese pythons: pay before pumping. *Journal of Experimental Biology* **198**: 1313–25.

15 Conley, K.E. & Linstedt, S.L. (1996) Minimal cost per twitch in rattlesnake tail muscle. *Nature* **383**: 71–2.

16 Frank, C.L. & Storey, K.B. (1995) The optimal depot fat composition for hibernation by golden-mantled ground squirrels (*Spermophilus lateralis*). *Journal of Comparative Physiology* **164B**: 536–42. Geiser, F. & Kenagy, G.J. (1987) Polyunsaturated lipid diet lengthens torpor and reduces body temperature in a hibernator. *American Journal of Physiology* **252**: R897-R901.

17 Sadler, D. (1998) The effects of ambient temperature and day length on the distribution, chemical composition and histological structure of adipose tissue in the dwarf hamster, *Phodopus campbelli*. Unpublished M.Phil thesis, The Open University.

6 The functions of fat

1 Pond, C.M. & Ramsay, M.A. (1992) Allometry of the distribution of adipose tissue in Carnivora. *Canadian Journal of Zoology* **70**: 342–7.

2 Worthy, G.A.J. & Edwards, E.F. (1990) Morphometric and biochemical factors affecting heat loss in a small temperate cetacean (*Phocaena phocaena*) and a small tropical cetacean (*Stenella attenuata*). *Physiological Zoology* **63**: 432–42.

3 Kradsheim, P.H., Folkow, L.P. & Blix, A.S. (1996) Thermal conductivity of minke whale blubber. *Journal of Thermal Biology* **21**: 123–8.

4 Käkelä, R. & Hyvärinen, H. (1996) Site-specific fatty acid composition in adipose tissues of several northern aquatic and terrestrial mammals. *Comparative Biochemistry and Physiology* **115B**: 501–14.

5 A single pup is also the rule among pinnipeds, but twins are very occasionally born to certain species of seals.

6 Bennett, M.B. & Ker, R. F. (1990) The mechanical properties of the human subcalcaneal fat pad. *Journal of Anatomy* **171**: 131–8. Ker, R.F. (1996) The time-dependent, mechanical properties of the human heel pad in the context of locomotion. *Journal of Experimental Biology* **199**: 1501–8.

7 Pond, C.M. & Mattacks, C.A. (1986) Allometry of the cellular structure of intra-orbital adipose tissue in eutherian mammals. *Journal of Zoology, London* **209**: 35–42. Mattacks, C.A. & Pond, C.M. (1985) The effects of dietary restriction and exercise on the volume of adipocytes in two intra-orbital depots in the guinea-pig. *British Journal of Nutrition* **53**: 207–13.

8 Hofer, H.O. (1972) On the corpus adiposum buccae (Bichat) in *Pan troglodytes*. *Folia Primatologica* **17**: 434–41.

9 Pond, C.M. & Mattacks, C.A. (1989) Biochemical correlates of the structural allometry and site-specific properties of mammalian adipose tissue. *Comparative Biochemistry and Physiology* **92A**: 455–63.

7 Fat people

1 Bradshaw, J.W.S., Goodwin, D., Legrand-Defrétin, V. & Nott, H.M.R. (1996) Food selection by the domestic cat, an obligate carnivore. *Comparative Biochemistry and Physiology* **114A**: 205–10.

2 The notes and drawings that Charles Darwin made on his travels in and around South America were published in 1839 as *The Voyage of the Beagle*. The book sold so well that Darwin revised it several times before the final edition appeared in 1860. Observations and ideas derived from that 5-year voyage around the world formed the basis for his most famous work, *The Origin of Species* (1859).

3 Sinclair, H. M. (1953) The diet of Canadian Indians and Eskimos. *Proceedings of the Nutrition Society* **12**: 69–82.

4 Kaplan, J.R., Manuck, S.B. & Shiveley, C.A. (1991) The effects of fat and cholesterol on social behavior in monkeys. *Psychosomatic Medicine* **53**: 634–42. Kaplan, J.R., Shiveley, C.A., Fontenot, M.B., Morgan, T.M., Howell, S.M., Manuck, S.B., Muldoon, M.F. & Mann, J.J. (1994) Demonstration of an association among dietary cholesterol, central serotonergic activity, and social behavior in monkeys. *Psychosomatic Medicine* **56**: 479–84. Kaplan, J.R., Fontenot, M.B., Manuck, S.B. & Muldoon, M.F. (1996) Influences of dietary lipids on agonistic and affiliative behavior in *Macaca fasicularis*. *American Journal of Primatology* **38**: 333–47.

5 Aiello, L.C. & Wheeler, P. (1995) The expensive-tissue hypothesis: the brain and digestive system in human and primate evolution. *Current Anthropology* **36**: 199–221.

6 Martin, R.D. (1996) Scaling of the mammalian brain: maternal energy hypothesis. *News in Physiology* **11**: 149–56.

7 Brown, P.J. & Konner, M. (1987) An anthropological perspective on obesity. *Annals of the New York Academy of Sciences* **499**: 29–46.

8 Pasquet, P., Brigant, L., Froment, A., Koppert, G.A., Bard, D., de Garine, I. & Apfelbaum, M. (1992) Massive overfeeding and energy balance in men: the *Guru Walla* model. *American Journal of Clinical Nutrition* **56**: 483–90.

9 Di Tomaso, E., Beltramo, M. & Piomelli, D. (1996) Brain cannabinoids in chocolate. *Nature* **382**: 677–78.

10 Laburn, H.P. (1996) How does the fetus cope with thermal challenges? *News in Physiological Sciences* **11**: 96–100.

11 Johnston, F.E., Cohen, S. & Beller, A. (1985) Body composition and temperature regulation in newborns. *Journal of Human Evolution* **14**: 341–5.

12 Mattacks, C.A. & Pond, C.M. (1988) Site-specific and sex differences in the rates of fatty acid/triacylglycerol substrate cycling in adipose tissue and muscle of sedentary and exercised dwarf hamsters (*Phodopus sungorus*). *International Journal of Obesity* **12**: 585–97.

13 Collins, S., Daniel, K.W., Petro, A.E. & Surwit, R.S. (1997) Strain-specific response to β_3-adrenergic receptor agonist treatment of diet-induced obesity in mice. *Endocrinology* **138**: 405–13.

14 Leppäluoto, J. & Hassi, J. (1991) Human physiological adaptation to the arctic climate. *Arctic* **44**: 139–45.

15 Hong, S.K., Rennie, D.W. & Park, Y.S. (1987) Humans can acclimatize to cold: a lesson from Korean women divers. *News in Physiological Sciences* **2**: 79–82.

16 Much of what we know about the Romans' attempts to conquer the indigenous German tribes between AD 9 and 17 comes from the account by Cornelius Tacitus, written about AD 98. The highly trained Roman armies suffered several humiliating defeats, as well as some famous victories, and Tacitus clearly had grudging respect for a people his countrymen regarded as savages.

17 Rebuffé-Scrive, M., Mårin, P. & Björntorp, P. (1991) Effects of testosterone on abdominal adipose tissue in men. *International Journal of Obesity* **15**: 791–5.

18 Pond, C.M. & Mattacks, C.A. (1987) The anatomy of adipose tissue in captive *Macaca* monkeys and its implications for human biology. *Folia Primatologica* **48**: 164–85.

19 Whitely, S. (1993) Predictors of milk production in lactating women. M.Phil. thesis. The Open University.

20 Prentice, A. & Prentice, A. (1988) Reproduction against the odds. *New Scientist* 118: 42–6. Prentice, A. M., Goldberg, G.R. & Prentice, A. (1994) Body mass index and lactation performance. *European Journal of Clinical Nutrition* **48** (suppl. 3): S78–S89.

21 Singh, D. (1993) Body shape and female attractiveness: the critical role of the waist-to-hip ratio (WHR). *Human Nature* **4**: 297–321. Singh, D. (1993) Adaptive significance of female physical attractiveness: the role of waist-to-hip ratio. *Journal of Personality and Social Psychology* **654**: 293–307.

22 Shattock, S.G. (1909) On normal and tumour-like formations of fat in Man and the lower animals. *Proceedings of the Royal Society of Medicine, Pathology Section* **2**: 207–70.

23 Krotkiewski, M., Björntorp, P., Sjöström, L. & Smith, U. (1983) Impact of obesity on metabolism in men and women. Importance of regional adipose tissue distribution. *Journal of Clinical Investigation* **72**: 1150–62.

24 Scarlett, J. M., Donoghue, S., Saidla, J. & Wills, J. (1994) Overweight cats: prevalence and risk factors. *International Journal of Obesity* **18** (suppl 1): S22–8.

25 Discussion of the course of human evolution is deliberately avoided: the taxonomy of extinct hominids is complicated and experts disagree upon exactly how species and subspecies should be defined, and upon when and where each lived.

26 Rice, P.C. (1981) Prehistoric venuses: symbols of motherhood or womanhood? *Journal of Anthropological Research* **37**: 402–14.

27 Most oriental people, including Japanese and Chinese did not traditionally eat dairy products as adults. Indeed, many are unable to digest whole milk satisfactorily. The Japanese kept few domestic mammals and preferred plant or whale oils to lard or butter for frying.

28 Leviticus, Chapter 11.

29 Exodus, Chapter 10.

30 In tombs such as those of Kagemni and Mereruka at Saqqara, built during the Old Kingdom around 4500 years ago.

31 *Theogony* line 535. Many of the epic poems attributed to Hesiod and Homer were based upon long-established myths and traditions that had hitherto been transmitted orally.

8 Fat and health

1 Wilske, J. & Arnbom, T. (1996) Seasonal variation in vitamin D metabolites in southern elephant seal (*Mirounga leonina*) females in South Georgia. *Comparative Biochemistry and Physiology* **114A**: 9–14.

2 Burr, G.O. & Burr, M.M. (1930) On the nature and role of the fatty acids essential to nutrition. *Journal of Biological Chemistry* **86**: 587–621.

3 Oils sold commercially are fairly constant in composition because each bottle is a mixture derived from many different plants that were cultivated under partially standardised conditions.

4 The Greek text of St Mark's gospel has 'ακριδες', which means locusts or grasshoppers (viz. Acrididae, the scientific name for the grasshopper family), but later editors altered it to refer to the pods of the carob tree (*Ceratonia siliqua*), also called the locust tree, a kind of acacia native to the Near East. Their aim may have been to make the saint's diet seem even more ascetic: the main local use of carob pods was as food for pigs, which were 'unclean' for Jews.

5 Gravatt, B.F., Prospero-Garcia, O., Siuzdak, G., Gilula, N.B., Henriksen, S.J., Boger, D.L. & Lerner, R.A. (1995) Chemical characterization of a family of brain lipids that induce sleep. *Science* **268**: 1506–11.

6 Macbeth so describes sleep in remarks to his wife following Duncan's murder (Shakespeare, W. (1606) Macbeth, Act 2, scene 2)

7 Calder, P.C., Bond, J.A., Bevan, S.J., Hunt, S.V. and Newsholme, E.A. (1991) Effect of fatty acids in the proliferation of concanavalin A-stimulated rat lymph node lymphocytes. *International Journal of Biochemistry* **23**: 579–88.

8 Calder, P.C. (1995) Fatty acids, dietary lipids and lymphocyte functions. *Biochemical Society Transactions* **23**: 302–9. Calder, P.C. (1996) Immunomodulatory and anti-inflammatory effects of *n*–3 polyunsaturated fatty acids. *Proceedings of the Nutrition Society* **55**: 737–74.

9 Fritsche, K.L., Shahbazian, L.M., Feng, C. & Berg, J.N. (1997) Dietary fish oil reduces survival and impairs bacterial clearance in C3H/Hen mice challenged with *Listeria monocytogenes*. *Clinical Science* **92**: 95–101.

10 Pond, C.M. & Mattacks, C.A. (1995) Interactions between adipose tissue around lymph nodes and lymphoid cells *in vitro*. *Journal of Lipid Research* **36**: 2219–31. Pond, C.M. (1996) Interactions between adipose tissue and the immune system. *Proceedings of the Nutrition Society* **55**: 111–26.

11 Mattacks, C.A. & Pond, C.M. (1997) The effects of feeding suet-enriched chow on site-specific differences in the composition of triacylglycerol fatty acids in adipose tissue and its interactions *in vitro* with lymphoid cells. *British Journal of Nutrition* **77**: 621–43.

12 Lajeunesse, T. A. & Peterson, R.O. (1993) Marrow and kidney fat as condition indices in gray wolves. *Wildlife Society Bulletin* 21: 87–90. Poullé, M.-L., Crête, M. & Huot, J. (1995) Seasonal variation in body mass and composition of eastern coyotes. *Canadian Journal of Zoology* **73**: 1625–33.

13 Plasma is the liquid portion of the blood from which the cells have been removed, usually by centrifugation, leaving lipoproteins and other dissolved material.

14 Finlayson, R. (1965) Spontaneous arterial disease in exotic animals. *Journal of Zoology, London* **147**: 239–343.

15 Most modern pigs are slaughtered for meat at less than a year old, before they are fully grown. Few breeding sows or boars are kept beyond the age of about 8 years.

16 Keys, A. & Keys, M. (1975) *How to eat well and stay well the Mediterranean way.* New York: Doubleday.

17 Sinclair, H.M. (1956) Deficiency of essential fatty acids and atherosclerosis etcetera. *The Lancet* i: 381–3.

18 In spite of its name, insulin is sometimes used to treat certain aspects of this disorder, especially episodes of coma that can occur when the body is under stress from some other infection or injury.

19 Friedman, J. M. (1997) The alphabet of weight control. *Nature* **385**: 119–20.

20 Montague, C.T., Prins, J.B., Sanders, L., Digby, J.E. & O'Rahilly, S. (1997) Depot- and sex-specific differences in human leptin mRNA expression. Implications for the control of regional fat distribution. *Diabetes* **46**: 324–47.

21 Flatt, P.J. (1995) Use and storage of carbohydrate and fat. *American Journal of Clinical Nutrition* **61**: 952S-9S.

22 Flemyng, M. (1760) *A discourse on the nature, causes and cure of corpulency.* London: Davis & Reymers. 28 pp.

23 Scheffel, U., Szabo, Z., Mathews, W.B., Findlay, P.A., Yuan, J., Callahan, B., Hatzidimitriou, G., Dannals, R.F., Ravert, H.T. & Ricaurte, G.A. (1996) Fenfluramine-induced loss of serotonin transporters in baboon brain visualized with PET. *Synapse* **24**: 395–98.

24 van der Kooy, K., Leenen, R., Siedell, J.C., Deurenberg, P., Droop, A. & Bakker, G.J.G. (1993) Waist-hip ratio is a poor predictor of changes in visceral fat. *American Journal of Clinical Nutrition* **57**: 327–33.

Epilogue

1 Tweedledee's poem, The Walrus and the Carpenter, from Lewis Carroll's *Through the Looking Glass* (1871). See also Chapters 6, 7, 1, 3 and 5.

Index